GEOLOGICAL RESOURCES AND GOOD GOVERNANCE IN SUB-SAHARAN AFRICA

Geological Resources and Good Governance in Sub-Saharan Africa

Holistic Approaches to Transparency and Sustainable Development in the Extractive Sector

Editors

Jürgen Runge
Department of Physical Geography, Goethe-University Frankfurt & Centre for Interdisciplinary Research on Africa (CIRA/ZIAF), Frankfurt am Main, Germany

James Shikwati
IREN, Inter Region Economic Network, Nairobi, Kenya

CRC Press
Taylor & Francis Group
Boca Raton London New York Leiden

CRC Press is an imprint of the
Taylor & Francis Group, an **informa** business
A BALKEMA BOOK

Front cover: Different aspects and impacts by mining and the extractive industries in Sub-Saharan Africa—from the informal, small-scale artisanal mining (bottom left), to mainly foreign enterprises dominated exploration (bottom middle) to large-scale, exploitation by open pit mining (bottom right) (Photos: J. Runge). Environmental degradation of forest ecosystems is linked to extractive industries (top) (Photo: J. Eisenberg).

Financially supported by CEMAC (Economic and Monetary Community of Central Africa) within the framework of the project "Strengthening governance in Central Africa's extractive sector" and by GIZ (German Society for International Cooperation/Deutsche Gesellschaft für Internationale Zusammenarbeit, www.giz.de)

First issued in paperback 2017

CRC Press/Balkema is an imprint of the Taylor & Francis Group, an informa business

Typeset by Vikatan Publishing Solutions (P) Ltd., Chennai, India

Published by: CRC Press/Balkema
P.O. Box 447, 2300 AK Leiden, The Netherlands
e-mail: Pub.NL@taylorandfrancis.com
www.crcpress.com – www.taylorandfrancis.co.uk – www.balkema.nl

ISBN 13: 978-1-138-11408-1 (pbk)
ISBN 13: 978-0-415-58267-4 (hbk)

Geological Resources and Good Governance in Sub-Saharan Africa – Runge & Shikwati (eds)
© 2011 Taylor & Francis Group, London, ISBN 978-0-415-58267-4

Table of contents

Geological Resources and Good Governance in Sub-Saharan Africa – Runge & Shikwati (eds)
© 2011 Taylor & Francis Group, London, ISBN 978-0-415-58267-4

Foreword

This book intends to give an update of the distribution, occurrences, potential and prospects of geological resources for sustainable development in Sub-Saharan Africa. By bringing together and publishing numerous different points of views, it will be holistic, interdisciplinary and scientific. Even if the procedure is academic, it addresses aside of researchers to stakeholders in the field including civil society, politicians and decision makers.

While the countries south of the Sahara are home to Africa's greatest concentration of wealth in geological resources, they often still count among the poorest nations on earth. The region is afflicted by crises and conflicts that severely harm national economies and enfeeble already weak governmental structures even further. Is the region's vast natural resource wealth really responsible for this situation? Or does it come down to individual political and social conditions which, within a single generation, have transformed certain countries' abundance in raw materials into a source of common wealth for all their citizens, while at the same time turning those same riches in other countries into a "resource curse" or the often cited "paradox of plenty"?

Today increased transparency in the extractive sector is widely recognised as crucial for achieving accountability, good governance and sustainable economic development. It offers benefits for all stakeholders in the sector—for resource-rich countries and their residents, industrialised or emerging countries, and both private and state-owned enterprises. Increased transparency and good governance are also in the interest of importers of raw materials as they increase confidence in the effectiveness of public institutions and political stability. They improve the investment climate and the security of the supply chain.

The idea for this book shaped up during an international conference held in September 2009 in Yaoundé, Cameroon, on 'Geological Resources and Good Governance in Central Africa'. Some 200 international experts from various fields like politics and science, the private sector and civil society came together and discussed the various demands being placed on good governance and transparency in Sub-Saharan's raw materials sector and the prerequisites that must be met, and how future challenges will be addressed.

On invitation, the Economic and Monetary Community of Central Africa CEMAC Commission, and representatives of all six CEMAC countries and international organisations helped to set the issues in the wider pan-African and global context.

The conference was organised by the CEMAC REMAP project (Renforcement de la governance dans le secteur des matières premières en Afrique Centrale) and AgenZ of the Deutsche Gesellschaft für Technische Zusammenarbeit (GTZ—German technical cooperation, which became in 2011 the GIZ, German international cooperation). The conference was generously supported by the German Federal Ministry for Economic Cooperation and Development (BMZ). It was also implemented by the German Federal Institute for Geosciences and Natural Resources (BGR).

Unfortunately, the originally planned 'big' conference volume subsequently received at the beginning of the editing process only a few properly drawn up papers that fulfilled the requirements for being considered in a scientific book. Therefore, the editors had to search for additional authors, and it took the Department of Physical Geography at Goethe-University Frankfurt, Germany, and IREN, Inter Region Economic Network, Nairobi, Kenya, another year to bring together the 25 manuscripts now presented in this edition. Of course these contributions are covering only a part of the complex situation facing the Sub-Saharan geological resources and its extractive sector. However, it is a promising beginning on the multi-stakeholder level.

The editors express their gratitude to all the contributors of the book and to the above mentioned organisations and institutions for their participation and support.

Jürgen Runge & James Shikwati
Bangui/Frankfurt/Nairobi, March 2011

Geological Resources and Good Governance in Sub-Saharan Africa – Runge & Shikwati (eds)
© 2011 Taylor & Francis Group, London, ISBN 978-0-415-58267-4

Opening words

The countries of Central Africa are among both the most resource-rich and the poorest on the continent, as they have so far been able to harness only little notable development impetus from their abundance of geological resources. The Commission of the Economic and Monetary Community of Central Africa (CEMAC) has placed improvement of the framework for economic development from natural resources on its agenda. In the meantime, five of the six member states (Cameroon, Central African Republic, Chad, Congo, and Gabon) have joined the Extractive Industries Transparency Initiative (EITI) in order to bring their government revenues from oil production and mining out into the open. In March 2011 the Central African Republic became the first compliant country within CEMAC. Other Sub-Saharan countries such as Liberia, Ghana, Nigeria and Niger have also succesfully established the EITI transparency standard which underlines the growing acceptance and the importance of the process.

The CEMAC Commission was given the task of advancing implementation of the EITI in the member states. In response to a request by the CEMAC Executive Secretariat in 2006, Germany has been advising since 2007 the regional organisation by the REMAP project (Strengthening governance in the extractive sector), and focusing on the following areas: 1) Increasing the value added from geological resources and improving their contribution to poverty reduction, 2) Promoting transparency and Good Governance, and 3) Intensifying international cooperation in the field of mining and energy in Central Africa.

The proceeds from the export of resources could play a key role in reducing poverty and securing the future of the Sub-Saharan region. Central Africa in particular and the CEMAC region almost 80% of export revenue comes from the extraction of oil and gas. The key question is how to make the most socially equitable use of this wealth from resource trade and ensure it benefits the countries and their people?

The support by the German Federal Ministry for Economic Cooperation and Development (BMZ), together with CEMAC's partners, supports the reforms in the member states that aim to make both the extraction of geological resources and the revenues from the extractive sector transparent, and to increase their development benefit. National implementation of the Extractive Industries Transparency Initiative in the member states has an important role to play in this connection.

With these issues in mind it is my pleasure to thank all the contributors to this book that will certainly assist to stimulate further discussions on the future potential of geological resources in the CEMAC member states and the wider Sub-Saharan region.

Dr. Bernard Zoba
Commissioner for Infrastructure and Sustainable Development CEMAC Commission
Bangui, in April 2011

 ISBN 978-0-415-58267-4

Acknowledgements

The editors like to express their gratitude to the following collaborators and colleagues at Goethe-University, Department of Physical Geography, Frankfurt, and at the Inter Region Economic Network (IREN), Nairobi, for assisting in the realization of this book: Eva Becker, Erik Hock, Josephat Juma and Purity Njeru for translations, corrections and proof-reading; Joachim Eisenberg for making the sketch of the book cover. Erik Hock for setting the text and correcting the contributions. Ursula Olbrich for cartographic and art work on numerous figures. Hilal Sertdere for assisting the editorial work; and Zaina Shisia for organizing James Shikwati's travel to Cameroon and for assisting in the presentation at the conference in Yaoundé in September 2009.

Geological Resources and Good Governance in Sub-Saharan Africa – Runge & Shikwati (eds)
© 2011 Taylor & Francis Group, London, ISBN 978-0-415-58267-4

Contributors

Ambomo, E.P.
EITI/BEAC, B.P. 1917 Yaoundé, Cameroon. Email: amepas2000@hotmail.fr, Phone: +237 77 05 93 16.

Asobie, H.
NEITI Chairman (Nigeria) and International EITI Board Member, Ruseløkkveien 26, 0251 Oslo, Norway. Email: hasobie@yahoo.com

Babies, H.G.
Federal Institute for Geosciences and Natural Resources (BGR), Stilleweg 2, 30655 Hannover, Germany. Email: HansGeorg.Babies@bgr.de

Bomba Fouda, A.
Expert, extractive industries data management, CEMAC SCTIIE Secretariat. Email: bombafo@yahoo.fr, Phone: +237 77 75 83 57.

Boukinda, I.
CEMAC, Department of Human Rights, Good Governance and Human and Social Development, B.P. 969, Bangui, Central African Republic. Email: boukinda_isabelle@yahoo.fr, Phone: +236 75 76 2156

Buchholz, P.
Geologist, Federal Institute for Geosciences and Natural Resources (BGR), Hannover, Germany. Email: peter.buchholz@bgr.de, Phone: +49 511 643 2518.

Dibeu, E.
Senior trade commissioner, Canadian High Commission, P.O. Box 572, Yaoundé, Cameroon, Email: eric.dibeu@international.gc.ca, Phone: +237 2223 2311.

Großmann, M.
Deutsche Gesellschaft für internationale Zusammenarbeit (GIZ), Dag-Hammarskjöld-Weg 1–5, 65726 Eschborn, Germany. Email: matthias.grossmann@giz.de, Phone: +49 6196 79 6353, Fax: +49 6196 79 806353.

Häßler, R.D.
Director, Product and Market Development, oekom research AG, Goethestr. 28, 80336 Munich, Germany. Email: rolf.haessler@oekom-research.de, Phone: +49 89 544184 57.

Hinton, J.
Small Scale Mining Consultant, P.O. Box 257, Entebbe, Uganda. Email: jenniferhinton@gmail.com, Phone: +256 782 913 315.

Ingram, V.
Centre for International Forestry Research (CIFOR) Central Africa Reginal Office, Yaoundé, Cameroon. Email: v.ingram@cgiar.org, Phone: +237 22 22 7451.

Kalenga, M.-A.
Regional Director, EITI International Secretariat, Ruseløkkveien 26, 0251 Oslo, Norway. Email: mkalenga@eiti.org, Phone: +47 222 42111.

Kema, K.
African Medical and Research Foundation. Email: info.tanzania@amref.org

Khamala, C.A.
Africa Nazarene University, Nairobi, Kenya. Email: chalekha@yahoo.co.uk, Phone: +254 (20) 0700 735 756.

Koyassé, F.-A.
Senior Program Officer, World Bank, Yaoundé, Cameroon. Email: fkoyasse@worldbank.org

Levin, E.
Estelle Levin Ltd., 25 Cockburn Street, Cambridge, CB1 3NB, UK. Email: estelle@estellelevin.com

Mtegha, H.
Senior Lecturer, School of Mining Engineering, Univeristy of Witwatersrand, Johannesburg, South Africa. Email: Hudson.Mtegha@wits.ac.za

Musingwini, C.
Senior Lecturer, School of Mining Engineering, Univeristy of Witwatersrand, Johannesburg, South Africa. Email: Cuthbert.Musingwini@wits.ac.za

Müller, M.
Researcher, Bonn International Center for Conversion (BICC), Bonn, Germany. Email: mueller@bicc.de, Phone: +49 228 911 9664.

Ndeki, L.
African Medical and Research Foundation, P.O. Box 10252 Mwanza, Tanzania. Email: lndeki@hotmail.com, Phone: +255 754 376946.

Ndikumagenge, C.
International Union for the Conservation of Nature (IUCN-PACO), Yaoundé, Cameroon. Email: cleto.ndikumagenge@pfbc-cbfp.org, Phone: +237 22 21 6496.

Ngarsandje, G.
Concordia University, Montreal, Quebec, Canada, 736 Rue Belanger, #402, Saint-Jerome, QC, Canada J7Z 1A5. Email: guelmbaye_ngarsandje@globalunification.com, Phone: (1) 450 275 7112, (2) 450-592-4927.

Ngakola, C.
Géomatitien REMAP, Renforcement de la gouvernance dans le secteur des matières premières en Afrique Centrale, CEMAC, B.P. 930, Bangui, Central African Republic, Email: ngakolaalpha@yahoo.fr, Phone: +236 75 32 03 75.

Oshokoya, O.
MSc Student, School of Mining Engineering, University of Witwatersrand, Johannesburg, South Africa. Email: tomi.oshokoya@gmail.com

Okedi, J.P.
Estelle Levin Ltd., 25 Cockburn Street, Cambridge, CB1 3NB, UK.

Pfeiffer, B.
Geologist, Dornröschenweg 3, 50997 Köln, Germany. Email: brittamoni@yahoo.de

Rungan, S.V.
Lecturer, School of Mining Engineering, Univeristy of Witwatersrand, Johannesburg, South Africa. Email: sivalingum.rungan@wits.ac.za

Runge, J.
Professor, Goethe-University Frankfurt, FB 11: Institute of Physical Geography, Altenhöferallee 1, 60438 Frankfurt, Germany. Email: j.runge@em.uni-frankfurt.de, Phone: +49 69 798 40160.

Schnell, M.
Deutsche Gesellschaft für internationale Zusammenarbeit (GIZ), Dag-Hammarskjöld-Weg 1–5, 65726 Eschborn, Germany. Email: maya.schnell@giz.de, Phone: +49 6196 79 6353, Fax: +49 6196 79 806353.

Schure, J.
Centre for International Forestry Research (CIFOR) Central Africa Reginal Office, Yaoundé, Cameroon. Email: j.schure@cgiar.org, Phone: +237 22 22 7449.

Sekule, P.
African Medical and Research Foundation. Email: info.tanzania@amref.org

Shikwati, J.
Director, IREN—InterRegionEconomicNetwork, Nyaku House, Argwings Kodhek Road, Hurlingham, P.O. Box 135 00100, Nairobi, Kenya. Email: james@irenkenya.org, Phone: +254 20 2731 497, +254 733 823 062.

Snook, S.
Senior Associate, Land Tenure and Property Rights, Tetra Tech ARD, 159 Bank Street, Suite 300, P.O. Box 1397, Burlington, VT 05402, USA. Email: stephen.snook@tetratech.com, Phone: + 1 802 658 3890

Stürmer, M.
Rheinische Friedrich-Wilhelms-University of Bonn, Bonn, Germany. Email: martin.stuermer@uni-bonn.de, Phone: +49 228 73 4032.

Surma, A.
Estelle Levin Ltd., 25 Cockburn Street, Cambridge, CB1 3NB, UK. Email: agatasurma@gmail.com

Temu, F.
African Medical and Research Foundation. Email: info.tanzania@amref.org

Tieguhong, J.C.
Technical Training and Research Centre for Development (TTRECED) Yaoundé, Cameroon. Email: chupezi@yahoo.co.uk, Phone: +237 22 23 7204.

Tourere, Z.
Researcher, Centre National d'Education (CNE), Ministere de la Recherche Scientifique et de l'Innovation, B.P. 1721, Yaoundé, Cameroon. Email: ze_na_bou@yahoo.fr, Phone: +237 99 65 46 19, +237 22 23 40 12, Fax: +237 22 20 19 72.

Villegas, C.
Estelle Levin Ltd., 25 Cockburn Street, Cambridge, CB1 3NB, UK.

Warner, N.
LEITI National Coordinator (Liberia) and International EITI Board Member, Ruseløkkveien 26, 0251 Oslo, Norway. Email: tnwarner@leiti.org.lr

Geological Resources and Good Governance in Sub-Saharan Africa – Runge & Shikwati (eds)
© 2011 Taylor & Francis Group, London, ISBN 978-0-415-58267-4

Holistic approaches to transparency and sustainable development in Sub-Saharan Africa: Background and objectives

J. Runge
Department of Physical Geography, Goethe-University Frankfurt & CIRA, Centre for Interdisciplinary Research on Africa, Frankfurt am Main, Germany

J. Shikwati
IREN, Inter Region Economic Network, Nairobi, Kenya

1 INTRODUCTION—WHY IS IT IMPORTANT?

The states of Sub-Saharan Africa in general and the Congo Basin region in particular, are among the world's most resource-rich regions. However, many of these countries are a long way from attaining their development potential; and some of them are among the least developed in the world. Paradoxically, often those countries that are most richly endowed with resources are the least developed. This phenomenon is exacerbated in many African countries by inadequate governance. And yet if the state is unable to provide basic services, social and environmental standards in the extractive and processing sectors will not be enforced.

In connection with the massive increase during the last 10–15 years in emerging countries' demand for raw materials, an unprecedented situation has been created. There is pressure on supplies and a sharp rise in metal prices, with an average, threefold increase since 2002. The market capitalisation of the major trans-national mining corporations has followed this trend. There is evidence that the rise in metal prices is likely on average to be a long-term one. The upward revaluation of mining assets held by companies and states increases the competition between operators to acquire rights to new mineral reserves and to secure transport routes. In Africa, this global competition generates tensions of various sorts—commercial, territorial and social.

Also mineral extraction has contributed to violent conflicts, and most of the countries in the region lack the political will or resources to ensure compliance with national and international norms and regulations. Conflicts have been fuelled by access to revenues from mineral resources, encouraging corruption, harmful environmental practices, and neglect of basic human rights.

In addition, high mineral prices are encouraging the development of mineral deposits, including the development of previously unviable deposits—companies are increasingly willing to invest substantial resources into developing mineral fields, including major infrastructure construction which will have further negative effects on environmental degradation.

Governments that rely on oil wealth may be facing shortages that encourage non-oil mineral development or pressures to increase logging or agricultural output as to develop income-generating alternatives to oil revenues. This means incursion into tropical forests or other sensitive areas that are rich in biodiversity.

Often it has been argued that there is an association between resource riches and poor economic performance (the 'paradox of plenty' or the 'resource curse'), and a huge amount of scientific literature from different disciplines has been published seeking to explain the relationships between resource abundance and economic performance. But obviously the resource curse is not inevitable. While there are many examples of poor management of resource wealth, a range of countries like Botswana, Canada, Chile, and Norway appear to have avoided these problems through prudent and transparent management practices. The key question for a large number of countries is how they can ensure that their abundance in resources remains a blessing.

According to the International Monetary Fund (IMF) today over 50 countries can be designated as rich in hydrocarbon and mineral resources. Many of these countries are located in Africa with a low and middle-income structure in which resource revenue accounts for over 50% of government revenue or export proceeds (for oil producing states). In addition to the possible adverse impact on growth, resource riches have been seen in many such cases as an important factor contributing to corruption and social unrest. In a number of countries, oil, diamonds (and timber) are associated with causing and financing civil war with its attendant social and economic costs. Given these potentially substantial costs of non-transparent practices, institutional strengthening to improve transparency in vulnerable resource-rich countries (e.g. by the Extractive Industries Transparency Initiative, EITI standard) should provide an ample pay-off for a relatively modest investment. Many analysts have emphasised the essential role played by fiscal transparency in improving resource revenue management. In the last several years, moreover, considerable agreement has been reached on a wide variety of good resource governance (particularly for oil and gas,) and revenue management practices.

2 MAJOR OBJECTIVES

This book publishes original manuscripts to give an update of the distribution, occurrences, potential and prospects of geological resources for sustainable development in Sub-Saharan Africa. By bringing together and discussing numerous different points of view, it follows a holistic, interdisciplinary and scientific approach. Even if the procedure is academic, this volume addresses turns to researchers and stakeholders in the field (including civil society), planners, politicians and decision makers.

Increased transparency in the extractive sector is widely recognised as crucial for achieving accountability, good governance and sustainable economic development. It benefits stakeholders in the sector—for resource-rich countries and their residents, the already industrialised or emerging countries, and for both private and state-owned enterprises. Increased transparency and good governance are also in the interest of importers of raw materials as they increase confidence in the effectiveness of public institutions and political stability. They improve the investment climate and the security of the supply chain.

The idea for this book shaped up during an international conference held for two days in September 2009 in Yaoundé, Cameroon, on 'Geological Resources and Good Governance in Central Africa' (www.yaounde2009.net). Some 200 international experts from various fields like politics and science, the private sector and civil society came together and discussed the various demands being placed on good governance and transparency in Sub-Saharan's raw materials sector and the prerequisites that must be met, and how future challenges will be addressed. New forms of intersectoral, transnational governance offer ways to take account of all the different stakeholder interests in the resources sector. Such novel approaches require institutionalised and constructive dialogues between governments, civil society and international companies operating in the resource sector. At invitation of the Economic and Monetary Community of Central Africa CEMAC Commission, representatives of all six CEMAC countries and of international organisations helped to set the issues in the wider pan-African and global context.

Some specific objectives of the book are: (1) to demonstrate the vast geological richness and potential of Sub-Saharan resources (as far as it is known to former and recent exploration; (2) to identify economic policy measures that can best turn these abundant natural resources into a motor for driving sustainable economic, social and environmental development; (3) to look at the economic and political conflicts that are brewing among the most powerful and weakest actors at local, national and international levels; including participation of civil society groups (e.g. NGOs); (4) to show how regional partnerships can implement the goals of good governance and thereby achieve transparent and efficient regulation, as it is proposed by the EITI and the Kimberley Certification Scheme (KCS); (5) to explain how local and equitable partnerships between indigenous communities and informal (artisanal mining) operators are indeed possible; and finally, (6) to define how the societal problem of the HIV/AIDS pandemic can be actively combated at the local level.

3 CONTRIBUTIONS TO THIS BOOK

Subsequently to the Yaoundé conference of September 2009, 25 papers which are different in scope and size in total were received. Not all aspects of geological resources and good governance were covered by them, for instance there is no contribution that tackles on the role of China in Africa's mineral sector. Furthermore, the environmental impact by mining activities on ecosystems has not been considered in such a prominent way as it was originally requested and planned by the editors.

Four major sections have been grouped within this book:

- Geological resources and governance;
- Case studies from different countries;
- Regional approaches and activities (including civil society groups and mining companies); and
- Artisanal mining, gender, and HIV/AIDS.

The first chapter contains a general and introductory discussion by James Shikwati, IREN, on 'How geological resources can aid Africa's development'. He outlines the historical dimension of mining in Africa and the colonial impact on the sector up to the requirements that are necessary to assist Africa's economic growth today.

Peter Buchholz and Michael Stürmer give an overview of known geological resources in Sub-Saharan Africa and they discuss the potential for tax revenues from the extractive sector by specific data. They highlight that to date, there is no medium and long-term projection on the difference the extractive sector can make from tax revenues, especially for non-major oil producing Sub-Saharan countries. They offer a methodology for estimating government revenues in Ghana, Zambia, Namibia and Mocambique from the extractive sector, illustrating that mining can have a significant impact to achieving the MDGs.

Alphonse Bomba Fouda from the CEMAC Central Bank and the extractive industries data management for the SCTII in Yaoundé focuses on the need for transparency along the value chain in Central Africa. He shows that the levelling-off of reserves in some CEMAC countries, the slowing down of oil revenue and the attention being paid to critics denouncing not only the industry's political, social, economic and environmental impact but also the opaque management of oil revenue, may encourage the oil-producing countries of Central Africa to move away from revenue-based economies.

Edwige Pascaline Ambomo contributes by reflections on 'Investments in mining industries for sustainable development'. She points out that the exploitation of the mineral wealth by mining companies should constitute an asset for the economic growth of the host countries. The question of 'where does the money generated by natural resources and extractive resources in particular (oil, gas, ore) go, mentioning the importance of the EITI mechanism that can offer a contribution to sustainable good governance.

A review article by Maya Schnell and Matthias Großmann from the Deutsche Gesellschaft für internationale Zusammenarbeit (GIZ) on 'International approaches to improve resource governance in Africa' sums up different elements of resource governance as highlighted by the World Bank Extractive Industries Value Chain, and present an overview of international and bilateral approaches to support resource governance in Africa. They emphasised that most support approaches focus on the regulation and monitoring of the extraction, processing and trade of resources as well as on the collection of taxes and royalties. Yet, resource governance should be understood and addressed in the broader context of good (democratic) governance and should focus on all steps of the Extractive Industries Value Chain.

The second main chapter of the book presents regional, country case studies. From the German Federal Institute for Geosciences and Natural Resources (BGR), Hans-Georg Babies' and Britta Pfeiffer's paper on 'Uganda's oil boom: Potential and risks'. This contribution analyses the development of the Ugandan petroleum sector, and it describes the geological potential, gives an evaluation of expectations relating to the benefits and characterises challenges of oil production.

Another emerging (and not undisputed) oil producer in Sub-Saharan Africa is Chad. It is discussed by Guelmbaye Ngarsandje in the paper 'World Bank's failure in the Chad-Cameroon oil project'. He describes the problems and conflicts that arose with World Bank when the Chadian government used the oil bonus to purchase arms to fight rebels at the Sudanese border. The author highlights and condemns how the Chadian government fooled all its partners in the project and made the people lose their benefits because of nepotism, clientelism and also by the incapacity of World Bank to coerce the government to respect its engagement.

Also strongly criticising the outcome of the EITI process in Nigeria (NEITI) is the topic of Marie Müller's paper titled: 'Turning the curse in a blessing: A convenient Illusion. Lessons learnt from the Nigeria EITI (NEITI) process'. She points out that the analysis of the conflict in the Niger Delta shows that revenue transparency is partially relevant to solving the conflict, although its origins involve issues that are beyond transparency. Her contribution delineates clear limitations to NEITI's potential, however, arising from its design and its being embedded in the political context of a 'petro-state'.

Jürgen Runge introduces measures of the technical cooperation to implement EITI structures, transparency and multi-stakeholder dialogues in the CEMAC sub-region. Besides, introducing the REMAP project, he points out by the comparative example of the quite heterogeneously shaped countries of Equatorial Guinea (professional oil exploitation) and the enclaved Central African Republic (informal diamond mining), some challenges facing the EITI standard and the limits of the process.

The case of the Tiomin titanium mine in Kenya is studied from a juridical perspective ('Whither Communities?') presented by Charles Khamala. He explains that the Kenyan government used its superior bargaining power to buy the relevant private land. In consequence, dissatisfied landowners omitted to pursue any public law remedies in appropriate forums. Instead, at successive proceedings not only were their collective sale contracts upheld, but also individual landowners were estopped from rescinding or invoking their constitutional rights against forced expropriation. Information scarcity deprives communities from participation.

He concludes that also environmental law requires common law reinforcement to attain good governance of polycentric natural resource disputes.

The paper by Rungan, Musingwini and Mtegha on 'Good governance, transparency and regulation in the extractive sector' is devoted to aspects of sustainable development. The authors examine the different initiatives directed towards the implementation of Sustainable Development (SD) internationally. Using South Africa's case study, they discuss the various legislative developments to regulate SD, such as the Mineral and Petroleum Resources Development Act, Mining Charter and environmental legislations. In conclusion, they emphasise the proactive role of the government that is essential.

Marie-Ange Kalenga, with colleagues from the EITI international board, evaluate first results of successful EITI implementation in Africa. Increased transparency also strengthens accountability and can promote greater economic and political stability. EITI training activities carried out by the World Bank and bilateral development partners, help to build management capacity in government institutions involved in the extractive sectors, as well as inform civil society groups about key aspects of the sector. These factors can enhance the investment returns from companies operating in EITI countries for investors, boosting a country's attractiveness as an investment destination.

Another contribution presented by Zenabou Tourere from Cameroon's government political science and planning division focuses from the political perspective, the role of investments done by extractive industries companies for sustainable development in the Central African region. Zenabou gives an engaged speech for the benefit of extractive industries to the people by expressing: 'the stakes in the sustainable development are the ones of the survival of the population of Central Africa even that of the humanity in the planet'.

Leaving this case-study oriented section of the book, the following topics address the roles regional political organisations as well as the private sector and civil society can play in strengthening good governance and transparency in Sub-Saharan Africa.

Hudson Mtegha et al. explain why regional organisations (e.g. EAC, ECOWAS, SADC) contributing to mining and exploitation are necessary and the kind they should be. They highlight that cross country approaches are essential and efforts have to be made by RECs to make an efford to develop common mining codes. Also, there is need to harmonise national policies, laws and regulations.

Another African regional organisation, CEMAC (Communauté Economique et Monetaire de l'Afrique Centrale), is introduced by Jürgen Runge on its role within the implementation of EITI in the six member states. He highlights that good resource governance depends strongly on ownership and political will that is expressed by professional political leaders.

The concept of Corporate Social Responsibility (CSR) in Cameroon supported by the Government of Canada and applied in the field by Canadian mining companies is illustrated by Erik Dibeu from the Canadian High Commission in Yaoundé, Cameroon. He presents a case study of a CSR initiative program provided by Alucam, Rio Tinto Alcan's subsidiary in Cameroon. Dibeu demonstrates that almost all CSR activities are taken as economic result-oriented measures and not as selfless philanthropy.

Another interesting approach on how to measure, evaluate and rate the social and environmental quality of commodities (gold) is presented by Rolf D. Häßler, oekom Research, Munich, Germany. By carrying out a risk-performance classification, a qualitative evaluation tool was set up and applied for different international and national mining companies. It can generally be stated that the social and environmental standards of large mining corporations are better than those of smaller local mine operators. However, this is not true for human rights standards.

The final sequence of this book focuses on artisanal mining (institutional, legal and fiscal regimes), which is often still the only way exploration of geological resources is carried out in the wider study area. Gender questions in the mining sector are also tackled, and finally the public health and HIV/AIDS conditions in informal mining communities are described and studied.

Alphonse Bomba Fouda discusses if artisanal mining is a benefit or a burden for sustainable development? He explains that this phenomenon of informal mining is, to a great extent, the result of the impoverishment of the population and/or economic crises. It has boomed within a context of economic liberalisation and globalisation.

Jolien Schure et al. (CIFOR, Yaoundé, Cameroon), examine 'Institutional aspects of artisanal mining in forest landscapes' in the environs of the high-biodiversity forest of the Congo Basin's Sangha Tri-National Park. From literature review, interviews and site visits, they conclude that diamonds and gold are an important but highly variable income source for at least 5% of the area's population. Environmental impacts are temporary and limited, mainly caused by mining inside the parks; overlaps between artisanal small-scale mining, large-scale mining, timber concessions, and national and trans-boundary protected areas have intensified competition for land resources.

Jennifer Hinton, Estelle Levin and Stephen Snook study different legal and fiscal regimes in the artisanal mining sector in 10 countries with regional focus on the Central African Republic (CAR). The study was carried out within a USAID project on the property rights of informal diamond miners (PRADD) in CAR. They conclude that other fiscal success factors include regional tax harmonisation, simple reporting procedures, stronger investment in artisanal diamond mining support services and tax collection at the point of export. Conducive fiscal provisions must nevertheless be coupled with legislation that sensibly recognises capacity constraints of miners and existing work arrangements on the ground. However, also within this study, the biggest challenge seems to lie in the institutional commitment needed to implement good governance in the resource sector.

The role of women in the artisanal mining sector is dealt with by Isabelle Boukinda from the CEMAC Commission in Bangui, Central African Republic and Jürgen Runge. Their paper explains against the background of the Millennium Development Goals (MDGs) why women are also so important in the extractive sector. Some reflections on CEMAC's responsibility and ownership in this important forward-looking economic field are highlighted and propositions are made on how women could be best supported by a project in small scale mining.

Finally, there are two papers focussing on the HIV/AIDS situation in commercial and informal small-scale mining. There are many reasons for the increased prevalence of HIV/AIDS in African mining areas, most of which are socio-economic in nature. Miners, the majority of whom are men between 18 and 49, often live apart from their families in order to work at mining sites. Prostitution therefore flourishes in mining areas—partly because by comparison with the rest of the population, miners are relatively well paid. The situation is exacerbated by limited knowledge on how the virus is spread and the miners' relatively low risk-awareness. Leonard Ndeki and colleagues from the African Medical and Resarch Foundation (AMREF), Mwanza, Tanzania, report on a National Multisectoral Strategic Framework (2008–2012) programme emphasizing the importance of scaling up of Public Private Partnerships (PPP) to complement the national HIV and AIDS response. By the example of the Lake Zone mine site of Anglo Ashanti Geita Gold mine there was evidence that such partnerships build sense of owned community resources supported by different HIV/AIDS workplace projects that brings results of impact to be sustainable on the long run.

The other HIV/AIDS study is by the REMAP CEMAC project (Jürgen Runge and Celestin Ngakola) on prevalence of artisanal miners in Central African Republic. It aimed the empirical and voluntarily testing of more than 3000 miners in the village of Bossui. Seroprevalence in Bossoui is high and varies between 8% and 23%. In consequence, the authors propose and recommend for the informal sector to set up more HIV/AIDS workplace programmes in rural mining areas and to make a better international support to the public health sector in CAR.

Many of the contributions in this book illustrate that mineral extraction can conflict with other development objectives such as the conservation and sustainable management of other natural resources. Therefore, a consensus needs to be achieved before mining starts. Support should be given for consultation and decision procedures among stakeholders (administrations, companies and civil society) and the studies required for informed decision-making that considers negative externalities and technical means for limiting or compensating them. Additionally, geological resource and the extractive industries sector does not only involve major companies, as it is often believed in a broader international public. In many regions of Sub-Saharan Africa artisanal mining is still the common way how geological resources are exploited.

REFERENCES AND FURTHER READING

African Union (AU) 2009. *Africa 2050 mining vision (preliminary document)*. p. 51.

Communauté Economique et Monetaire de l'Afrique Centrale (CEMAC) 2010. *Ressources géologiques et bonne gouvernance en Afrique Centrale, 24–25 September 2009, Yaoundé, Cameroon*, p. 37.

Interministerial Committee for International Cooperation and Development (CICID) 2008. *Mineral resources and development in Africa. Strategic guideline document*. CESMAT, Paris, p. 47.

International Monetary Fund (IMF) 2005. *Guide on resource revenue transparency*. Washington, DC, USA, p. 77.

Reed, E. & Miranda, M. 2007. *Assessment of the mining sector and infrastructure development in the Congo Basin region*. WWF Macroeconomics for Sustainable Development Program Office, Washington, DC, USA, p. 27.

Spitz, K. & Trudinger, J. 2008. *Mining and the Environment*. Routledge/Taylor & Francis Publishers, p. 900.

Geological resources and governance

Geological Resources and Good Governance in Sub-Saharan Africa – Runge & Shikwati (eds)
 ISBN 978-0-415-58267-4

How geological resources can aid Africa's development

J. Shikwati
IREN, Inter Region Economic Network, Nairobi, Kenya

ABSTRACT: Mining played a key role in African economies as early as 2000 BC. West Africa and Zimbabwe produced gold that propped up the monetary and commercial systems of Western Europe and Indian Ocean trading zones. Recent exploitation of geological resources in Africa has been characterised by little or no accountability by governments during pre-colonial, colonial and post colonial Africa until the late 1990s. Insufficient information on mapped resources and data regarding the explored Sub-Saharan Africa's natural resources has restricted countries to poverty amidst vast geological wealth. Geological resources can help drive up economic growth on the continent through development corridors that inter link major cities and mining sites and enhanced bargaining skills with foreign investors. The surge in demand for and increase in commodity prices calls for an urgent review of governance, economic and mining policies to draw in more indigenous players into this sector.

1 INTRODUCTION

Mining of iron, copper, tin, gold and salt played a key role in sections of African economies as early as 2000 to 3000 years BC. According to Zeleza (2003), West Africa and Zimbabwe produced the world's gold supply that propped up the monetary and commercial systems of Western Europe and the Indian Ocean trading zones. Utilisation of geological resources and agricultural activities drove African ancient economies as evidenced by a series of archeological discoveries of furnaces and implements produced out of minerals. Recent exploitation of geological resources in Africa has been characterised by little or no accountability by governments during pre-colonial, colonial and post colonial Africa until the late 1990s, observes Frick (2002).

Insufficient information on mapped resources and data regarding the exploitation of Sub-Saharan Africa's natural resources has restricted countries to poverty amidst vast geological wealth. Quashie (1996) observes that colonial governments put in place skewed mining leases that ensured that they retained mineral rights and tenure to the disadvantage of Sub-Saharan countries. The mining sector was used to develop the economies of western nations and the situation did not change even in the post colonial era (United Nations Economic and Social Council and Economic Commission for Africa 2009). Africans have largely been confined to the informal business sector due to reasons such as colonial legacy, post colonial paternalistic and ethnic political economy and the high cost of formalising business notes Kadenge (undated). Increased global commodity prices do not capture the limited returns to African people due to the suffocating effect of mining agreements signed between African governments with foreign investors.

Africa experienced a surge in investment in commercial exploration for mineral resources that rose from US$575 million (14% of World's commercial exploration) in 2000 to US$1.7 billion in 2007 (Raw Material Group 2009). Exploration interests have grown in line with increase in commodity prices and demand for geological resources especially by the emerging and re-remerging economies such as China and India. The renewed interest in geological resources can help drive up economic growth on the continent through access to relevant data from mineral explorers; creating special development corridors that inter link major cities and mining sites, and enhanced bargaining skills with foreign investors.

Geological resources have the potential to drive economic development in Africa through reviewed global market systems that free resources to enable value addition on the continent. African governance system that currently focuses more on establishing a suitable environment for external investors, whilst relegating its citizenry to informality also has to be reviewed.

Transformation of virgin geological resources into valuable products requires intellectual capital, financial resources and long term focus. Africa ought to take advantage of the competition between developed, emerging and re-emerging economies to invest in her people and establish effective financial systems to transform geological resources into key drivers to prosperity. The surge in demand for and increase in commodity prices calls for an urgent review of economic and mining policies to draw in more indigenous players into this sector. The continent's governance system has to focus on the long term because, as Atlas (2010) observes mining projects can span up to 10 years or more and financing of explorations are driven by demand.

2 BACKGROUND

Geological resources have been synonymous to Sub-Saharan Africa's curse of underdevelopment. Weak political governance and institutional structures on the continent transformed the vast valuable geological resources into a nightmare for the citizenry. The fraudulent nature, with which European nations seated in Berlin initiated the birth of the current African nations in 1885, has continued to impact negatively on how the African elites manage treaties and concessions for access to geological resources.

European powers when creating African nations, were keen to exploit the continent's resources. At independence of African nations; European powers negotiated their way into installing governments that would safeguard their interests—African populations did not have an avenue to engage and create a legitimate social contract with the new governance systems (since African negotiators for independence had been oriented to colonisers education system and worldview that conflicted with expectations of those they represented).

Africans are still grappling with the question of the legitimacy of their governments and governance systems. This quest has continued to be overshadowed by interests of both developed and emerging economies that take advantage to ensure easy access to resources. As developed nations jostle to control sources of valuable geological resources; they have stationed foreign armed forces, military bases and peace keeping forces in the region. The increase in demand for natural resources by emerging economies is likely to sustain old commercial, territorial and social tensions.

It therefore does not come as a surprise that according to FAO more than 40% of people in Sub-Saharan Africa live on less than US$1 a day; while more than 70% live on less than US$2 a day. Contrast this with the fact that Sub-Saharan Africa governments preside over vast geological resources. Beri and Uttam (2009) observe that the continent has 9.7% of world proven oil reserves, 7.8% of world's total natural gas and 6% of world's proven coal reserves. Africa has about 30% of the planet's mineral reserves including 40% of gold, 60% cobalt, 40% of hydroelectric power and 90% of World's Platinoid Group of Metal reserves. Africa's contribution to the world's major metals (copper, lead and zinc) is less than 7%; Angola alone is said to have 11% of the world's known diamond reserves (Ayittey 1998).

Sub-Saharan Africa is faced with a great challenge to transform governance institutions originally designed to safeguard colonial interests to those that will address the citizenry's interests (see also App. 1). Such a transformation will provide a platform for geological resources to play a key role in the development of the region. Without governance and institutional transformation, the region finds itself bedeviled with political instability and conflict. An urgent paradigm shift is required to move away from focusing on short term electoral politics as a gauge for democracy and good governance, to politics that focus on the long term and one that promotes a knowledge driven participation in investment and wealth creation in their countries.

Extraction and eventual transformation of virgin rock and natural fluids into items of high value requires huge capital investment, technological know-how and huge environmental impact. Unfortunately, Africa's knowledge of geological resources on the continent is scanty. While seeking out African authors to contribute to this topic, I was shocked at how the energies and expertise of the continent's geologists has turned into writing about human rights, forestry and tourism. Over reliance on funding from foreign countries to determine research topics has literally compromised the continent's ability to assess and strategise on its geological wealth. Africa is invisible on world research map, accounting for only 1% of the world's research and development effort (UNECA 2001). The common trend is for trained African geologists to work for international corporations as an in country brain drain. There is little effort to build the continent's capacity to identify, map and transform geological resources into wealth. The general policy and financing infrastructure favours growth of the mining industry in wealthy nations, while sustaining Africans largely in the informal and artisanal mining ventures.

3 GEOLOGICAL RESOURCES AND THE STATE OF AFFAIRS IN SUB-SAHARAN AFRICA

Sub-Saharan Africa has experienced over 70 wars since independence. In the past 25 years, the continent has experienced ten high intensity conflicts that have directly or indirectly affected an astounding 155 million people (Lind & Sturman 2002). The quest to access and control valuable natural resources has been a crucial factor in the occurrence of violent conflicts across the continent. Conflict in Sudan has led to an estimated 2 million deaths since 1983. The Democratic Republic of Congo witnessed Africa's ever "First World War" that resulted to an estimated 1.5 million deaths in 1998 alone and 5.4 million by 2010, observes Shah (2010). Natural resources in Africa breed conflict, poor governance and push populations to extreme poverty, while powering Western country and re-emerging markets' industrial might.

As illustrated in Figure 1, the quest for geological resources has made human life so cheap to cost human blood to power the industries of the developed and emerging economies. Only South Africa, Botswana, Cameroon and Ghana record a relatively stable political situation, while the rest of Sub-Saharan Africa oscillates from active and severe tensions to zones of political instability and internal confrontations.

4 THE "GANDHIAN TRAP" AND INFORMALITY IN AFRICA

Africans remain bogged down by what I refer to as the "Gandhian Trap." The global market system has restricted Africa to exporting raw materials without value addition. Importing countries consequently grow their job markets, knowledge industry, and financial sector at the expense of Africa.

In *The New World Encyclopedia*, Mahatma Gandhi addressed a similar predicament in early 1900s in India, where Britain banned processed cotton cloth from India preferring to import raw cotton. Africa's focus on exporting raw materials to developed and emerging economies places it in the situation that India faced with her cotton exports; Indians were forced into subsistence whilst the industrial, financial and knowledge sectors of the United Kingdom expanded.

Africans thus export raw geological resources (like Indians did for cotton in the days of Gandhi) and import electronic and engineering products

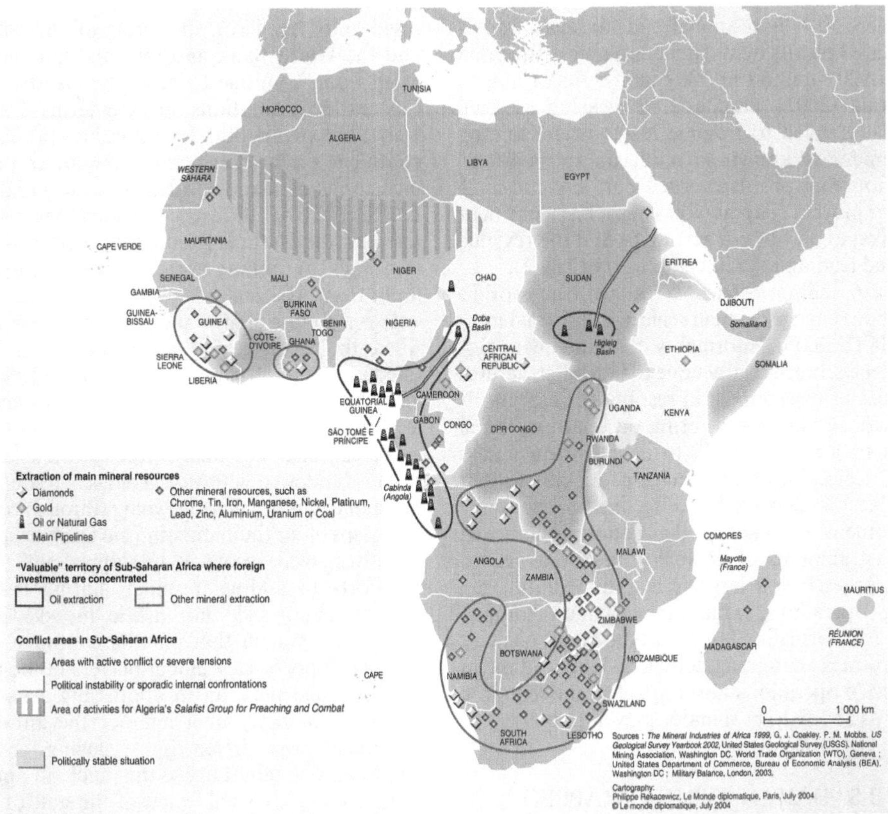

Figure 1. Mineral resources and political instability in Sub-Saharan Africa (Source: UNEP/GRID) (For a colour version see Colour Plate 1).

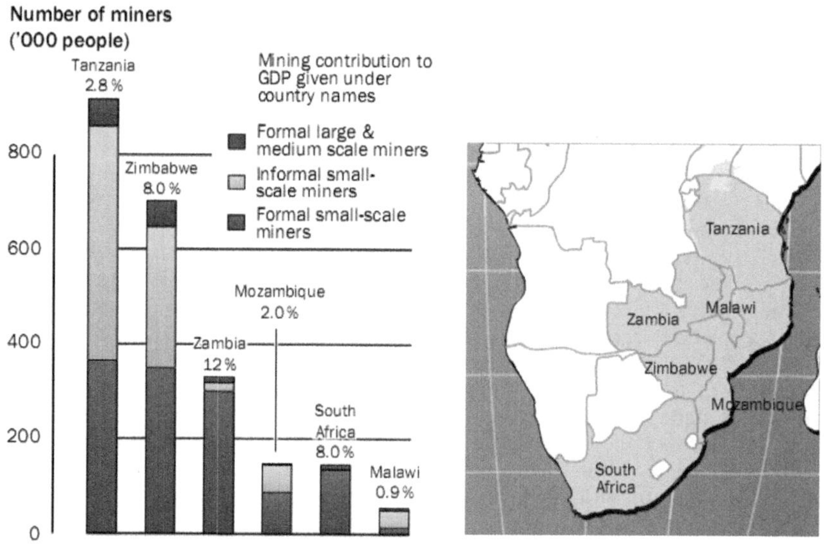

Figure 2. Number of people in the mining sector in selected countries (UNEP/GRID).

from developed and emerging economies. Worse still, Jomo (2005) observes that the continent looses over US$150 billion annually in tax evasion and undeclared profits by foreign companies; Tanzania lost US$207 million in 2007 alone. According to Mugarula (2001), emphasis in pleasing external investors has led to revenue losses where mining companies are known to remit up to 54.4% of taxes not from profits accrued but deductions of their employees' pay as you earn. The continent has leakages on both its resources and the revenue collected leading to artificial poverty (Fig. 2).

Over 6 million Africans (out of the world's 20 million) engage in small scale and informal mining (CICID 2005). Informality combined with global interests, breed conflict due to lack of structures and rules of engagement in resource utilisation. It is in African's interest to bring on board informal miners to formal systems to enable them access capital, technology and management skills so that natural resources contribute effectively to economic development. It's a shame that countries endowed with vast amounts of geological resources record minimal presence of large-scale and medium-sized mining operations, whilst they remain trapped largely in informality. Informality makes it difficult to keep track of mining activities. Uncoordinated small-scale operations also impact negatively on a country's ability to be a major player globally.

5 SUB-SAHARAN AFRICA TRAPPED IN HISTORY

Sub-Saharan Africa countries' history stretching to say 250 years back, paints a very horrifying scenario. Populations were enslaved and freed on the terms of their slaver; communities were colonised and "freed" on the terms of the coloniser; and the World Bank and the International Monetary Fund continue to prescribe disastrous policies and offer solutions on their terms. Countries whose companies plunder natural resources on the continent offer transparency initiatives on their own terms as demonstrated in such initiatives as the Extractive Industries Transparency Initiative and the Natural Resources Charter. Both are responses by mining sector players with limited input from Africans.

It is a picture akin to the water boarding torture, where the agent seeking information from the tortured uses water to simulate drowning and allows the victim to gasp for air (in this case both monetary and technical aid), because the intention is not to kill but get intelligence. For many African countries, more so those endowed with vast geological resources, the technique is to simulate drowning (through conflicts) and gasps of air (humanitarian aid and transparency initiatives), while resources are shipped out.

Efforts to address revenue sharing from raw material exports do not tackle the skewed global market system that pin the continent to raw material exports. Key global players in the mining industry take the lead in addressing evils of the extractive industry in a manner the anti-slavery movement was driven by enslaving countries. Although one might argue that such an approach ended slavery; a casual review of the state of Africa and former slave populated countries indicates a system that mutated to economic slavery instead. The same manner in which a slave's confidence was destroyed to make them subservient, plays out today on how donor countries push their policies onto Sub-Saharan countries in return for cheap access to geological resources.

6 HOW GEOLOGICAL RESOURCES CAN ASSIST SUB-SAHARAN AFRICA'S DEVELOPMENT

Any discussion on how geological resources ought to drive economic development must actively review existing unfair leases and mineral concessions. It has also to factor in reform of governance structure and clearly define role of state in geological resource exploitation. African must initiate a deliberate push to scale up small and medium sized enterprises on the continent. It is imperative that the African people re-examine the legitimacy of their governments and put in place constitutional mechanisms to ensure genuine people participation. Policy makers must also re-examine the legal framework that is currently skewed in favour of foreign investors at the detriment of indigenous people.

According to Page (2009), geological resources can be effective drivers of growth and poverty reduction in Africa if (a) the continent invests in building institutions that address the needs of the populace. Such systems have to limit the rampant discretionary powers enjoyed by the political elite and develop effective mechanisms of engaging external interests of both the developed and the re-emerging economies; (b) countries' use revenues accrued to diversify their economies and allow people to participate in and benefit from the growth; and (c) the country gets a fair price for its resources, has high transparency and accountability; saves in good times in anticipation of bad times, strengthens its public financial management and accountability and monitors and evaluates outcomes and report results. He adds that for a country to enjoy growth driven by natural resources, it should encourage:

- high savings,
- long term investments,
- continuous improvements in organisation, technology and management,
- investing in highly visible wealth sharing mechanisms, for example, universal primary education, rural development and basic healthcare.

Mining disrupts communities' subsistence economic activities and can be a threat in the long run when resources run out. Countries have to however guard against the fact that natural resource exports drive down exports from other sectors as countries focus on revenues from mining exports. The tendency to forget normal economic activity in favour of revenues from natural resources leads to lost growth (Shah 2009). Focus on excess revenues over all costs including normal profit margins from natural resource ventures leads to resource rent. For Africa to tap effectively into its geological resources, strategies must be put in place to escape the "natural resource trap." The trap is described by Collier (2007) in terms of exports that increase value of an exporting country's currency against other currencies; which make other export activities from the same country to be uncompetitive, consequently leading to abandoning of its tradable products in favour of revenues from geological resources. Surplus from natural resources exports reduces growth and excess revenues over all costs, leading to natural resource "rent" which drives up corrupt activities in countries with weak governance institutions.

To escape the trap, Africa ought to consider use of development corridors that link intra continent cities and mining nodes to spur development. Thomas (2009) describes development corridors are described as transport (or trade) corridors with underutilised economic potential which can spur growth of inter-related infrastructure leading to economic growth. Negotiations for mineral concessions ought to factor in possibilities of infrastructure supporting adjoining economic activities in order to spur growth on the continent.

Revenues from geological resources should be used to diversify Africa's economic activities during boom time. Sub-Saharan Africa countries ought to go beyond revenue sharing, diversification and include reform governance structures that facilitate Africans to consciously sign onto a social contract with governments that serve their interests and desires. A revamped governance structure will ensure that citizenry of resource rich countries are well equipped with geological knowledge and knowhow.

Investment in research is the key to enable the continent's population create efficient means of profitable exploitation and responsive organisational structures that ensure delivery of public goods. African governments must equip their populations with education that facilitates talent and knowledge expansion; accountable to their people and one that embraces an open global market system (not the current system that relegates Africa to raw material exports).

Africa must adopt a strategy of value addition to transform virgin geological resources into valuable items before they are exported to wealthy markets. Such strategies include reorientation of the education system to ensure that knowledge in science and technology is not simply geared towards churning out experts who repair machines for developed country industries but rather those who can map and exploit resources on the continent. Technological advances on the continent devoid of

Africans' own engagement and input will simply drive down living standards to subsistence while driving up population growth (Clark 2007).

Through adoption of a knowledge driven economy, Sub-Saharan Africa can take advantage of the commodity prices boom to push for diversification of their economies. The continent ought to scale up small scale mining operations to medium sized and push for building of intra Africa infrastructure from natural resource revenues. An interconnected continent will give more value to diversified country economies since they will be able to establish markets with Africa's one billion people. It is also crucial that the financial sector be revamped to ensure a robust infrastructure for natural resource exploitation.

According to UNECA (2010), prices of non-energy sector have surged after 30 years of price stagnation. Driven by commodity exports, Africa's GDP is projected to grow from an average of 4.5% in 2010 to 5.2% in 2011 by Africa Economic Outlook. Major international mining companies' profits soared with cumulative net profits of three global mining groups (Rio Tinto, BHP Billiton and Anglo American) hitting US$26.9 billion in 2006 up from US$4.3 billion in 2002. Copper recorded a sharp price increase from US$2000/t in 2002 to US$5500/t with a peak at US$8000/t in 2007. Prices of Nickel rose from US$7000/t in 2002 to US$30,000/t in 2008 with a peak at US$50,000/t. Gold prices rose from US$300/oz in 2002 to US$1000/oz in 2008. Scaling up and formalisation of mining operations will enable governments keep track of benefits accrued from the surge in profits. As global giants swim in profits, informality restricts benefits to top government functionaries who supervise revenue from such conglomerates.

7 CONCLUSION

It ought to be clear in the Africans' mind that jostling for sphere of influence by European powers led to partitioning of the continent, massive land grabs and treaties that impede people from exploitation of resources to their benefit. The surge in commodity prices and competition among developed and emerging economies is likely to restrict Africa's quest to have geological resources play a significant role in transforming their economies for the better. It is in the continent's interest to reform internally and equip her people with necessary skills, general understanding and the know how on the operations of global economy.

Geological resources can assist drive Africa's economic growth through harmonisation of both in-country, continent and international mining codes. African countries on their part have to ensure participatory governance, long term focus and systems that efficiently respond to the needs of their constituents. The countries must facilitate the formalisation of the artisanal and informal mining infrastructures to spur a growth path. Small and Medium sized mining enterprises ought to be supported by clear policies to enable them secure financing to invest in value added operations that draw from natural resources from the continent.

The best way to assist somebody under systemic torture is to stop it and allow the person to come to his senses. Geological resources can help a country develop only if people are given the freedom to participate in building institutions that manage the exploitation of such resources. It is up to a country's people and its leaders to evaluate global systems as regards geological resources for purposes of creating benefits as opposed to "curses" from natural resources. It is people who Utilise the human mind—the ultimate capital—to transform such resources into valuable products, hence the need to promote the concept of *A Free Human Mind is the Ultimate Capital* in Sub-Saharan Africa.

REFERENCES

African Economic Outlook (undated). *African Economies will gain strength in 2010 and 2011* http://www.africaneconomicoutlook.org/en/outlook/macroeconomic-situation-and-prospects/african-economies-will-gain-strength/(accessed January 23, 2011).

Atlas Copco 2010. *Exploration Drilling: There is Always A Better Way*, www.podshop.se/Content/10/opensearchresult.aspx?file...L.pdf (accessed February 24, 2011).

Ayittey, B.N. 1998. *Africa in Chaos*, p. 5. St. Martin's Press, New York.

Beri, R. & Sinha, U.K. 2009. *Africa and Energy Security: Global Issues, Local Responses*, p. 51. Academic Foundation, New Delhi.

CICID 2005. *Mineral Resources and Development in Africa: Strategic Guideline* www.diplomatie.gouv.fr/fr/IMG/pdf/mineral_resources_DOS_2010.pdf (accessed January 23, 2011).

Clark, G. 2007. *A farewell to Alms: a brief economic history of the world*, p. 3. Princeton University Press, New Jersey.

Collier, P. 2007. *The Bottom Billion: why the poorest countries are failing and what can be done about it*, p. 38. Oxford University Press, NY.

FAO, *Mapping poverty, water and agriculture in Sub-Saharan Africa* ftp://ftp.fao.org/docrep/fao/010/i0132e/i0132e03a. pdf (accessed February 7, 2011).

Frick, C. *Direct Foreign Investment and the environment: African Mining Sector* http://www.oecd.org/dataoecd/46/24/2074862.pdf (accessed February 7, 2011).

Jomo, F. (2008) "Africa demands a larger slice of mining revenues," http://www.mineweb.co.za/mineweb/view/

mineweb/en/page68?oid=46891&sn=Detail (accessed February 11, 2011).
Kadenge, J. *Barriers of the Sub-Saharan Miracle: the Kenyan experience* http://depot.gdnet.org/newkb/fulltext/esuha.pdf (accessed February 7, 2011).
Lind, J. & Sturman, K. (editors) 2002. *Scarcity and Surfeit: The Ecology of Africa's Conflicts*, Pretoria, Institute for Security Studies.
Mugarula, F. 2011. Bulk of Mining Taxes Come from Workers, *The Citizen*, http://www.thecitizen.co.tz/component/content/article/37-tanzania-top-news-story/8217-bulk-of-mining-taxes-comes-from-workers.html (accessed February 15, 2011).
New World Encyclopedia, *Cotton* http://www.newworldencyclopedia.org/entry/Cotton (accessed January 23, 2011).
Page, J. 2009. Africa's Growth Turnaround: from fewer mistakes to sustained growth. *Commission on Growth and Development; Working Paper* 54 http://www.growthcommission.org/storage/cgdev/documents/gcwp054web.pdf (accessed January 23, 2011).
Quashie, A.K.L. *The case for mineral resources management and development in Sub-Saharan Africa* http://unu.edu/unupress/unupbooks/80918e/80918E0w.htm (accessed February 7, 2011).
Raw Material Group 2009. *Overview for SGU: Botswana, Namibia and South Africa*, Aug., www.meetingpoints-mining.net/pdf/NamibiaBotswanaOverview.pdf (accessed February 24, 2011).
Shah, A. 2010. *The Democratic Republic of Congo* http://www.globalissues.org/article/87/the-democratic-republic-of-congo (accessed February 7, 2011).
Thomas, H.R. 2009. *Development Corridors and Spatial Development Initiatives* available at http://www.fdi.net/documents/WorldBank/databases/africa_infrastructure/Thomas_SDI_paper_lowres.pdf (accessed February 15, 2011).
UNECA 2001. *Science and Technology and Competitiveness of Natural Resources in Africa* (accessed February 7, 2011).
UNEP/GRID-Arendal(undated). *UNEP/GRID-Arendal Maps and Graphics Library*, http://maps.grida.no/go/graphic/sub-saharan-africa-mineral-resources-and-political-instability (Accessed January 23, 2011).
UNESC & ECA 2009. *Summary of Africa Review Report on Mining*, available at www.uneca.org/csd/csd6/AficanReviewReport-on-MiningSummary.pdf (accessed February 15, 2011).
Zeleza, T.P. 2003. A Modern Economic History of Africa: The Nineteenth Century," *CODESRIA Book Series*, 1, Dakar.

APPENDIX 1

Summary of The Resolutions from The 6th Africa Resource Bank meeting held on November 2008 on "How Africa can Utilise its natural resources to raise the living standards of its people" held in Mombasa Kenya; hosted by the Inter Region Economic Network (IREN).

Delegates drawn from Kenya government geology departments, legislators from Uganda and Tanzania, private sector representatives, scholars and media resolved as follows:
African governments and its people must ensure:

- Strategic planning on how revenues from minerals and oil will be used (people involvement);
- Open access to the public contracts negotiated between business and government in the mining industry (transparency);
- Independent audit on governments use of current and future revenues from mineral resources (transparency);
- Establishment of an African Commodity Stock Market (to address the skewed global market system).

Approaches to Managing Natural Resources:

- Identify qualified and experienced competent African negotiators to ensure good deals/contracts;
- Make education responsive to African challenges and their environment/sensitize people about benefits of geological resources;
- Royalties ought to be factored to pay communities on whose land minerals are found;
- Consult mining experts to value and quantify mineral wealth to help determine investment relationships in Africa;
- Promote public/private partnerships between indigenous and international investors in exploration and exploitation of minerals;
- Develop sound policies that regulate and facilitate exploration/exploitation of mineral resources;
- Project the business acumen of the African people (to correct the impression that Africans are "genetically corrupt" and are only good in working for "NGOs");
- Develop processing abilities in the countries of mineral origin;
- Develop regional and continental institutions to manage geological wealth in a harmonised/standardised manner;
- Continuous environmental rehabilitation of mining areas.

Tax exemption to mineral exploration companies that disclose their findings to government and relevant agencies answerable to the people.

Geological Resources and Good Governance in Sub-Saharan Africa – Runge & Shikwati (eds)

An overview of geological resources in Sub-Saharan Africa: Potential and opportunities for tax revenue from the extractive sector

P. Buchholz
Federal Institute for Geosciences and Natural Resources (BGR), Hannover, Germany

M. Stürmer
Rheinische Friedrich-Wilhelms-University of Bonn, Bonn, Germany

ABSTRACT: Sub-Saharan African countries are among the world's largest producers of mineral commodities. Due to its geological nature, the region has an enormous potential for new discoveries in the long term. The recent boom in mineral commodity prices has raised the awareness that revenue from the mining sector could enable resource-rich countries in Africa to mobilise additional domestic funds towards improving their economy and achieving the Millennium Development Goals (MDGs). However, there are a number of economic hurdles to overcome in order to transform geological endowments into government income. Indeed, one could say that tax revenue drawn from the extractive sector is not easy money. To date, there is also a lack of medium and long-term projections as to the extent the extractive sector can make a difference for tax revenue, especially for non-major oil producing Sub-Saharan countries. Here, we present a scenario-based method for estimating tax revenue from the extractive sector. Our results show that in the cases of Ghana, Zambia and Namibia, the extractive sector could make an important contribution to achieving the MDGs.

1 INTRODUCTION

In the past few decades, the extractive sector, notably mining as well as gas and oil extraction, has been regarded as an impediment to economic development by several authors. This includes the Prebisch-Singer hypothesis of falling terms of trade for commodity exporters (Prebisch 1950, Singer 1950) and the debate on the so called "resource curse" for a number of countries (Sachs & Warner 1995).

The economic rise of China, India and other emerging economies has altered international commodity markets. During the commodity boom from 2003 to 2008, prices increased significantly and terms of trade turned in favour of commodiy exporting countries in Sub-Saharan Africa. At the same time, China and India have become important players with respect to investments in the extractive sector. Although the world economic turmoil has depressed some mineral commodity prices partly to near pre-boom levels, this phenomenon has taken the extractive sector high on the development agenda as it could provide these countries with the opportunity to raise domestic funds for achieving the Millennium Development Goals (MDGs). It may help to finance public goods such as infrastructure, education and basic health (Sachs 2007). This would provide the basis for further economic development. At the G8 Summit in L'Aquila, Italy, 2009, the Heads of State and Governments reiterated their support for responsible leadership and increased transparency, notably through the Extractive Industries Transparency Initiative (EITI). Furthermore, the World Bank and the United Nations Development Programme (UNDP) focus on establishing "Good Governance" all along the value chains of the extractive sector and provide capacity building in negotiating contracts. Other initiatives related to the mining sector include the Kimberley Process, the Revenue Watch Institute, Publish What You Pay, the Communities and small-scale mining or the German pilot project Certified Trading Chains (CTC).

The aim of this article is to give an overview of the geological potential of Africa and to show the opportunities for additional tax revenue from the extractive sector that could contribute to further financing the sustainable development of Sub-Saharan Africa. We first provide an introduction to the actual economic importance of the extractive sector in Sub-Saharan Africa. We will then show the geological potential of Sub-Saharan

African countries. Subsequently, we analyse the main hurdles to further develop the sector and to generate tax revenue. Then, we introduce a methodology to estimate potential tax revenue and their contribution to financing the MDGs. Finally, we present the results for the cases of Zambia, Ghana, Mozambique and Namibia.

2 THE EXTRACTIVE SECTOR—AN OPPORTUNITY FOR GROWTH?

African countries are among the world's largest producers of mineral commodities, with Sub-Saharan Africa providing 54% of global diamond production, 63% of global platinum group metals (PGM) production (mainly from South Africa and Zimbabwe), and 55% of the global cobalt production from the Zambian-Congolese Copperbelt. Sub-Saharan Africa also provides important stocks of chromite, vanadium, manganese and gold, and about 12% of global petroleum production with Nigeria and Angola being the largest petroleum producers in the region. The region also provides an important share of global production of a number of further mineral commodities, natural gas, hard coal and uranium (Fig. 1 and Appx. 1).

For many Sub-Saharan African countries, minerals and fuels are important export products. Metals, metal ores and concentrates provide more than 50% of total merchandise exports, e.g. in Mali (80%), Zambia (79%), the DRC (78%), Guinea (73%), Niger (71%), Mauritania (69%), Liberia (65%), Mocambique (64%), and Tanzania (54%) (Fig. 2a). Fuels provide an important share of total merchandise exports in Angola, Nigeria, and Chad. Mining exports also provide more than 10% of the GDP, e.g. in Zambia (31.5%), Mozambique (21.6%) and Namibia (11.6%) (Fig. 2b). This clearly shows that mining has an important impact on many Sub-Saharan African economies.

Many of the Sub-Saharan African countries have experienced rising export revenue from the extractive sector due to high world market prices for mineral commodities in recent years. During the commodity boom from 2003 to 2008, annual export revenue nearly tripled to US$139 billion in 2008. Sub-Saharan African countries have exported minerals and fuels worth a total of US$482 billion for the time from 2003 to 2008 (UNSTAT 2010, World Bank 2010).

While earlier studies (Sachs & Warner 1995) supported the conclusion that natural resources had a negative effect on economic growth, recent

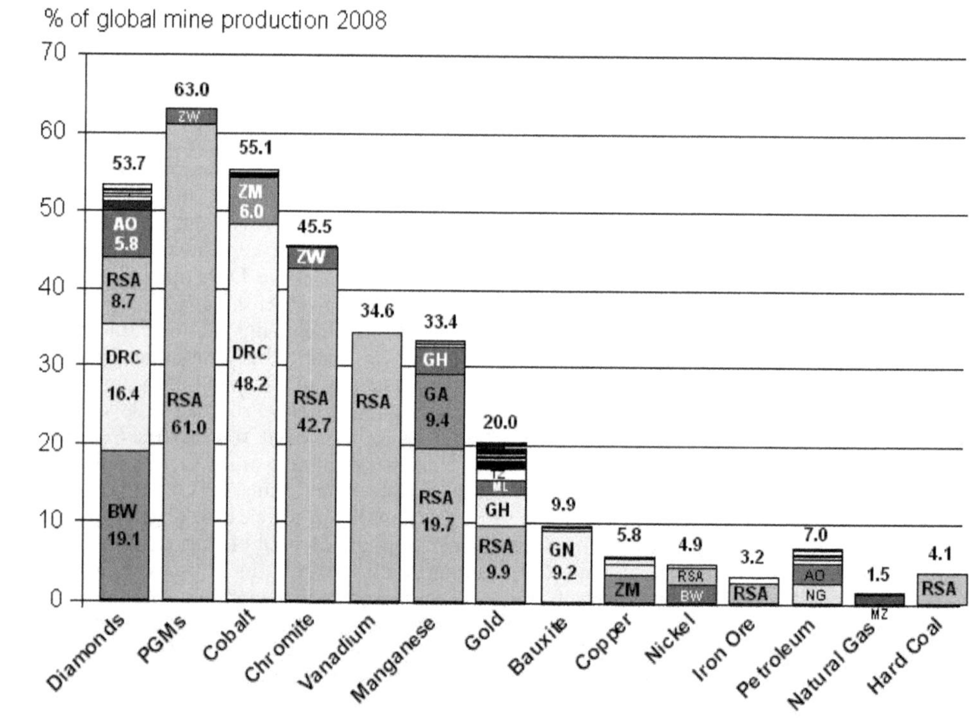

Figure 1. Sub-Saharan production of mineral and energy commodities as a share of global mine production 2008. (Source: BGR database).

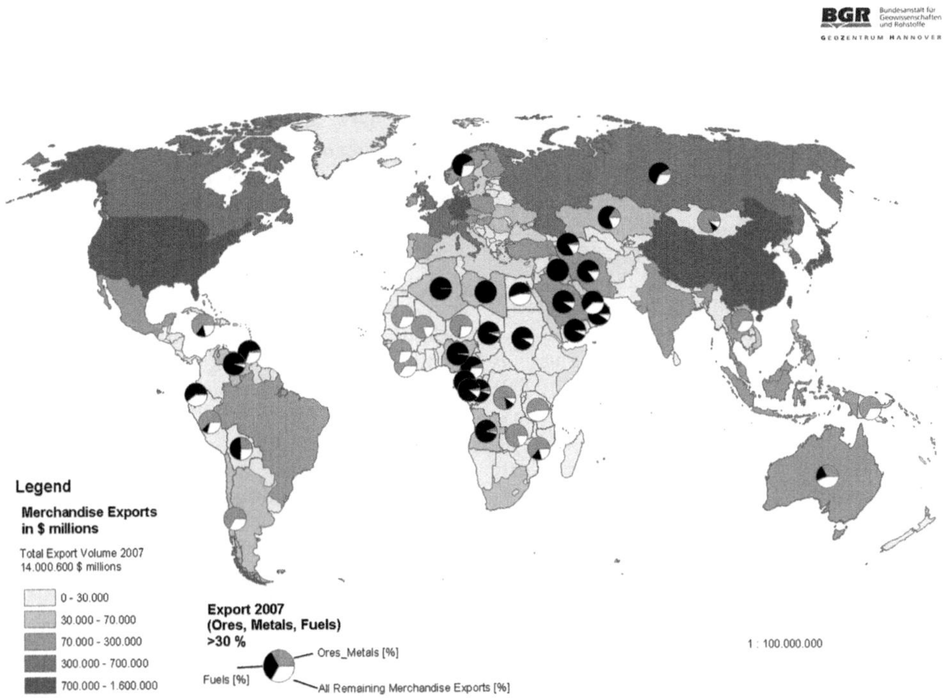

Figure 2a. Total merchandise export volume (global and national) and share of fuels, metals, ores, and concentrates. (Data source: World Bank 2010).

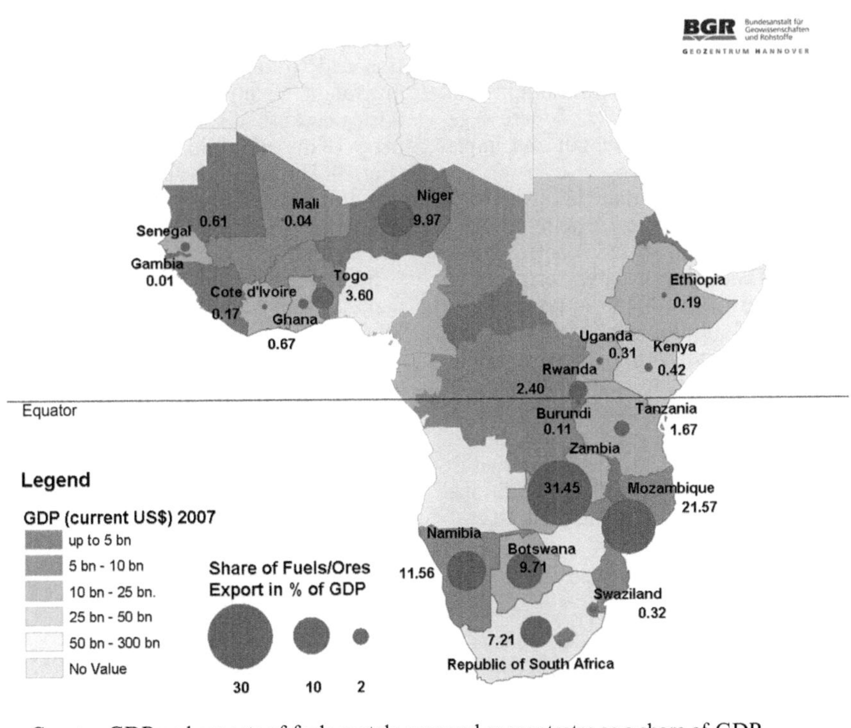

Figure 2b. Country GDP and exports of fuels, metals, ores and concentrates as a share of GDP. (Data source: World Bank 2010; limited availability of data).

studies have challenged these findings. For example, Lambrechts (2009) concludes in a report for several development organisations that it is the quality of national legislative and policy processes, state institutions and individual political leadership that will determine whether potential wealth goes towards financing national development or lining the pockets of political and business elites. Pedro (2005) of the United Nations Economic Commission for Africa writes in a report that it is "the quality of institutions that determines the development gains from mining" and that "weak governance institutions have been at the root of Africa's 'resource curse', and not mining activity itself". Also according to Béland and Tiagi (2009), the quality of management of the natural resources sector has an immediate effect on whether the resource wealth can turn the "resource curse" into a "resource blessing".

Despite the rich endowment of mineral and energy resources, all Sub-Sahara African countries except for South Africa and the fuel producing countries of Angola and Nigeria rank among the countries with the lowest GDP in the world. The total GDP of Sub-Saharan African nations was only US$993 Bn in 2008, which is 1.64% of global GDP. And for most Sub-Saharan countries the share of global GDP is way below 0.4%. This shows that exporting mining products is not the single solution to sustained growth.

The extractive sector also has negative impacts on local communities and the environment including the loss of land for farming, soil and water contamination, air pollution, deforestation, and damage to dwellings. The intensity of these negative impacts depends on the strength and implementation of regulation.

At the same time, fuels and minerals are a major potential source of tax income for many resource dependent countries and could contribute to finance the MDGs. The extractive sector could provide these countries with the opportunity to raise domestic funds for financing public goods such as infrastructure, education and basic health, which could induce further economic development (Sachs 2007). This would also enable investments in higher value-added products which need a minimum of public goods such as infrastructure or trained staff. Overall, the benefits of additional tax revenue should be weighed against the costs to communities and the environment.

3 AFRICA'S MINERAL RESOURCES—A FINITE SOURCE OF WEALTH?

Due to its geological nature and its relatively low degree of exploration, Africa has an enormous potential for new raw material discoveries. About 88% of today's known global PGM reserve base lies in Africa, predominantly in South Africa and Zimbabwe. South Africa and Ghana account for almost 80% of the manganese reserve base. The largest prospective diamond deposits are in West Africa, the DRC, Angola, Botwana, South Africa and Namibia, and the largest prospective global cobalt deposits lie in the Zambian-Congolese Copperbelt. For energy resources, Africa accounts for up to 5% of the global reserve base (Fig. 3). Many areas such as the Congolese basin and the Kalahari Desert are covered by soil and sediments and the potential for discovering new mineral deposits is huge. There is no doubt that Sub-Saharan Africa will remain in the focus of business interests for the coming decades.

But what defines the mineral potential of a country and how do mineral occurrences become part of officially declared reserves or resources? It is not only the geological potential, but also the attractiveness of a country for investment that defines the mineral potential.

The geological endowments of a country are described by its reserves and resources. The rock formations of a country have their own geological characteristics due to plate tectonics, volcanic activity and past environmental changes such as sea level rise and fall. For example, the "Karoo group" is a rock formation that is found throughout South-Central Africa. It is known for its coal deposits due to the extensive biomass production and sedimentation in the region. Diamonds are found mainly in Southern Africa and West Africa because of the frequent occurrences of primary (Kimberley pipes) and secondary deposits (Fig. 3). However, for all mineral commodities the rule applies that there is a large quantity of occurrences and deposits and very few large or giant deposits.

So what makes mineral occurrences appear as deposits with respective reserves and resources in official statistics? Reserves are largely defined as the known mineral occurrences that can be mined economically with today's technology. Resources are defined as a mineral occurrence of intrinsic economic interest in such form, quality and quantity that there are reasonable prospects for eventual economic extraction.

This means that—in contrast to intuition—reserves are not stock but a flow variable. Economical costs and technological progress are its major determinants. For example, the trend of haul truck size leads to rationalisation effects through economics of scale in large open pits. This results in a growth of reserves (Tilton 2002). Reserves and resources do not fully describe the geological potential of a country as there might

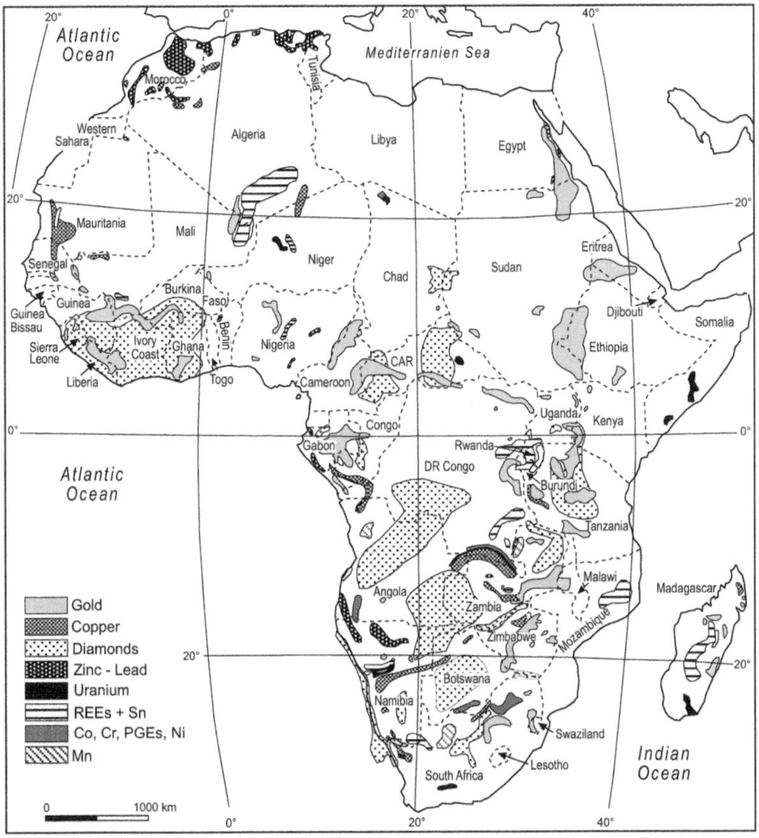

Share of global reserve base 2008.

	%		%		%		%
PGMs	88	Chromite**	43	Bauxite	23	Petroleum*	5
Manganese	79	Gold	34	Copper	3	Natural Gas*	3
Diamonds	56	Vanadium	32	Iron ore	1	Hard Coal*	4
Cobalt	41	Nickel	9				

Figure 3. Mineral provinces in Africa with enormous potential for new discoveries. Source of map: with kind permission of CSA Group Ltd. 2005; Share of global reserve base 2008 calculated from USGS data, 2009 (*2007; **% of reserve base in the United States of America, India, Kazakhstan and South Africa.; Definition: The reserve base is the in place demonstrated (measured plus indicated) resource from which reserves are estimated. It may encompass those parts of the resources that have a reasonable potential for becoming economically available within planning horizons beyond those that assume proven technology and current economics. The reserve base includes those resources that are currently economic (reserves), marginally economic (marginal reserves), and some of those that are currently subeconomic (subeconomic resources)).

still be many mineable deposits that are simply unknown.

An increase in prices or technological inventions in mining and processing leads to the conversion of resources into reserves (Fig. 4). Improving infrastructure in remote areas might also transform formerly marginal resources into reserves since mining these ore deposits becomes profitable. This relation also applies to other factors of accessibility such as war, civil unrest, misgovernance, or a general lack of economic development. Increasing the attractiveness of investment conditions may not only increase the economics of the reserves as such, but also may lead to new investments and mineral discoveries.

Overall, Africa is far less explored in terms of geology than for example Canada. An average of US$16 per km^2 was spent on exploration in Africa

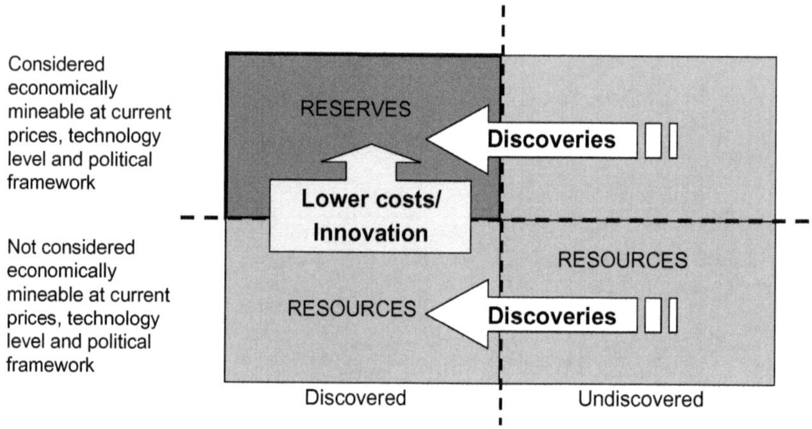

Figure 4. The flow of resources to reserves over time.

whereas the corresponding figure for Canada was US\$55 per km^2. Accessible basic geological data is often not adequate. Exploration in 2009 was highest in Latin America (25% of total exploration investment) followed by Canada (19%), Africa (15%) and Australia (14%) (Metals Economics Group, 2008). Despite the price boom of recent years the share of total exploration budgets changed little for Africa as compared to 2002.

4 GET THE PRODUCTION GOING: INVESTMENT CONDITIONS AND INFRASTRUCTURE

To transform geological endowments to production and manufacturing, infrastructure and other investment conditions are important factors. In general, Africa is far behind its peers such as Canada or Australia in terms of exports. In 2008 Sub-Saharan Africa exported fuels and minerals worth US\$5700 per km^2, whereas Australia exported over US\$16,000 per km^2 (Computations based on data from UNSTAT 2010 and World Bank 2010).

Investments in the extractive sector are capital intensive and are done over the long term. The assessment of risk plays a major role in investment decisions. Therefore, sound and stable public administration and political stability are essential for the long term development of the sector.

The Fraser Institute for example, rates the attractiveness of a country to mining and exploration investments (2009). The institute inquires each year international mining managers about the investment conditions in 65 mining countries and lists them up in a "Policy Potential Index". The Quebec Province in Canada for example ranks top on the Fraser index as it has a high mineral potential and provides a sound legislation and taxation system for mining. Although for many Sub-Saharan countries, the mineral potential is rated high. The attractiveness of exploration investment differs widely, with Botswana being most attractive, Ghana lies in the upper middle and the DR Congo is one of the least attractive countries for investment. Zimbabwe ranks number 67 of 68 entities surveyed due to political instability and corruption (Fig. 5). The most important points are the clear grant of mining titles, continuity of mining legislation, the ownership of permits, predictable taxation, political stability, geological data, and security issues.

Infrastructure such as transport and electricity are of particular importance for developing the extractive sector. Many African countries currently lack the infrastructure for an expansion of the sector. Most mineral commodities except precious and some rare metals as well as gemstones require bulk transport facilities such as railways, roads, inland waterway transportation and adequate harbor facilities. The same is true for electricity which is a major bottleneck for developing additional production.

Thus, the geological potential can only be sufficiently used if investment flows into the countries. This heavily depends on the overall investment conditions, which strongly depend on the country's investment ratings. Botswana, South Africa, Mauritius and Namibia lead in Sub-Saharan Africa with A to BBB ratings.

For other countries there is a lot of room for improvement (Fig. 6). Good ratings mean lower costs for borrowed money, bad ratings mean high premiums and therefore higher costs for financing investment projects in all sectors. Because of the

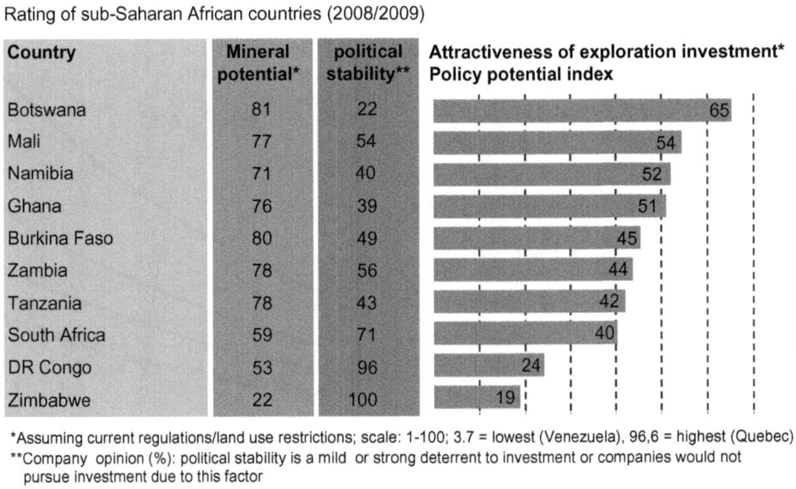

Figure 5. Rated attractiveness of exploration investment in Sub-Saharan African countries. (Fraser Institute 2009).

	Standard & Poor's		Moody's		Fitch	
Country	Rating	Date	Rating	Date	Rating	Date
Botswana	A	Dec 2009	A2	Mar 2009		
South Africa	BBB+	Nov 2008	A3	Jul 2009	BBB+	Nov 2008
Mauritius			Baa2	Dec 2007		
Namibia					BBB–	Dec 2005
Lesotho					BB–	Sep 2006
Gabon	BB–	Nov 2007			BB–	Oct 2007
Nigeria	B+	Aug 2009			BB–	May 2008
Cape Verde					B+	Jun 2009
Ghana	B+	Mar 2009			B+	Mar 2009
Kenya	B	Aug 2008			B+	Jan 2009
Senegal	B+	Dec 2009				
Seychelles	CCC	Aug 2008				
Cameroon	B	Feb 2007			B	Mar 2007
Benin	B	Dec 2007			B	Sep 2004
Burkina Faso	B	Aug 2008				
Madagascar	B–	Mar 2009				
Mozambique	B+	Dec 2007			B	Jul 2003
Uganda	B+	Dec 2008			B	Aug 2009
Mali	B	Nov 2005			B–	Apr 2004
Malawi					B–	Mar 2007
Gambia					CCC	Dec 2005

Figure 6. Foreign currency ratings (long term) of Sub-Saharan African countries.

financial crisis, the costs for loans are now high for countries with bad ratings, if they are made available at all (Fig. 7).

Although the ratings for Sub-Saharan African countries are rather weak, a total of 28 Sub-Saharan countries implemented 58 reforms from 2007 to 2008, a continuous upward trend (World Bank 2009). Business reforms in many of the Sub-Saharan African countries reflect a sustained commitment to improving competitiveness. Thus for Africa, foreign direct investment has already increased from US$57 bn to US$87 bn.

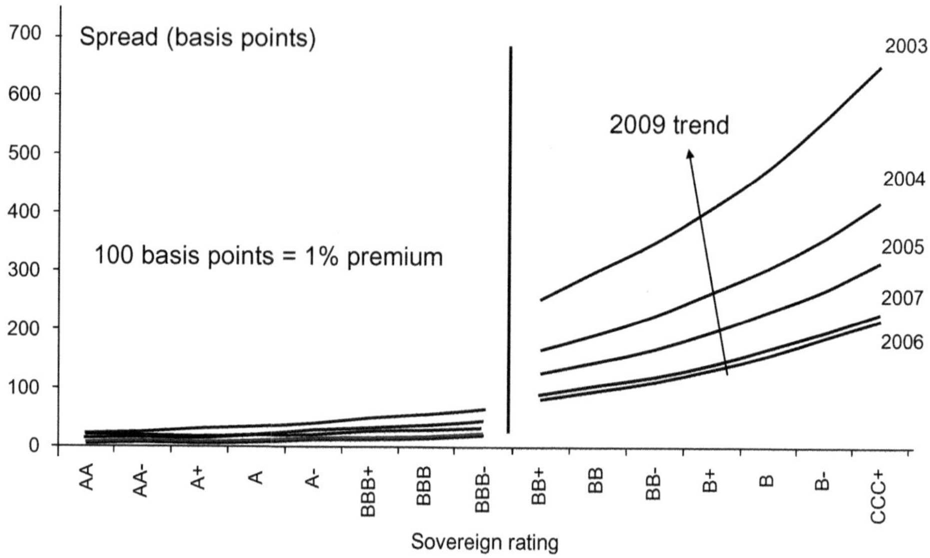

Figure 7. Sovereign Bond Launch Spreads & Ratings. (adapted from PricewaterhouseCoopers and Worldbank August 2009).

5 THE EXTRACTIVE SECTOR— A POTENTIAL FOR TAX REVENUE TO FINANCE THE MILLENNIUM DEVELOPMENT GOALS?

As demonstrated above, Africa has the geological potential to further develop its extractive industry. In this chapter we examine to what extent tax revenue from the extractive sector could contribute to financing the MDGs.

We will present a short introduction into taxing in the extractive sector, introduce a methodology to estimate potential tax revenue and finally show the potential revenue from the extractive sector and compare them to the costs of achieving the MDGs up to 2015.

We rely heavily on a study which we have conducted at the German Federal Institute for Geosciences and Natural Resources (BGR; Stürmer & Buchholz 2009). The case study includes Namibia, Ghana, Mozambique and Zambia. The study covers different mineral resources and economic importance of the extractive sector and aims to bring into focus the potential of the non-energy minerals sector.

5.1 *Taxing the extractive sector*

The extractive sector has a couple of special features, which makes a distinct taxation system usual. These include the size and timescale of investments, high sector specific risks, as well as the instability of world market prices. African mining tax regimes are often a mix of discretionary agreements, which grant tax holidays or an individual tax regime to the respective company as well as tax laws enacted through parliament.

In most countries the fiscal regime consists of a mix of royalties, windfall profit taxes, corporate income taxes, free carried interest/government shares, and concession charges.

Royalties have historically been the most common instrument for taxing the extractive sector and are widely used. They normally tax the fiscal dues on the basis of either volume ("unit" royalty) or the value ("ad valorem" royalty) of production or exports. As the definition of value varies, there are many different forms for the exact calculation of "ad valorem" royalties. Royalties have the advantage of easy assessment and application even though calculation can become complicated if the value is adjusted to subtract costs for transportation, handling etc. (Otto et al. 2006). They also ensure a relatively stable revenue stream to the government since production and sales normally vary much less than profits (Radetzki 2008). To producers, royalties constitute additional costs that have to be paid irrespective of profit levels. They can wipe out the entire profit margin or even result losses, when world prices and pre-tax profits are low. Very high royalties are therefore a major deterrent to investments, especially for minerals with highly cyclical prices or for less outstanding resource endowments with only normal profitability. They increase the economic cut-off rate of a mine and therefore reduce the economic lifetime of a project (Tordo 2007, Otto et al. 2006, Radetzki 2008).

Windfall profit taxes cream off a substantial proportion of profits that are considered "above" normal. Normally, it depends on a certain threshold e.g. a certain world market price level. There are a broad range of different applications of windfall profits taxes. For example, in 2006 the Mongolian Parliament enacted a windfall profits tax. The windfall profits tax called for a 68% tax on sales of gold when world market prices would rise above US$500 per ounce and US$2600 per tonne of copper. As a result, gold was smuggled out of the country so that the official gold production decreased by several tons.

The corporate income tax extracts fiscal dues on the basis of profits, i.e. taxes are only due when annual revenue exceeds some measure of costs and allowances. Key variations of corporate income taxes are the specification of allowable costs, the definition of taxable income, and the applied rate. In its proportional formulation (a fixed tax rate), corporate income taxes are relatively regressive, as their burden in percentage terms remains the same at different levels of profitability. Corporate income taxes avoid the problem of royalties where companies have to pay taxes even if they make losses. For the government, corporate income taxes are much more difficult to compute and to impose as profits have to be assessed. Furthermore, corporate income taxes yield a far greater fluctuation of public revenue than royalties since profits fluctuate much more than volume of output or sales. This is especially true when a progressive rate is applied (Tordo 2007, Otto et al. 2006, Radetzki 2008).

Another measure for fiscal extraction is government participation for free or on concessional terms. The government acquires a carried interest and pays for its share out of future earnings of the project, or it demands a minority equity share for free at the time of the original investment decision (called free carried interest). Public ownership is not always easy to transform into a fiscal income flow as it may expose the government to the costs involved in reinvestments and expansions. There may also be a conflict of interest in the government's role as equity holder and regulator overseeing environmental and social impacts. Overall, government participation represents a cost to the investor. On concessional terms it also reduces cash flow and raises the risk profile of an investment (Tordo 2007, Radetzki 2008).

Finally, there are different taxes and charges that add to the tax burden of the extractive sector. These include concession charges, duties for imported equipment, payroll taxes, value added taxes, and environmental taxes. Other important sources of tax revenue from the extractive sector are direct grants or credits from foreign government in exchange for access to resources.

Many states lack the capacity to develop sound taxation systems and to negotiate favourable agreements with enterprises. For example, Zambia negotiated a number of "development agreements" when it privatised its copper sector in the late 1990s. These agreements lowered the applied royalty rate to 0.6% in order to attract foreign investment. As a consequence, tax revenue have been quite low in the past few years even though copper prices reached record levels from 2004 to 2008.

It is difficult to strike the right balance between a fair share of income for the state and a fair level of taxation for the sector. The strong price cycles of mineral commodity markets make it especially challenging to strike this balance. Several African countries are currently reforming their mineral taxation systems (e.g. DR Congo, Sierra Leone and Zambia) with a view to obtaining a higher government share. At the same time, the rapid change in tax systems in combination with easing world market prices may chase away potential investors who fear a lack of regulatory stability.

There are many problems with the execution and implementation of tax regimes. Many countries lack a sound tax and mining administration for effective tax collection. Due to a lack of knowledge on geology and mining operations, the administration is often not able to check tax statements. Corruption is often widespread. Competences between the different state authorities are unclearly distributed. An additional problem is the pricing of minerals, which leads to variations in the computation of royalty payments. Mining companies also apply different exchange rate regimes for the payment of mineral royalties. Some taxes are often simply not paid (Government of the Republic of Ghana/Ministry of Finance and Economic Planning 2006). As a result, effective tax rates are often much lower than nominal tax rates. In addition, high corruption is a major obstacle to collecting and distributing tax revenue even if the legislative framework has been improved. Unfortunately, the corruption perception index for many Sub-Saharan African countries is still high (Transparency International 2009).

To conclude, the extractive sector has a distinct taxation system and there is no 'one size fits all' tax regime. It should keep the balance between government intakes and investment incentives. Most importantly, the tax regime must be well executed.

5.2 *Estimating potential tax revenue from mining*

Our scenario-based methodology to estimate potential tax revenue from the extractive sector through 2015 consists of the following main parts:

Production is defined as the sum of all mineral and energy commodities produced in a respective country. For describing the total future production, future projects have been added to the current production level through 2015. As mining projects have a lead time of 5 to 10 years, the range of possible production levels up to 2015 has been determined by today's information on new projects. Many mining projects are not realised or the start of production is delayed due to the decreased in world market prices between 2008 and 2009 and complications with infrastructure or equipment. We use different parameters for the realisation of these projects (Tab. 1).

Gross revenue is defined as the total revenue from the sales of all mineral and energy commodities that are produced in a country. To compute gross revenue, transportation costs had to be subtracted from world market prices. World market prices are either in f.o.b. or c.i.f. F.o.b. means "free on board" and implies that the producer delivers the product for free to a given point of sale, normally on board of a ship in a harbour of the producer country. Therefore, the freight costs from the producer country to the consumer country are paid for by the purchaser. C.i.f. stands for "cost, insurance, freight" and implies that costs such as customs, documentation, freight and insurance during the shipment to a consumer country, are paid by the seller.

In the case of f.o.b. prices, the transport costs to the point of sale such as the harbour have been subtracted from the world market price. This is especially important for countries such as Zambia. As it is a land locked country, copper and cobalt have to be transported to the coast on train. To compute gross revenue from c.i.f. world market prices, freight costs to the consumer country have been deducted. Freight costs have been calculated based on rules of thumb from Wellmer et al. (2008).

We define tax revenue as all direct government intakes from the extractive sector including royalties, corporate income tax and free carried interests. They are by far the most important contributors to tax revenue in the extractive sector (compare e.g. Australian tax revenue from the extractive sector: Minerals Council of Australia 2008).

The method does not include indirect taxes and taxes levied on others as it is hard to retrieve the necessary data to compute e.g. personal income taxes, license fees, VAT, withholding taxes, and other local duties. As these indirect taxes are a component of operating expenses, they are also not visible in the financial statements of the respective companies. All tax revenue is computed in constant US$ prices.

Tax rates have been taken from the respective tax codes as well as from different journal articles and information provided by the World Bank and the IMF. Unfortunately, there is a lack of transparency with regard to individual production and tax agreements between government and mining and oil companies. As these agreements are not freely available, it is hard to estimate the effective tax rate. The nominal tax rate might deviate widely from the effective tax rate, because of special agreements between governments and companies with respect to depreciation, loss carryforward and ring fencing. Ineffective tax collection and administration are other reasons. To show the magnitude of these effects, we have also computed the corporate income tax on the basis of national effective tax rates from the "Paying Taxes 2008" report by the World Bank and PricewaterhouseCoopers. Unfortunately, these effective tax rates are based on normal business and are not specific for the extractive sector (PricewaterhouseCoopers and World Bank 2008). Furthermore, we have used the world average effective corporate income tax rate for mining companies as a comparison. This rate has been derived from the PricewaterhouseCoopers mining survey (PricewaterhouseCoopers 2008). This survey analyses the operations of the top 40 mining companies.

The share of taxable income has also been derived from the PricewaterhouseCoopers mining survey. In 2007, the average share of taxable income was 36%. In scenario 2 and 3 the share of taxable income develops on the basis of a logistic sigmoid function in parallel to world market prices down to 6.3%. This was the rate in 2002 before the recent price hike.

5.3 *Estimating the MDG costs*

The MDG costing describes the funds needed to achieve the MDGs by 2015. There is no broadly accepted definition of what should be included in a costing and which spending should be counted as relevant for achieving the MDGs. The estimates are meant to give guidance on the overall volume of aid that will be needed to achieve the goals, but they should not be confused with the detailed costing that will have to be done at the country level.

Table 1. Estimating tax revenue from the extractive sector of a given country.

Parameter	Formula
Production	1. $Q_{i,t} = HQ_i + Q_{i,t-1}(1+a) + NQ_{i,t}b$
	Q = Production HQ = Historic production level a = Parameter for the development of the general production level NQ = New production projects b = Parameter for the realization of future projects i = Subscript for the respective mineral commodity t = Subscript for the respective year
Gross revenue	2. $R_{i,t} = P_{i,t}Q_{i,t} - F_iQ_{i,t} - RC_iQ_{i,t} - TC_iQ_{i,t}$
	2a. $F_i = D \cdot SF_i$
	2b. $P_{i,t} = P_{i,2007} + (P_{i,2007} - P_{i,2015})\left(1 - \frac{1}{1+e^{-\alpha(t-2007-T_0)}}\right)$
	R = Gross revenue P = Commodity price; in scenario 1 the price is assumed to be constant over the period 2008 to 2015; in scenario 2 and 3 the price development is computed on the basis of 2b with different parameter values. RC = Refining charge (where applicable) TC = Treatment charge (where applicable) F = Freight costs D = Distance (where applicable) SF = Commodity and mean of transport specific freight costs per mile T_0 = Parameter to shift the timeline of the curve α = Parameter to control the acceleration of the curve
Tax revenue	3. $T = \sum_{i=1}^{I} \sum_{t=2008}^{2015} R_{i,t}Rr_i + R_{i,t}\,\mathrm{Pr}_t\,Cr + R_{i,t}\,\mathrm{Pr}_t\,DPFr_i + R_{i,t}Wr_i$
	3a. $\mathrm{Pr}_t = \mathrm{Pr}_{2007} + (\mathrm{Pr}_{2007} - \mathrm{Pr}_{2015})\left(1 - \frac{1}{1+e^{-\alpha(t-2007-T_0)}}\right)$
	T = Total tax revenue of a given country Rr = Royalty rate Cr = Corporate income tax rate Wr = Windfall profits tax rate (where applicable) Fr = Rate of free carried interest (where applicable) DP = Share of dividend payments (where applicable) Pr = Share of taxable income from gross revenues; in scenario 1 the share is assumed to be constant over the period 2008 to 2015; in scenario 2 and 3 the development of the share is computed on the basis of 2b with different parameter values. i = Respective mineral commodity I = Total number of mineral commodities included T_0 = Parameter to shift the timeline of the curve α = Parameter to control the acceleration of the curve Pr_{2007} = 0.36 Pr_{2015} = 0.63

(*Continued*)

Table 1. (*Continued*)

Parameter	Formula
MDG costing	$MDG = \sum_{t=2008}^{2015} PCC_{t,s}N_t - PPCHC_tN_t - CPA_t$
	MDG = Total MDG financing gap that needs to be covered by government expenditures
	PCC = Per capita costs for achieving the MDGs
	N = Population
	$PPCHC$ = Private per capita household contributions to financing the MDGs
	CPA = Country programmable aid

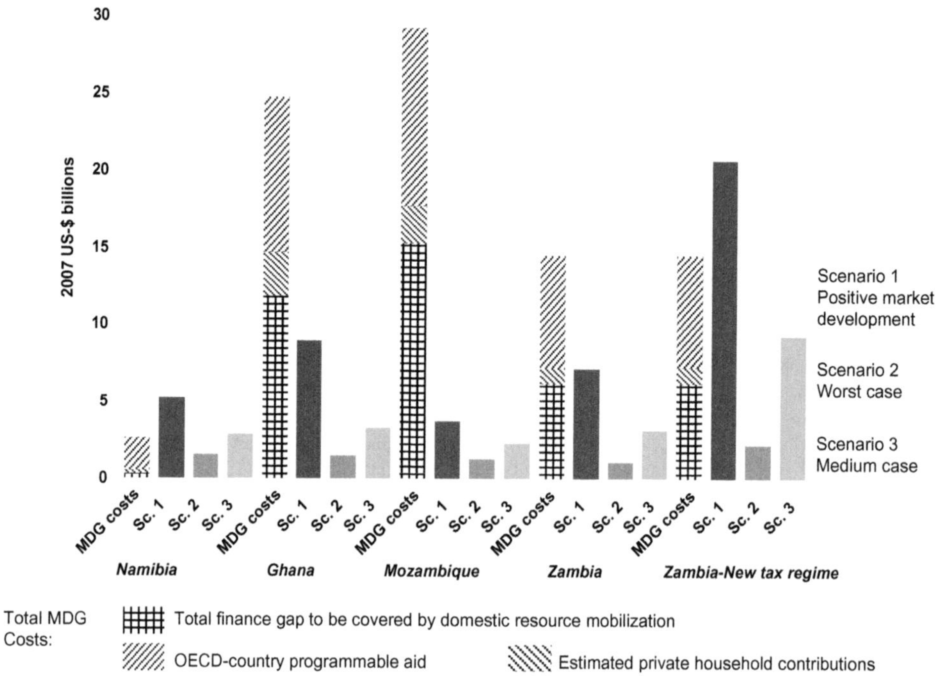

Figure 8. Total MDG costs and government revenues from the extractive sector according to the different scenarios over the period 2008 to 2015 for Ghana, Mocambique, Namibia and Zambia (in constant US$, base year: 2007) (Stürmer & Buchholz 2008).

The UN Millennium Project has done a MDG costing for Ghana and four other developing countries (United Nations Millennium Project 2005). Based on this methodology the University of Zambia in Lusaka has computed the MDG costs for Zambia (Mphuka 2005). Unfortunately, there are no such costings available for Mozambique and Namibia. To estimate their MDG costs, we have compared key indicators of the respective countries with those of the five case study countries from the report of the UN Millennium Project (Tab. 1). We have assigned the per capita costs for each indicator to the corresponding indicator in the case study. The per capita costs were converted to 2007 US$ and summed up for all sectors over the period from 2008 to 2015. Similar to the UN Millennium Project, we assume that the annual costs increase over time due to the need for building capacities to absorb the financial inflows. Underlying these estimates is the assumption that the scaling up of investment goes hand in hand with optimising current public expenditures using best practices.

After computing the total costs, we have subtracted the contributions from private household similar to the UN Millennium Project. We subtract the Country Programmable Aid (CPA) from bilateral and multilateral donors to determine the financial gap that has to be filled by the government.

Additionally, we have developed three different scenarios which include assumptions on future mine production, prices, and shares of taxable income (Fig. 8).

5.4 *Results*

The results for the case study countries show that the extractive sector could be a significant source of future income in SSA, and it could contribute to financing the Millennium Development Goal (MDG) costs.

In Zambia tax revenue from the extractive sector, dominated by copper and cobalt production, had been negligible for a long time. "Development agreements" with mining companies were negotiated at the end of the ninenties when copper prices were very low and the Zambian government was trying to reactivate the depressed copper sector by privatising and reforming state owned mining companies. In order to attract foreign investors the "development agreements" reduced royalty rates to 0.6%. As a result, mining taxes from the sector were only about US$10 m in 2005 at a time when world copper prices had already doubled and Zambia exported copper worth US$1.5 bn. Even in 2007 when world market prices had risen six-fold since the 1990s and copper exports were US$3.4 bn, mining taxes summed up to only around US$160 m. However, from April 2008, the government introduced a new tax regime with applied royalty rates of 3%, a corporate income tax of 35%, and a variable windfall tax at 25 to 75%. Even under a pessimistic scenario, based on low world market prices and low production increases, tax revenue could provide around 15% of the total MDG financing and more than 30% of the finance gap that is left to the national government. The construction of infrastructure, especially with regard to electricity, is an important factor for the further development of the sector.

In contrast to long standing mining countries such as Zambia, the outlook for financing the MDGs through the extractive sector in Mozambique is quite low (for further discussion see Stürmer & Buchholz 2009).

In Namibia the situation is more advantageous. The MDG costs are moderate and tax revenue from the extractive sector has been high for the last few years. Diamonds, which make up about 30% of Namibian export revenue, contribute heavily to most of tax revenue. At the same time, uranium and zinc resources have become the focus of international mining companies and could generate extra tax revenue through 2015. Overall, the potential revenue could outweigh the MDG gap to be filled by the government in all three scenarios for the period 2008 to 2015. However, environmental standards especially for uranium mining are important for sustainable development in this sector.

Ghana is a stronghold of gold mining in West Africa and is the second largest producer of this metal in Africa. There is also high potential for extracting hydrocarbons offshore. Ghana shows that oil production really makes a difference for tax revenue. Past revenue from corporate taxes on mining companies is moderate due to accelerated depreciation and the carry forward of loss making concessions. Most companies pay a minimum of 3% in royalties. According to the Ghana Chamber of Mines, corporate income tax revenue from mining companies amounted to about US$2.8 m and about US$18 m in royalties in 2002. These incomes increased to US$11 m and US$24 m respectively in 2004. Under the current tax regime, the potential tax revenue from the oil sector would make up about 75% of the total tax revenue from the extractive sector. The extraction of oil alone could add US$1.6 bn to government coffers in the period 2008–2015 for the medium scenario. Ghana could thus finance about one sixth of its MDG needs from the extractive sector.

Overall, these case study countries serve as a starting point. They can only provide rough estimates based on strong assumptions. More specific data on the mining operations in each case study country would be necessary for more exact outcomes. Furthermore, secondary effects of the mining industry on tax revenue from the construction and service sector are not included in this analysis due to a lack of data. For the same reason, it was also not possible to consider indirect taxes such as payroll taxes. Further studies should be done in close cooperation with the respective country officials and local research institutions.

6 CONCLUSION

The extractive sector can be a major source of tax income in Sub-Saharan Africa and it can contribute to financing the MDG costs. If well managed, Sub-Saharan Africa has enormous potential for the discovery of new mineral and energy resources and additional mineral production. Major cornerstones for reasonable management of the sector are stable and enforced mining legislation including high and well implemented social and environmental standards, transparency of financial flows and trade, a reasonable taxation adapted to the fluctuating commodity markets and the local situation, infrastructure measures and improvements on the geological knowledge base. Better management and governance of the sector will be a fundamental base for to fully utilising the geological potential of countries in a sustainable way, with less negative social and environmental impacts. It is therefore important to weight the benefits of additional tax revenue against the costs to communities and the environment. Hence, tax revenue from the extractive sector "no low hanging fruits" and need to be carefully invested in the welfare and economic systems for achieving the MDGs.

REFERENCES

CSA Group Ltd. 2005. Now SLR Consulting Ireland, Presentation.

Fraser Institute 2009. *Fraser Institute annual survey of mining companies*. 2008/2009. Download from http://www.fraserinstitute.org/commerce.web/product_files/MiningSurvey20082009_Cdn.pdf on 20.01.2010.

Government of the Republic of Ghana/ Ministry of Finance and Economic Planning 2006. *Inception report on the aggregation of payments and receipts of mining benefits in Ghana*. Download from http://www.geiti.gov.gh/downloads/inception.doc on 01/12/2008.

Lambrechts, K. (Ed.) 2009. *Breaking the Curse: How Transparent Taxation and Fair Taxes can Turn Africa's Mineral Wealth into Development*. Report published by Open Society Institute of Southern Africa, Johannesburg; Third World Network Africa, Accra; Tax Justice Network Africa, Nairobi; Action Aid International, Johannesburg; Christian Aid, London. Download from http://www.taxjustice.net/cms/upload/pdf/TJN4Africa_0903_breaking_the_curse_final_text.pdf on 20.01.2010.

Metals Economics Group 2008. Overview of worldwide exploration budgets. *Strategic report*, 21, 6: 1–11.

Minerals Council of Australia 2008. *Minerals Industry Survey Report 2007*. Download from http://www.minerals.org.au/__data/assets/pdf_file/0012/25311/79374_Minerals_Industry_Surveyv-FINAL_ebook.pdf on 15/09/2008.

Minerals Council of Australia 2009. *Minerals Industry Survey Report 2007*. Download from http://www.minerals.org.au/__data/assets/pdf_file/0012/25311/79374_Minerals_Industry_Surveyv-FINAL_ebook.pdf on 15/09/2008.

Mphuka, C. 2005. *The cost of meeting the MDGs in Zambia*. Download from http://www.debtireland.org/resources/ZambiaMDGsReport05.pdf on 05/10/2008.

Otto, J., Andrews, C., Cawood, F., Doggett, M., Guj, P., Stermole, F., Stermole, J. & Tilton, J.E. 2006. *Mining royalties. A global study of their impact on investors, government, and civil society*. Washington DC: World Bank. Download from http://siteresources.worldbank.org/INTOGMC/Resources/336099-1156955107170/miningroyaltiespublication.pdf on 20.01.2010.

Pedro, A. 2005. *Mainstreaming mineral wealth in growth and poverty reduction strategies*. Download from http://www.dundee.ac.uk/cepmlp/journal/html/Vol16/Vol16_5.pdf on 30.01.2010.

Prebisch, R. 1950. *The Economic Development of Latin America and its Principal Problems*. New York, UN.

PricewaterhouseCoopers 2008. Kursiv. *As good as it gets? Review of the global trends in the mining industry 2008*. Download from http://www.pwc.com/extweb/pwcpublications.nsf/docid/4E436FE336691868852 5746B005C8D9E/$File/Mine_2008_v7_Final.pdf 06.09.2008.

PricewaterhouseCoopers/World Bank 2008. *Paying taxes 2008. The global picture*. Download from http://www.doingbusiness.org/documents/Paying_Taxes_2008.pdf on 10.09.2008.

Radetzki, M. 2008. *A handbook of primary commodities in the global economy*. Cambridge, Cambridge University Press.

Ratha, D., De, P. & Mohapatra, S. 2007. Shadow sovereign ratings for unrated developing countries. World *Bank Policy Research Working Paper*, 4269. Download from http://www-wds.worldbank.org/external/default/WDSContentServer/IW3P/IB/2007/06/21/000016406_20070621154413/Rendered/PDF/wps4269.pdf

Sachs, J.D. 2007. How to handle the macroeconomics of oil wealth. In: Humphreys, M. / J.D. Sachs / J.E. Stiglitz, *Escaping the resource curse*. New York: Columbia University Press.

Sachs, J.D. & Warner, A.W. 1995. Natural resource abundance and economic growth. *NBER Working Papers*, 5398. Download from http://www.nber.org/papers/w5398.pdf on 15.01.2010.

Singer, H. 1950. The Distribution of Gains between Investing and Borrowing Countries. *American Economic Review*, 40: 473–485.

Stürmer, M. & Buchholz, P. 2008. *Government revenues from the extractive sector in Africa. A potential for funding the Millennium Development Goals*? Hannover: Federal Institute for Geosciences and Natural Resources. Download from http://www.die-gdi.de/CMS-Homepage/openwebcms3.nsf/(ynDK_contentByKey)/ANES-7LTG2N/$FILE/BP%209.2008.pdf on 20.01.2010.

Tilton, J.E. 2002. *On borrowed time? Assessing the threat of mineral depletion*. Download from http://www.mines.edu/Fac_staff/senate/dist_lecture/tilton_text.pdf on 20/07/2008.

Tordo, S. 2007. Fiscal systems for hydrocarbons. Design issues. *World Bank Working Paper*, 123. Download fromhttp://siteresources.worldbank.org/INTOGMC/Resources/fiscal_systems_for_hydrocarbons.pdf on 20/06/2008.

Transparency International 2009. *Global corruption barometer 2009*. Download from http://www.transparency.org/content/download/43788/701097/ on 30.01.2010.

United Nations Millennium Project 2005. Investing in Development. A practical plan to achieve the Millennium Development Goals. *Report to the UN-General Secretary*. Download from http://www.unmillenniumproject.org/documents/MainReportComplete-lowres.pdf on 19/02/2008.

United States Geological Survey 2009. *Minerals Information*, http://minerals.usgs.gov/minerals/

UNSTAT = United Nations Statistics Division 2010. *United Nation Commodity Trade Statistics database. UN comtrade*. Download from http://comtrade.un.org/db/dqBasicQueryResults.aspx? on 06.01.2010.

Wellmer, F.-W., Dalheimer, M. & Wagner, M. 2008. *Economic evaluations in exploration*. Berlin/ Heidelberg/ New York, Springer.

World Bank 2010. *World Development Indicators*. Download from http://databank.worldbank.org/ddp/home.do

8 APPENDIX

Appendix 1. Sub-Saharan Africa: Mine production in 2008 (*2007 data) of selected mineral commodities and their share in the world's total production (United States Geological Survey 2009; BGR database).

Commodity/Unit	Country	Production	% of World	Reserve base	% of World
Iron Ore*	RSA	41,559	2.5	2,300,000	0.7
1000 t	Mauritania	11,917	0.7	1,500,000	0.4
			3.2		1.1
PGMs*	RSA	304,031	61.0	70,000,000	87.5
kg cont.	Zimbabwe	9870	2.0	n.a.	n.a.
			63.0		87.5
Vanadium*	RSA	23,500	34.6	12,000,000	31.6
t cont.			34.6		31.6
Bauxite	Guinea	19,296	9.2	8,600,000	22.6
1000t	Sierra Leone	945	0.4	n.a.	n.a.
	Ghana	574	0.3	n.a.	n.a.
			9.9		22.6
Chromite	RSA	10,300	42.7	150,000,000	42.8**
t	Zimbabwe	650	2.7	n.a.	n.a.
	Sudan	32	0.1	n.a.	n.a.
			45.5		n.a.
Cobalt	DR Congo	31,000	48.2	4,700,000	36.1
t cont.	Zambia	3,841	6.0	680,000	5.2
	Botswana	337	0.5	n.a.	n.a.
	RSA	244	0.4	n.a.	n.a.
			55.1		41.3
Copper	Zambia	545	3.5	35,000	3.5
1000t cont.	DR Congo	214	1.4	n.a.	n.a.
	RSA	109	0.7	n.a.	n.a.
	Botswana	29	0.2	n.a.	n.a.
			5.8		3.5
Gold	RSA	220,100	9.9	31,000,000	31.0
kg cont.	Ghana	86,996	3.9	2,700,000	2.7
	Mali	40,700	1.8	n.a.	n.a.
	Tanzania	36,600	1.6	n.a.	n.a.
	Guinea	19,900	0.9	n.a.	n.a.
	Zimbabwe	6400	0.3	n.a.	n.a.
	Mauritania	6300	0.3	n.a.	n.a.
	Burkina Faso	4030	0.2	n.a.	n.a.
	Burundi	4000	0.2	n.a.	n.a.
	Ethiopia	3300	0.2	n.a.	n.a.
	Botswana	3100	0.1	n.a.	n.a.
	Kenya	3000	0.1	n.a.	n.a.
	Ivory Coast	2700	0.1	n.a.	n.a.
	Niger	2600	0.1	n.a.	n.a.
	Uganda	2500	0.1	n.a.	n.a.
	Sudan	2300	0.1	n.a.	n.a.
	Namibia	2126	0.1	n.a.	n.a.
			20.0		33.7
Manganese	RSA	6807	19.7	4,000,000	76.9
1000t	Gabon	3250	9.4	90,000	1,7
	Ghana	1261	3.6	n.a.	n.a.
	Ivory Coast	150	0.4	n.a.	n.a.
	Namibia	90	0.3	n.a.	n.a.
			33.4		78.6
Nickel	Botswana	34,900	2.3	920,000	0.6
t cont.	RSA	31,700	2.1	12,000,000	8.0

(*Continued*)

Appendix 1 (*Continued*)

Commodity/Unit	Country	Production	% of World	Reserve base	% of World
	Zimbabwe	7900	0.5	260,000	0.2
			4.9		8.8
Diamonds 1000ct	Botswana	33,000	19.1	230,000	17.7
	DR Congo	28,400	16.4	350,000	26.9
	RSA	15,100	8.7	150,000	11.5
	Angola	10,000	5.8	n.a.	n.a.
	Namibia	2132	1.2	n.a.	n.a.
	Guinea	1100	0.6	n.a.	n.a.
	Ghana	720	0.4	n.a.	n.a.
	Zimbabwe	690	0.4	n.a.	n.a.
	Sierra Leone	600	0.3	n.a.	n.a.
	Central Afr. Rep.	470	0.3	n.a.	n.a.
	Lesotho	450	0.3	n.a.	n.a.
	Tansania	230	0.1	n.a.	n.a.
	Ivory Coast	210	0.1	n.a.	n.a.
			53.7		56.1
Petroleum Mt	Nigeria	105.3	2.7	4928	3.1
	Angola	92.2	2.4	1,837	1.1
	Sudan	23.7	0.6	912	0.6
	Equatorial Guinea	17.9	0.4	231	0.1
	Republic of the Congo	12.9	0.3	259	0.2
	Gabon	11.8	0.3	272	0.2
	Chad	6.7	0.2	204	0.1
	Cameroon	4.3	0.1	27	0
			7.0		5.4
Natural Gas Bcm	Nigeria	35.0	1.1	5215	2.8
	Equatorial Guinea	6.6	0.2	37	0
	Mozambique	2.5	0.1	119	0.1
	RSA	2.0	0.1	20	0
			1.5		2.9
Hard Coal Mt	RSA	235.8	4.1	31,022	4.3
			4.1		4.3

**% of reserve base in the United States of America, India, Kazakhstan and South Africa.

Geological Resources and Good Governance in Sub-Saharan Africa – Runge & Shikwati (eds)
ISBN 978-0-415-58267-4

Transparency and the value chain in extractive industries in Central Africa

A. Bomba Fouda
Extractive industries data management, SCTIIE-CEMAC, Yaoundé, Cameroon

ABSTRACT: The oil economies of Central African countries are predominantly revenue-based economies characterised by an opaque management of oil revenue, widespread corruption and a poor development track record. The extractive sector is characterised by a certain lack of clarity and confusion in the way it operates. The confusion relates, for example, to the destination of the oil, gas and mineral products extracted in Central African countries, and to who really benefits from the revenue generated. This confusion can impact on social order, leading to conflict, resentment and disruption to businesses and public services. In the worst cases, it can result in physical damage to installations and equipment, and even the intimidation of company employees. In Central Africa, oil revenue has provoked or fuelled civil wars and repeated human rights violations, as well as increasing poverty. Transparency could help improve poor economic performance, reduce corruption and poverty, promote social order and remove conflicts that are not infrequently based on oil and mineral resources.

1 INTRODUCTION

In the vast majority of Central African countries, the oil, gas and mining sectors are characterised by state-owned resources that form part of a national heritage. The extractive sector is able to generate significant, highly concentrated revenue that may be siphoned off by an élite class instead of being used for the public good. In Central Africa, the oil sector is often controlled by the head of state, i.e. the President of the Republic, via a state-owned company. The sector has for a long time been dominated by major international oil companies that extract crude oil within the terms of licence agreements, some of which were obtained during the colonial period or shortly after independence. The levelling-off of reserves in some countries, the slowing down of oil revenue and the attention being paid to critics denouncing not only the industry's political, social, economic and environmental impact but also the opaque management of oil revenue may encourage the oil-producing countries of Central Africa to move away from revenue-based economies focusing solely on production and towards revenue-based economies that also take account of wealth distribution and social well-being.

2 THE EXTRACTIVE INDUSTRY VALUE CHAIN

Companies extract raw materials, which then need to be processed so that the end customer can make use of them. Apart from the process of extraction, the various processing phases that create added value are significant in terms of job creation and development. Several African countries that are major oil exporters find themselves having to import refined products as they have insufficient refining capacity themselves. In Central Africa, there are three distinct types of actor involved in the upstream oil processes (exploration and production)—national companies, major international companies and more or less independent companies.

National companies have recently been having problems financing their share of investment in joint ventures (or associations), and now prefer production sharing agreements. Moreover, their management has been highlighted by donors when granting loans to states. These national companies include Societé National des Pétroles au Congo (SNPC) in Congo, Societé National des Hydrocarbures (SNH) in Cameroon and Sociedade Nacional de Petróleos de Angola (SONANGOL) in Angola.

The major international companies include ExxonMobil, Shell, ChevronTexaco, Agip and TotalFinaElf, and have a presence in Angola, Cameroon, Gabon, Nigeria, Congo, etc. In African countries, these companies express their power in terms of their financial capacity and technical expertise. They have extensive investment capabilities and represent an easy source of credit for governments. Profit sharing between these companies and governments is essentially based on future revenue, hence the emergence of production sharing agreements.

'Independent' companies of varying sizes pursue a policy of positioning themselves in specific locations across various African countries. For example, the countries around the Gulf of Guinea provide opportunities for small independent companies seeking to geographically diversify their activities.

Oil and gas exploration and production are activities that are largely independent from the other economic activities within the extractive industry value chain. Their main characteristic is that they currently generate few jobs.

Oil in the form of fuel constitutes the main source of energy in Central Africa, with the exception of traditional forms of energy such as wood, charcoal and animal and vegetable waste. Fuel is distributed via service stations controlled by international and/or national companies. Central Africa has limited capacity for refining oil products in Cameroon (Societé Nationale des Raffinage, SONARA at Limbé) and the Congo (a small refinery at Pointe-Noire), and others. This limited refining capacity is insufficient to meet local demand. The countries of Central Africa therefore have to import fuel. Investment in both upstream and downstream extraction activities is limited in Africa.

The extractive industry value chain can be represented by five links (Fig. 1). The first link in the chain is the negotiation of agreements. The agreements basically relate to production.

In Cameroon, there are two types of agreement in the oil sector—licence agreements and production sharing agreements. Licence agreements are oil agreements linked to a hydrocarbon exploration licence and, if need be, one or more extraction licences. Production sharing agreements are oil agreements that give the holder payment in kind in the form of a share in production.

The second link represents the monitoring of operations. This involves the actual production and processing of the materials extracted (refining). The third link is the collection of taxes and fees. The fourth link is the distribution of revenue and the fifth link is the use of this revenue in sustainable projects.

The Extractive Industries Transparency Initiative (EITI) is only involved in one link of the extractive industry value chain, namely the fourth link—the distribution of revenue. The stakeholders involved here are states or governments, civil society and the private sector (businesses). The EITI does not guarantee that countries will develop, but it does check actual flows of money and the quality of governance in countries benefiting from the revenue. It aims to increase transparency and the accountability of those involved in extractive industries by publishing and checking full details of payments made by companies and revenue received by governments. It is a process by which government revenue generated by extractive industries is published in independently verified reports. These reports are based on information regarding payments made by companies and revenue received by governments.

The EITI is able to promote a transparent commercial environment and help reduce corruption and poverty. This transparency may lead to an influx of additional investment, helping to improve economic and social conditions in countries and promoting sustainable development. The fact that civil society is informed of government revenue from extractive industries may help reduce social tension and lead to improved public accountability and political stability. Transparency therefore contributes towards creating a culture of public accountability, and of public trust between civil society, companies and governments. A lack of transparency often undermines public trust by questioning the legitimacy of those who control mineral resources. It is therefore important to achieve transparency in revenue and financial management in order to make the authorities accountable, gain the public's confidence and reduce corruption and poverty.

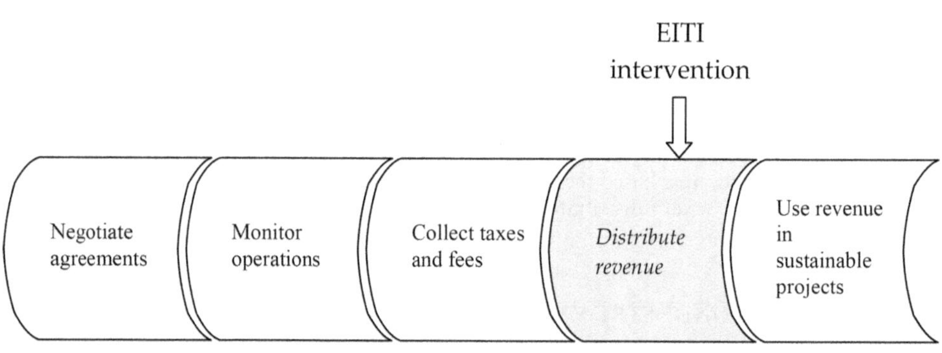

Figure 1. The extractive industry value chain can be represented by five links.

3 WAYS OF ACHIEVING TRANSPARENCY IN THE EXTRACTIVE INDUSTRY VALUE CHAIN

The extractive industry value chain consists of several links to which the EITI principles could be applied. Revenue is involved in every link of the value chain. Exploration eats away at the revenue that was anticipated when the agreements were negotiated. Oil agreements can only be concluded with a single oil company or jointly with several commercial partnerships, at least one of which is an oil company. Negotiating oil agreements is often not a transparent process, for example if carried out in accordance with Article 9 of the Cameroon Oil Code. This article stipulates that the government will handle proposals for oil agreements and requests for authorisation at its absolute discretion. If proposals or requests are rejected, either absolutely or conditionally, this does not give the applicant any right of appeal or entitlement to compensation whatsoever.

The 'monitor operations' link relates to production. The majority of oil agreements in Central Africa are production sharing agreements. Much information regarding production is still not available to the public. Congo remains the only country in Central Africa that has attempted to publish information on its oil agreements.

4 WAYS OF ACHIEVING TRANSPARENCY IN THE PROCESS OF ISSUING LICENCES

In upstream extraction activities, transparency could be applied to various links in the value chain. For example, it could apply to the process of negotiating oil agreements in the first link, and to information on 'cost oil' and 'profit oil' in the second link. Note that 'cost oil', or the production required in order to recover costs, represents that proportion of total hydrocarbon production earmarked to reimburse the costs that are effectively borne by the licence holder. 'Profit oil' is the proportion shared between the State and the licence holder in line with the production sharing agreement. The agreement negotiation process is still opaque, as is the issuing of associated licences for exploration and extraction.

5 CONCLUSION

The EITI could lay down stricter criteria regarding the quality and consistency of reports. The CEMAC CTIIE secretariat (Commission Technique Interministerielle sur l'Information Elaboré—Interministerial Technical Commission for Information) could also play this role within CEMAC by defining stricter criteria.

- Extractive industries would therefore have to publish the market value of their payments, not only the total amount of the payments made.
- EITI member states should investigate the entire value chain thoroughly, from the granting of licences and the negotiation of agreements through to the way governments use the resulting revenue.
- Subnational payments should also be made public. A subnational EITI report could be drawn up by agencies or experts from the CEMAC CTIIE secretariat.

The major challenge facing Central African oil states today is their transition from revenue-based economies focusing solely on production, moving towards revenue-based economies focusing on both production and revenue-sharing. This represents a change from an unbalanced, opaque situation involving two parties (governments and oil companies) that is unjust due to its remoteness from the people, moving towards a tripartite situation (involving governments, the private sector and civil society) that favours sustainable development.

REFERENCES

BIT 1999. Problèmes sociaux et de travail dans les petites exploitations minières. Bureau International du Travail (BIT), Genève, 1999.

Ernst et al. 2007. *Rapport sur les revenus pétroliers et miniers de la République Gabonaise pour l'année 2005*. ITIE—Gabon, 2007.

Ernst et al. 2008. *Rapport sur les revenus pétroliers et miniers de la République Gabonaise pour l'année 2006*. ITIE—Gabon, 2007.

Foute, R.-J. 2008. Exploitation minière—le Cameroun attire. *Cameroon Tribune*, 07.02.2008.

Gendron, C. 2006. Le développement durable comme compromis. *PUQ*, 2006.

Harsh, E. 2005. Les investisseurs se tournent vers l'Afrique. *Afrique Renouveau*, 2005.

Mazars/Hart Group 2006. *Rapport de conciliation des chiffres et des volumes dans le cadre de l'ITIE au Cameroun pour les années 2001–2004*. ITIE—Cameroun, 2006.

Mazars/Hart Group 2007. *Rapport de conciliation des chiffres et des volumes dans le cadre de l'ITIE au Cameroun pour l'année 2005*. ITIE—Cameroun, 2007.

MINIMIDT 2001. *Code minier du Cameroun. Loi N° 001 du 16 avril 2001 portant code minier*. Ministère de l'Industrie, des Mines et du Développement Technologique (MINIMIDT), 16.04.2001.

Ndanga Ndinga, B. 2008. Mieux rentabiliser les ressources minières du Cameroun. *Cameroon Tribune*, 25.01.2008.

Theys, J. 2005. *Le développement durable: une innovation sous-exploitée*. Dalloz, Paris 2005.
Touna, R. 2008. Production de cobalt dès 2010 au Cameroun. *Jeune Afrique*, 30.06.2008.
UNCTAD 1997. *Environmental management of small-scale and artisanal mining sites in developing countries*. L.R. Blinker. UNCTAD Consultancy Report.

Geological Resources and Good Governance in Sub-Saharan Africa – Runge & Shikwati (eds)
© 2011 Taylor & Francis Group, London, ISBN 978-0-415-58267-4

Investments in the mining industries for sustainable development

E.P. Ambomo
EITI / BEAC, Yaoundé, Cameroon

ABSTRACT: The paradox of mineral resources in Central Africa was scientifically proved. The exploitation of the mineral resources by the mining industries should constitute an asset for the economic growth, and the question is where does the wealth generated by natural resources go? The Initiative of Transparency in the Mining Industries (ITIE) came up with principles and criteria for the countries which adhered to the initiative and were worried about good governance such as publishing the payments made by companies exploring and exploiting minerals for better management of the wealth. There are various definitions for sustainable development. For decades, this concept has been westernised and implies the ecological, social and economic durability. The environment becomes a political key stake constituting a real brake for a sustainable economic growth of the African countries in the process of development and those of the world generally. A conference on the global warming in Copenhagen, Denmark put together about approximately 193 countries with the objective, of reducing of 2 °C of the greenhouse gas, so as to reduce pollution in the world. All these steps aim at maintaining the economic growth, hence reducing poverty.

1 INTRODUCTION

The investments in the mining industries for the well-being of the populations were always a marigold for the countries of Central Africa. The recipes of the major part of these states rich in natural resources result from mineral wealth. However, these countries are among the poorest in the world, yet the resources should nevertheless contribute to the economic growth of these states.

To mitigate this paradox, experts, scientists and specialists raised the question of good governance in exploitation and distribution of wealth. A lack of transparency leads to disparities entailing various internal and state conflicts, corruption and various crises.

So, the exploration and the export of natural resources by the mining industries (oil, gases and ores) should constitute an asset for a sustainable economic growth, better, that leads to sustainable development which is a conception meant to reconcile the ecological, social and politico-economic factors, and which recommends use of the resources, the investments, the technologies and the institutional development in a manner that does not affect the health and the well-being of future generations. Sustainable development should ensure that the survival of human beings is achieved so as to reduce poverty.

2 SUSTAINABLE DEVELOPMENT

There are various definitions for sustainable development. In 1979, the economist Howe gave two definitions to this concept.

1. Sustainable development is an example of economic, social and cultural transformations which optimises benefactions social and economic, accessible at the present moment, without compromising the similar benefactions in the future. One of the essential objectives of the sustainable development being to reach an acceptable level (however defined) and fair of the distributions of the economic well-being, which can be constantly immortalised for a large number of generations.
2. Sustainable development implies the use of the renewable natural resources in a way which degrades them or does not eliminate them, or which does not decrease their potential use by the future generations. Sustainable development also implies depletion of renewable energy resources in a very reduced speed to assure a methodical social transition towards renewable sources of energy.

For decades, the notion of sustainable development is typically western, and does not make sense in most of the African countries where human beings live in accordance with nature. In Central

Africa, the fight against the failing management of recipes resulting from mining industries is on the right track and the questions on sustainable development keep arising. Sustainable development hast to do with ecological durability, social durability and economic durability which allows the establishment of the link between the investments in the mining industries and the sustainable development. These three factors do not work in isolation but together to form a whole.

2.1 *Ecological durability*

The ecological durability is a science of the biology. It studies connections between the geography and the politics of States. It is also the capacity for a company to create the business model which can contribute and benefit from big environmental tendencies. Ecological durability is the foundation of sustainable development and it recognises that sustainable economic growth is not practicable without considering data and environmental impacts. It can be possible only if we protect more the nature for a better balance between the human being and its environment.

The environment is a key political stake with which the governments are confronted. Reduction of pollution is a crucial objective for companies proposing environment solutions. Ecological durability aims at regeneration of the nature.

To protect the environment and use in the long term what nature offers, awareness by the human being is necessary. The current exploitation of natural resources which does not take into account the regeneration of the nature, and capacity of ecological load is a danger for future generations.

2.2 *The social durability*

The social durability was defined during the conference of United Nations for the Environment and the Development (UNCED) in 1992 in Rio as follows:

1. Right of human life for all,
2. Equity generative, intra generative and international inter,
3. Participation of all the groups of social actors.

It is a process of social development and social justice without which we cannot arrive at the environmental durability. It is one of the fundamental constituents of the sustainable development supported by scientists, economists, specialists, politicians, whose main objective is the harmonisation of social relationships.

Previously, human needs (the social, economic and political obstacles) which confronted individuals were not taken into account. Only particular or selfish interest counted for the leaders in the power. However, due to degradation of the environment and all the problems connected to the social progress in the world, more particularly in Africa, measures are taken by the governments to find adequate solutions to satisfy the fundamental human needs, as community facilities, public utilities, housing, health, education and employment.

In a socially sustainable approach, the question is on equity and equality of the opportunity. Income stemming from mining industries in Central Africa is not still distributed in a fair way with regard to the priorities of sustainable development. They are generally diverted by a handful of people leading to diverse conflicts.

To guarantee social cohesion, organisations in charge of the management of this income should take into account all the sectors of society of the community and ensure good governance, job creation and transparency in the redistribution of recipes.

2.3 *Economic durability*

Economic durability is the preservation of the economic growth and the well-being of the individuals, its objective is to reduce the disparities between various sectors of society.

For economic durability to be possible in Central Africa, its development should be assured directly by states. These states which set up services and measures of support for economic security offer possibilities of employment to marginalised citizens in the society, by assuring better usage of technologies connected to the processes of production and by minimising negative effects on the environment.

3 INVESTMENTS IN THE MINING INDUSTRIES

Mining industries indicate that they explore and extract minerals under natural, underground, on-surface shape, by digging a well or exploiting the sea bed.

These exploitations include the production of the crude oil and the gas by the extraction of hydrocarbons, the subterranean exploitation, and by the extraction of ores, etc. A better productivity in the industries requires certain number of investments like placed funds, every average human beings and used materials which help to improve the quality of the work and constitute an asset for sustainable economic growth. So one of the objectives is the reduction of the poverty by sustainable and sure means of support which minimise the depletion of resources, the degradation of the environment, the cultural break and the social instability.

How can the investments in the mining industries be translated by a sustainable development? The economic growth in the sector of the mining industries can be determined by the quality of the means set up for better productivity by protecting spaces, resources, natural circles, sites and landscapes, botanical and animal species. Their exploitation, their development, their restitution and their management should be for the people's interest.

4 EXAMPLE OF AN INVESTMENT IN THE MINING INDUSTRIES FOR A SUSTAINABLE DEVELOPMENT

The case of the pipeline Chad-Cameroon built from 2001 till 2003 for a duration of 25 years is a convincing example of an investment in the mining industries in Central Africa. All the activities connected to the construction of this oil pipeline that is 1070 km long between Doba (Chad) and Kribi (Cameroon) required enormous human, material and financial investments.

Indeed, the discovery of the quantities of the oil to Doba in the South of the Chad in 1969 and the possibility of development due to this oil, led to the origin of the Chad-Cameroon pipeline. Chad being a landlocked country, Cameroon was chosen to evacuate the Chadian oil for export.

The Chad-Cameroon pipeline assures the transport of the Chadian crude oil and across five ecological zones between Chad and Cameroon. The exploitation of this discovery was not realised without impact on the ecological, social and economic factors.

On the ecological plan and for the resolution of the environmental problems such as the forest, the project "CAPECE" had been retained to protect the environment without which the well-being of the future populations would be affected. The project was to substitute forest resources destroyed through for example reforestation.

On the social plan, job creation thanks to the various activities in the mining sector, the standard of living of the populations changed considerably. Several families were compensated, and changes of cultures were observed, etc.

On the economic plan, the construction of this oil pipeline brought recipes and currencies as well in the Chad as in Cameroon. A stability of recipes during at least 25 years and an economic preservation for these two countries is an experience, but especially for the Chad.

5 CONCLUSION

Central Africa's rich in natural resources have no other considerable sources of income. It can make sure that sustainable development is maintained, thanks to recipes generated by this wealth. Good governance, fair redistribution, and transparency play a key role in sustainable development and should not be ignored.

Do these African countries arrange for means that allow them to exploit (run) all these natural resources which can take out of underdevelopment? It is important to indicate that slowness in the implementation of the fight against the pollution in the world was observed. It is very urgent to act on the degradation parameters like water, ground, air, forest, climate, health, of the massive population growth and the mismanagement of the resources.

REFERENCES

Assidon, E. 2002. *The economic theories of the development. Collection Marks*, 3rd edition. The Discovery, p. 122.

Faucheux, S. & Joumni, H. 2009. *Economy and politics of climate change*. Ed. The Discovery.

Gauchon, P. 2005. *Geopolitics of the sustainable development*. PUF.

Gendron, C. 2006. *The development sustainable as compromise*. PUQ. p. 276.

Gendron, C. & Reveret, J.-P. The sustainable development. *Development, growth and progress*, 37, p. 14.

Godard, O. 1994. The sustainable development: intellectual landscape. *Natures, Sciences and Companies. Flight (theft)* 2, 4: 309–322.

Mancebo, F. 2008. *The sustainable development*. Ed. Armand Colin.

Tellenne, C. 2006. *Sustainable development in the heart of the company*, Collective, Dunod.

Theys, J. 2005. *The sustainable development: an under exploited innovation*: 108–119. In Smouts, Dalloz, Paris.

Geological Resources and Good Governance in Sub-Saharan Africa – Runge & Shikwati (eds)
 ISBN 978-0-415-58267-4

International approaches to improve resource governance in Africa*

M. Schnell & M. Großmann
Economic Section of the Africa Region, Gesellschaft für Internationale Zusammenarbeit (GIZ) GmbH, Eschborn, Germany

ABSTRACT: Revenues resulting from the extraction, processing and export of mineral resources can potentially contribute significantly towards the financing of Africa's development needs. Different steps, ranging from the award of contracts and licences, regulation and monitoring of extraction activities, to the allocation of revenues are required to put a country's resource endowment to good use. This paper will analyse in more detail the different elements of resource governance as highlighted by the World Bank Extractive Industries Value Chain and present an overview of international and bilateral approaches to support resource governance in Africa. Until today, most support approaches focus on the regulation and monitoring of the extraction, processing and trade of resources as well as on the collection of taxes and royalties. Yet, resource governance should be understood and addressed in the broader context of good (democratic) governance and should focus on all steps of the Extractive Industries Value Chain.

1 INTRODUCTION

Africa is rich in mineral resources such as oil, gold or copper. Many African countries like Nigeria and Angola (oil), or Botswana and South Africa (diamonds) became major global suppliers of mineral resources. If used appropriately, the revenues resulting from the extraction and export of resources might contribute substantially towards the social and economic development of the continent. In 2008, the export revenues from African fuel and mining products were US$394 billion (Africa Progress Panel 2010). This amount is a multiple of the Official Development Assistance (ODA) that was provided to Africa in 2008. According to a study conducted by the Overseas Development Institute (ODI), revenues from the resource sector could fund a large part of the financing gap that exists for the successful implementation of the MDGs by 2015 (Warner & Kyle 2005).

However, despite these potentially good conditions, Africa's resource endowment was often a curse rather than a blessing in the past. In the majority of resource-rich African countries, poverty rates remain high and inequalities are increasing. Resource endowment has rather impeded sustainable development of those countries.

This phenomenon is also known under the terms "paradox of plenty" or "resource curse." Bad governance and the strong dependency on revenues from the resource sector in many countries led to corruption, weak development orientation of national policies and strategies and a higher conflict potential in general (El Bakri 2009). Only if countries adopt the principles of good governance, strengthen their institutions and help reducing corruption, might Africa's resource endowment actually contribute towards sustainable development.

The perception that improved resource governance is a major pre-condition for sustainable development has motivated more and more donors and Non-Governmental Organisations (NGOs) to increase their engagement in this area. A major focus of this engagement lies on the Extractive Industries Transparency Initiate (EITI), which aims at improving the transparency of public revenues emanating from the resource sector. Although the transparency of revenues is a major pre-condition for ensuring sustainable development, resource governance comprises more than just transparent revenues. In this sense, it is also important to take into account the award of contracts and licences, the regulation and monitoring of extractive operations (including the processing and trading of resources), the collection of taxes and royalties, and the revenue management and allocation.

The aim of this paper is to analyse in more detail the different elements and preconditions of resource governance as highlighted by the

*This paper reflects the authors' opinion and does not necessarily reflect the position of GIZ.

World Bank Extractive Industries Value Chain (Alba 2009) and to present existing international approaches to support Resource Governance in Africa. In addition, we will also provide examples that are being implemented by the German bilateral development cooperation to illustrate different support approaches. While we hereby focus on GIZ examples in the context of technical development cooperation, which we do know best, neither do we intend to provide a complete picture of existing approaches nor do we represent the official position of German development cooperation or GIZ. The discussion includes an evaluation of the approaches' major strengths and weaknesses and an analysis of their relevance for sustainable development. For this purpose, the major concepts and definitions will be presented first in section two, including an overview of the different elements of the Extractive Industries Value Chain and their relevance for sustainable development. This is followed by an overview and analysis of major international and bilateral approaches in section three. The last section summarises the main arguments and suggests a number of recommendations for practical development cooperation in this area.

2 RESOURCE GOVERNANCE AND THE EXTRATIVE INDUSTRIES VALUE CHAIN

2.1 *Definitions*

According to the definition of the Bonn International Center for Conversion (BICC), the term resource governance mainly focuses on government actions (BICC 2007). Resource governance describes how governments or government bodies regulate the extraction of extractive resources and the distribution of the resulting benefits and costs. This definition covers all steps of the resource extraction process for sustainable development, which will be explained in greater detail below with the help of the Extractive Industries Value Chain.

In this context, we also argue for a broader term or definition of "resource governance" that includes the actions and behaviours of all relevant stakeholders in the resource sector. By focusing the term "resource governance" on government actions, other crucial stakeholders such as private businesses that extract resources might not be sufficiently taken into account. As will be explained in more detail in section three below, it might be useful to strengthen resource governance via the private sector, parliaments and non-governmental organisations as well.

The term "extractive resources" defines non-renewable mineral and energetic resources that are extracted or mined. This definition does not include agricultural resources as they are renewable and thus require different approaches and policies than non-renewable resources.

2.2 *The Extractive Industries Value Chain*

Figure 1 shows the major elements of the World Bank Extractive Industries Value Chain that are relevant for making use of extractive resources for sustainable development (adapted from the model suggested by Alba 2009).

The value chain comprises the following elements: 1) award of contract and licences; 2) regulation and monitoring of extraction, processing and trade; 3) collection of taxes and royalties; and 4) management and allocation of revenues for sustainable development strategies and programmes. For the different elements to be in line with the principles of good governance, sufficient institutional capacity, transparency, participation of stakeholders and the parliament are necessary.

The Extractive Industries Value Chain was a model developed by the World Bank to show the different steps that are needed to make use of extractive resources for sustainable development (Alba 2009). The model is not exhaustive and can be complemented by additional steps, like the surveying of resources by the state. The elements presented here cover the major steps in the extraction process of resources and the usage of the resulting revenues for sustainable development.

2.2.1 *The award of contracts and licences*

The extent to which the extraction of resources might contribute towards sustainable development is already influenced at the stage of awarding contracts and licences. As a consequence, a number of conditions should be met at this stage of the extraction process. First, contracts concluded between the respective government and the extracting company should be comprehensible and implementable. This includes in particular the implementation capacities of local institutions: only contracts that take into account the existing capacities can be successfully implemented and monitored. In many African countries, this criterion is often not met: the implementation of complex contracts with several hundreds of pages is more difficult and prone to mismanagement in countries with weak capacities and limited technical know-how.

Secondly, contracts should ensure a fair balance of interests between the extracting company and the respective host country. This also means that contracts should include provisions and rules that describe the company's duties and engagement,

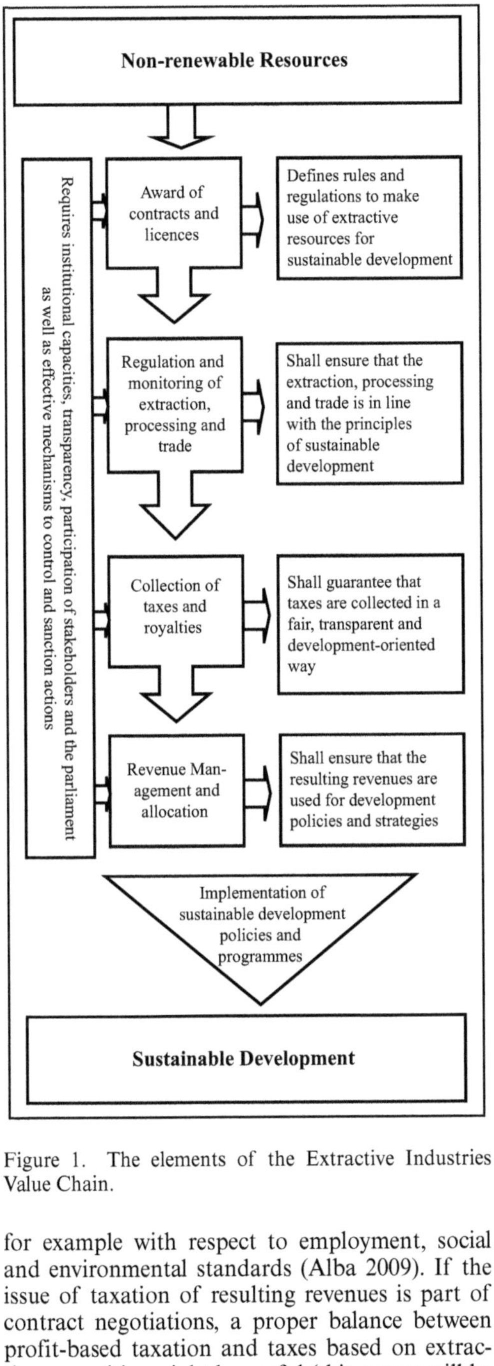

Figure 1. The elements of the Extractive Industries Value Chain.

for example with respect to employment, social and environmental standards (Alba 2009). If the issue of taxation of resulting revenues is part of contract negotiations, a proper balance between profit-based taxation and taxes based on extraction quantities might be useful (this aspect will be explained in greater detail in section three below).

Thirdly, whenever possible and sensible, the process of awarding contracts and licences as well as their content should be published in order to limit potential incentives for corruption or unfavourable conditions for the respective host country. Only if the public is informed about the details of the contracts and licences, can politicians and companies be held accountable. In many African countries, these principles are not yet fully taken into account as neither the contract awarding process nor the contracts themselves are transparent. They often contain secret clauses and are not subject to parliamentary control or scrutiny by the civil society (Unmüßig & Fuhr 2007). For those cases, where secret clauses might be necessary due to competition reasons, an independent organisation should monitor that minimum standards are kept and should inform the public accordingly.

Before licences are awarded, detailed analyses of the expected economic, social and environmental consequences should be made compulsory in order to assess and address the potentially positive and negative impacts of resource extraction.

In general, the negotiation of contracts is often more difficult for governments of developing countries than for international companies that often have considerable negotiation advantages due to their year-long expertise, sector knowledge as well as a pool of specialised lawyers.

2.2.2 *Regulation and monitoring of extraction activities, processing and trade*

The regulation and monitoring of the resource extraction process does not only require efficient lawmaking procedures, but also the effective implementation of adopted rules and standards. In this context, the public sector administration plays a key role. In order for the administration to effectively carry out its duties, its rights and responsibilities need to be clearly defined. In many African countries, the implementation of rules and standards often fails due to weak sector administrations.

Certification systems are important elements of an effective supervisory mechanism for the extraction, processing and trade in resources. They should also be designed to such an extent as to ensure that social and ecological standards are met (e.g. preventing trade in so-called "conflicts diamonds", certification of timber or other rare resources).

2.2.3 *Collection of taxes and royalties*

The extraction of resources is subject to a complex taxation system that includes taxes and other charges like licence fees and royalties collected by the government. For many countries, these taxes and fees are important sources of financing that can contribute towards poverty reduction and economic development.

The way companies that extract resources are taxed, depends partly on the characteristics of the existing taxation system and the type of resources (Land 2009). In principal, taxation can occur in two different ways. First, companies can be subjected to individual taxation on the basis of licensing agreements. As a result of such a system, each company is subjected to individual taxation principles with different regulations.

Contrary to this approach, the taxation of the resource sector can also occur through the general tax system. In this case, all resource extracting companies are subjected to a common and transparent tax regime that is no longer part of individual negotiations. In many African countries, a hybrid system is being used: although general tax systems exist that were legitimated by parliaments and that contain fixed tax rates for resource extracting companies, tax rates are still often regulated individually through separate agreements and contracts.

One additional aspect of the taxation of resources is the issue of profit-based taxes versus taxes based on turnovers. Profit-based taxes provide attractive investment incentives for resource extracting companies but might lack the legitimacy of the public. This could be due to the fact that companies can write off their initial investments, thus reducing profits during the first couple of years. Profit-based taxes should therefore be balanced carefully with taxes based on the amount of resources that are extracted and the resulting company turnover.

Irrespective of how resources are taxed in a specific country, the transparency of public revenues from the resource sector plays an important role. In many African countries, the legislature and civil society have only limited information about the amount and use of revenues emanating from the resource sector and thus have difficulties in fulfilling their accountability function. In countries with governance deficits, this might create additional incentives for the misappropriation of funds, leading to corruption, embezzlement and mismanagement. In this context, the disclosure of payments is an important condition for better resource governance—only if parliaments, civil society and the public at large are informed about the payments, will they be able to fulfil their supervisory role and—for the case of parliamentarians—execute appropriate budget control.

2.2.4 *Revenue management and allocation*

The central aspect of resource governance relates to the question of how and for what purposes the revenues emanating from the resource sector are used. A country's resource endowments can only contribute towards development if the revenues are used for development objectives. Therefore, institutions and mechanisms that ensure the proper use of those funds are crucial. The biggest challenge in this context is to improve governance and institutional capacities in the respective countries.

Numerous studies revealed a negative relationship between high revenues from the resource sector and governance (Isham et al. 2006, Damania & Bulte 2003, Collier & Hoeffler 2005). This is due to the fact that substantial revenues from the resource sector reduce governments' dependency on other tax revenues (and thus also reduces the dependency on the tax payers themselves), which might create incentives for corruption and clientelism. In order to prevent such a development, one priority should be to strengthen and increase the transparency of the public financial system and to strengthen parliaments' supervisory and accountability functions.

Closely linked to the quality of the public financial system is the question whether funds are invested in a sustainable manner. In line with the fact that mineral resources are finite and deposits will be depleted sooner or later, revenues resulting from these resources are also finite and non-sustainable in the long run. The proper investment of these resources is important for the post-resource era. One focus of the investments should be on the diversification of the economy, e.g. through strengthening promising industries, the agriculture or tourism sectors. Moreover, governments should also invest in measures such as education and training, health and environmental programmes or projects aimed at the local public in the affected mining areas. The example of Botswana demonstrates that it is possible to put a country's resource sector to good use for sustainable development (Adler 2006).

Another option is to invest parts of the revenues from the resource sector in special funds. One example often cited is Norway that pays part of its oil and gas revenues into a public pension fund administered by the central bank. A maximum of 4% (more than US$300 billion have already been paid into the fund) can be used yearly to fund innovations and projects in the context of the state budget. The majority of funds are reserved for the period after the end of the resource boom (Williams 2007).

In order to prevent macroeconomic imbalances like rising inflation and currency appreciations (also known as "Dutch Disease"), economic policies need to be adjusted accordingly. This includes, for example, investment policies and programmes that are aligned with the amount of revenues received from the resource sector. Moreover, reserves for critical times might contribute towards the economy's long-term stabilisation.

3 OVERVIEW AND EVALUATION OF INTERNATIONAL AND BILATERAL APPROACHES TO IMPROVE RESOURCE GOVERNANCE IN AFRICA

As the following examples demonstrate, there are currently a number of international and bilateral approaches and measures to support resource governance that focus indirectly or directly on the different steps of the value chain. Figure 2 provides an overview of the different approaches. They will be explained and evaluated in more detail in the following sections.

3.1 *Award of contracts and licences*

One established approach to support developing countries already at the stage of negotiating with international resource extracting companies is the provision of so-called negotiation advisors. These advisors support developing countries with their negotiation process with companies in the resource sector. One crucial issue often not covered by advisors, however, is whether the contracts are implementable and comprehensive. In the context of limited implementation capacity in many developing countries, it is of particular importance that the negotiated contracts are clear and also take into account the existing capacities of local institutions and processes. One possibility to address this issue is to use "Model Development Contracts." The model contracts can be complemented by the different parties during the negotiation process and have the advantage that the structure of the contract and ready-formulated paragraphs are already provided. This might improve the negotiation and coherence of such contracts. One way of promoting the introduction of model contracts might be through an international body (e.g. at the UN level). It could support developing countries with the preparation of contracts and might also serve as a knowledge manager in this area. Contracts might then be screened before entering into force and governments could be advised accordingly. One example here is the support provided to developing countries in the context of WTO negotiations, where developing countries are equally challenged to negotiate rules and regulations, often without having the proper negotiation capacities (Lal Das 2009).

Furthermore, there is currently a lack of approaches that aim at improving the transparency of negotiation processes and contracts. As was discussed above, neither are most contracts concerning the resource sector publicly available, nor are the negotiation processes fully transparent. Future support approaches could also be part of projects and programmes in the areas of good governance (e.g. public procurement) or private sector development (e.g. investment promotion) that could be extended to cover issues relevant for the resource sector.

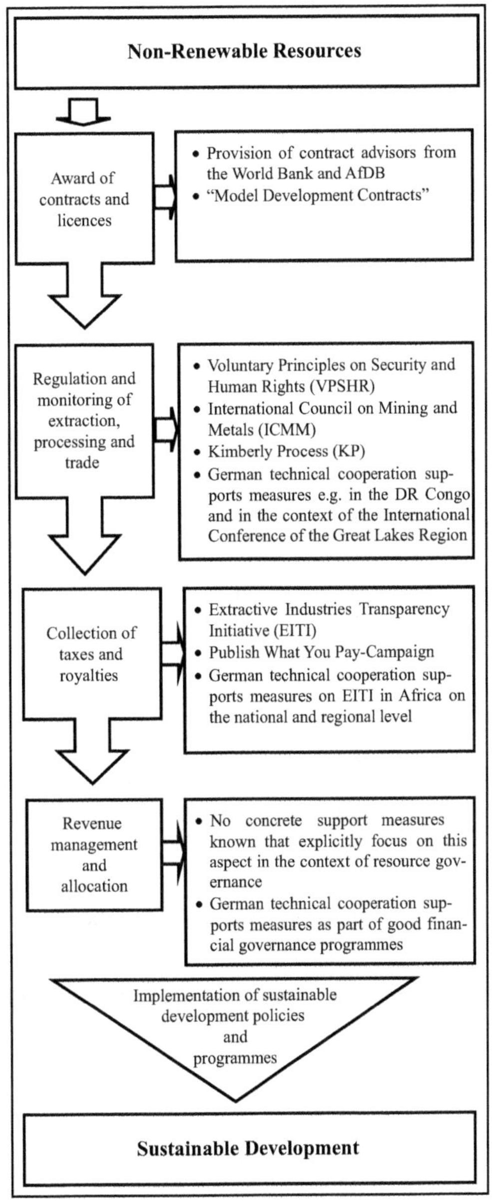

Figure 2. International and bilateral approaches to support resource governance at the different levels of the value chain.

3.2 *Regulation and monitoring of extraction activities, processing and trade*

In the context of regulating the extraction, processing and trade in resources, a number of international approaches exist. First, there are a number of voluntary principles and behavioural norms, especially with regard to human rights and social and ecological standards that corporations adopted. The most important ones are the Voluntary Principles on Security and Human Rights and the voluntary standards of the International Council on Mining and Metals (ICMM).

3.2.1 *Voluntary Principles on Security and Human Rights*

The Voluntary Principles on Security and Human Rights (VPSHR) are voluntary principles adopted by a number of companies in the oil and mining sector (Voluntary Principles 2000). They provide guidelines for dealing with state or private security forces. Furthermore, they aim at reducing the risk of human rights violations in the case of interventions by security forces. The principles were jointly developed by governments, resource extraction companies and NGOs.

3.2.2 *International Council on Mining and Metals*

The International Council on Mining and Metals (ICMM) that consists of 19 mining companies and 30 mining associations developed 10 principles for sustainable development in the mining industry (ICMM 2010). They aim at improving social and ecological standards and contribute to the solution of conflicts between resource extraction and the preservation of nature.

Although these standards represent a step in the right direction to ensure social and environmental sustainability and to reduce conflicts, they are based on voluntary principles and thus are only relevant to businesses that intend to adhere to these principles anyways. Corporations that are subject to less public scrutiny might be inclined not to adhere to them. Moreover, the standards with regard to human rights often comprise unclear rules, and the requirements to fulfil the standards are not always entirely clear.

3.2.3 *Kimberley Process*

In addition to standards and rules, a number of certification systems for different resources exist. The best-known certification system was established as part of the Kimberley-Process (Kimberley Process 2002). It was established in the context of the issue of "blood diamonds" resulting from the civil wars in Liberia, Sierra Leone and Angola in the 1990s and 2000s to prevent trade in conflict diamonds with the help of a complex certification system. Currently, 47 countries—including all of Africa's diamond producers—participate in the process. An international body is tasked to assess whether countries correctly issue certificates.

Although the Kimberley-Process was certainly successful in many areas, it is also subject to increasing criticism by human rights advocates, scientists and NGOs. The latest report of *Partnership Africa Canada* (*PAC*) criticises above all governments' lacking willingness to effectively prevent smuggling in diamonds, human rights abuses and the lack of sanctioning mechanisms (PAC 2010). PAC provides the example of the authoritatively governed Zimbabwe, where the political elite assumedly engaged extensively in illegal dealings with diamonds; however the participants of the Kimberley-Process could not agree on effective sanctions at one of their last meetings in November 2009 (PAC 2010). The German aid organisation Medico International furthermore criticises weak control mechanisms (Medico International 2007).

The Kimberley-Process is complemented by a number of other certification processes for extractive resources. These also serve to guarantee quality standards in global trade and to document that standards were met. The German Federal Institute for Geosciences and Natural Resources (BGR), for examples, supports the development of a certification system for the tantalum ore "*Coltan*". For this purpose, the scientists at BGR developed a forensic test that allows identifying the origins of Coltan (BGR 2010).

One example for bilateral support approaches is the certification of resources: in the context of the regulation and monitoring of operations, German technical cooperation, for example, supports the certification of extractive resources in the DR Congo as part of its support to the International Conference of the Great Lakes Region (Schnell & Theissen 2010).

3.3 *Collection of taxes and royalties*

One important precondition for the sustainable use of extractive resources at this level of the value chain is the publication of the public revenues from the resource sector and the strengthening of local tax authorities.

3.3.1 *Extractive Industries Transparency Initiative (EITI)*

Internationally the most important initiative is the Extractive Industries Transparency Initiative (EITI) that was established in 2003 (Müller 2009, EITI 2003). The aim of this international, multistakeholder initiative is to increase the transparency of public revenues emanating from the resource sector. About half of the 53 countries worldwide with considerable extractive resource endowments committed themselves to the implementation of the EITI. One important incentive for a membership are positive external effects that result from better transparency in the resource sector—especially in terms of international prestige and attractiveness to foreign investors.

For the EITI implementation to be successful, member countries need to go through a validation process after two years the latest. Members will only receive full membership status (compliant status) if this validation is successful. In total, six criteria need to be met (EITI 2003). First, it is necessary that all payments resulting from the resource extraction that are made by the corporations as well as the revenues received by the government need to be published and validated through an independent auditing process (criterion 1 and 2). Furthermore, an independent administrator needs to compare the amounts that were spent with the amounts received and is required to document any discrepancies (criterion 3). The process needs to cover all public and private businesses in the resource sector (criterion 4). In addition, the civil society needs to participate actively in the design and monitoring of these processes (criterion 5). Criterion 6 is the development of a public and financially sound work plan for the implementation of these processes (including measurable target indicators and an implementation time frame) by the government. Countries that meet these criteria are re-evaluated every two years. Currently, only five out of a total of 35 countries passed the validation (these are Azerbaijan, Ghana, Liberia, Mongolia and Timor-Leste).

Internationally, EITI is the most prominent approach to improve transparency in the resource sector. The international community supports the implementation on the bilateral level and via the EITI Multi Donor Trust Fund managed by the World Bank (EITI 2010). Highlighting examples from bilateral development cooperation, one major focus is equally on EITI: six support measures that are currently being implemented by German development cooperation in Africa focus on EITI at the national and regional (i.e. at the level of Regional Economic Communities) levels (Schnell & Theissen 2009).

Due to the regular publication of revenues emanating from the resource sector and the dialogue with different stakeholders at the local level, the EITI has contributed significantly towards improved transparency in the resource sector since 2003. However, despite the progress made so far, the instrument also has weaknesses, in particular with regard to the incentive system. For one, the initiative is voluntary and, until today, it is not binding for all resource rich countries. Moreover, there is no effective sanctioning mechanism in case of a breach of standards and regulations. In 2008, the validation process was introduced as a means of quality guarantee. It is yet too early to assess, whether and to what extent the validation represents an effective sanctioning mechanism. Only five countries have passed the validation so far. For another 20 countries, the deadline for submitting a validation report ends in October 2010. It remains to be seen to what extent the EITI board intends to apply the validation rules and how it weighs the criterion of a free civil society. According to a study conducted by the German Heinrich-Böll Foundation, civil society organisations still face numerous challenges in many EITI member countries that include, for example, travel bans or arrests (Heinrich Böll Stiftung 2005). The EITI initiative is at a crossing point: only if countries that deliberately breach the rules and regulations of the EITI are sanctioned, can the EITI maintain its reliability, making the validation process an effective sanctioning mechanism.

3.3.2 *Publish What You Pay Campaign*

Another international approach to improve the transparency in the resource sector is the Publish What You Pay Campaign (PWYP 2010). PWYP is a global coalition of the civil society that supports citizens to hold their respective governments accountable for the management of the resource sector. Compared to the EITI that aims at the voluntary publication of all revenues from the resource sector, PWYP requests the compulsory publication of all payments and revenues from the resource sector and a transparent and responsible expenditure management system. In addition, PWYP lobbies for the publication of all licences and agreements between the governments and resource extracting companies. Today, PWYP comprises 70 countries and 26 national coalitions attached to PWYP. The campaign is based on a comprehensive public relations/ awareness rising agenda. Moreover, PWYP also provides inputs to the EITI (the campaign is represented on the EITI board through civil society groups) and carries out advocacy work with regard to the rules for stock exchange listings and accounting standards. PWYP members also support various capacity building programmes, e.g. in the form of training for local civil society representatives on EITI processes, contract processes, tax regimes, audit and monitoring processes, credit rules of international financial institutions (IFIs) and the tracking of government expenditures (Müller 2009).

Another important aspect in the context of collecting taxes and royalties in the resource sector is the design of a sustainable and development-oriented tax system for extractive resources. Another possibility to address the issue of natural resource taxation is in the context of good financial governance approaches that focus, for example, on strengthening local tax systems and authorities, or the accountability of financial management systems.

3.4 *Revenue management and allocation*

Whether a country's resource endowments can be put to use for sustainable development depends above all on the quality of the country's governance and institutions. Sustainable development will only be possible if the revenues received from the resource sector are used for development purposes.

Internationally, different approaches and discussions exist in the context of the management and allocation of these revenues. Many donors currently support measures to strengthen good governance in the partner countries. In the context of German bilateral cooperation, numerous measures with a good governance focus are being implemented that aim, for example, at strengthening public accountability, national and local administrative processes or civil society bodies. Support is provided in countries like Ethiopia, Mauretania, Rwanda, Zambia or Tanzania. Focusing on the key stakeholders in the respective decision-making and supervisory processes, e.g. in the form of strengthening parliaments or civil society groups, might contribute towards better resource governance.

To highlight another example for bilateral approaches, German technical cooperation supports the strengthening of public financial systems as part of good governance programmes. These approaches that focus, for example, on budgetary processes, the management of public finances, the taxation system and authorities can equally contribute towards strengthening resource governance.

Concrete measures in the context of the management and allocation of revenues from the resource sector are currently discussed by governments and donors. One example is Ghana, where the option of implementing a special trust fund (similar to the Norwegian model) is currently being evaluated (Das et al. 2009).

3.5 *Overarching approaches*

There are also overarching approaches to strengthen resource governance like the Natural Resource Charter, the United Nations Global Compact and the OECD Guidelines for Multinational Enterprises. They are a step in the direction of not only considering resource governance as the domain of government behaviour and actions; they also take into account the responsibility that private businesses have for resource governance. It would be important, though, to also establish binding rules and introduce sanctioning mechanisms that substantiate the role and behaviour of enterprises vis-à-vis the government and other stakeholders.

3.5.1 *Natural Resource Charter*

The Natural Resource Charter is a global initiative for the voluntary improvement of the management of the resource sector and its revenues (NRC 2010). The charter is supported by a number of high profile experts from science and politics and was initiated by Paul Collier. The Charter consists of 12 principles that cover aspects like ownership of the extractive resources, transparency, supervision, competition, sustainability or the development orientation of the management of the resource sector. According to these principles, for example, licences for the extraction of resources should be auctioned off through a competitive and transparent process. Revenues from the resource sector should be paid into a sovereign wealth fund (example: Norway), and an excessive expansion of expenditures should be avoided, if possible. The key message is to use the revenues above all for investment purposes. The long-term goal of the charter is to establish an international convention on resource governance that is binding for governments and corporations. Until now, the charter serves only as an orientation and is not (yet) binding.

Additional approaches that do not have an explicit or unique focus on resource governance are the United Nations Global Compact and the OECD Guidelines for Multinational Enterprises.

3.5.2 *United Nations Global Compact*

The United Nations Global Compact represents an agreement between corporations and the United Nations to align the behaviour of enterprises with social and economic sustainability principles (UNGC 2010). In order to implement the Global Compact, companies need to adhere to 10 social and ecological minimum standards. Until 2009, 5,000 enterprises, numerous NGOs, research institutes as well as business and employee associations joined the Compact. The Compact is well-suited as an advocacy tool to promote good behaviour. However, like the EITI, it does not include any sanctioning mechanisms in case standards are not adhered to.

3.5.3 *OECD Guidelines for Multinational Enterprises*

The OECD Guidelines for Multinational Enterprises contain a number of principles on meeting social and environmental standards, adhering to laws and tax rules, or on fighting corruption

(OECD 2000). The guidelines are voluntary and only concern multinational enterprises from member countries (most of them being OECD member countries) so that a large part of internationally operating firms are not covered. Although a formal complaint mechanism exists, there are no sanctioning mechanisms.

4 CONCLUSIONS AND RECOMMENDATIONS

The Extractive Industries Value Chain shows that sustainable development through the resource sector comprises several steps and is a complex process ranging from the award of contracts and licences for exploring and extracting extractive resources to the allocation and management of the resulting revenues. Accordingly, there are numerous potentially positive and negative effects at the different levels of the value chain. The complexity of the value chain also requires broad strategic resource governance approaches that address the different levels accordingly. Of key importance is the effective integration of the different approaches into a comprehensive strategy.

The overview of international approaches and measures supported by bilateral development cooperation as highlight by examples from GIZ demonstrates that the majority of support measures are focusing on steps two and three of the Extractive Industries Value Chain—the regulation and monitoring of the extraction, processing and trade in resources and the collection of taxes and royalties. One key focus is the improvement of transparency in the context of the Extractive Industries Transparency Initiative (EITI).

In addition to this, a number of voluntary principles and guidelines exist, especially with regard to human rights and social and ecological standards that enterprises in the resource sector adhere to. One weakness is that most of these principles and guidelines are voluntary, rather than binding and do not include effective supervisory and sanctioning mechanisms.

Fewer approaches exist for the aspects of awarding contracts and licences or the management and allocation of revenues. There are support measures to improve governance and the management of public finances in general. However, the characteristics of the resource sector have ramifications for the taxation system, for example, with regard to the degree of differentiation of taxes according to different types of resources or with regard to the question to what extent revenues should be collected through licences or taxes. These aspects often go beyond the issues that are addressed within the typical support measures implemented in the context of good financial governance approaches. It requires that support measures which are designed for resource-rich countries, should always take into account the relevance and effects that they might have for the resource sector as well. This is to ensure a comprehensive and cross-cutting approach to strengthen resource governance.

A comprehensive approach to strengthen resource governance should also include measures focusing on the other steps of the value chain. This includes, for example, the award of exploration and extraction licences. One approach discussed in this paper is to support partner countries with the development and introduction of "Model Development Contracts." Moreover, the issue of managing and allocating revenues from the resource sector should be integrated into a comprehensive strategy as well. One symptom of the "*resource curse*" is that revenues from the resource sector are often misappropriated by the elites to ensure their own stay in power, instead of being invested for the public good. Resource governance thus should always be understood and supported in the context of broader good governance measures. The strengthening of supervisory mechanisms and structures of parliaments and civil society are playing an important role.

In this context, the term and definition of resource governance as primarily being in the public domain (as being influenced by the actions of governments) should be extended to also include the behaviours and inter-actions between all relevant stakeholders. Enterprises in the resource sector equally influence—through their behaviours and actions—whether and to what extent a country's resource endowments have positive or negative effects on development. Binding standards and rules that can be supervised and that could also be sanctioned, if necessary, would be useful in this context.

Based on this analysis, the following recommendations are suggested:

Resource governance should be considered part of a broader good governance agenda. In order to improve resource governance, governments and donors should also strengthen key stakeholders in the respective decision-making and supervisory processes (e.g. parliaments, audit courts and civil society groups).

For the development and implementation of approaches to strengthen resource governance, the different steps of the value chain should be taken into account. Furthermore, support measures in other areas (e.g. sustainable economic development, land and natural resource management, or environment and climate) should comprise different aspects of resource governance as well.

In addition, the transparency of revenues collected from the resource sector and the development orientation of these revenues could be strengthened in the context of bi-and multilateral policy dialogue, including a stronger dialogue among donors.

Generally, binding rules that contain provisions for supervision and sanctioning are necessary for generating proper incentives for sustainable development. Ideally, this should be done via an international convention. In reality, however, such a convention might be difficult to agree upon in the short or medium-term. The establishment of proper incentive systems and rules can also be supported in the context of programmes and projects in a number of areas, e.g. in the form of strengthening parliaments or accountability systems. Similarly, in the context of supporting national development strategies, it would be possible—through a participative process—to design and agree upon behavioural rules to be implemented for the resource sector.

If these different challenges in the resource sector are addressed effectively, chances are good that the "resource curse" might actually change into a "resource blessing" for sustainable development.

REFERENCES

Adler, M. 2006. *Paradox des 'schwarzen Goldes' oder neue Hoffnung für die Armen? Chancen und Risiken des afrikanischen Ölbooms*, April 2006. Frankfurt am Main: KfW.

Africa Progress Panel 2010. *Africa Progress Report 2010*. Genva: Africa Progress Panel.

Alba, E.M. 2009. *Extractive Industries Value Chain: A Comprehensive Integrated Approach to Developing Extractive Industries*. Washington D.C.: Oil, Gas, and Mining Policy Division, World Bank.

BGR 2010. *Zertifizierte Handelsketten im Bereich mineralischer Rohstoffe*. Hannover: Bundesanstalt für Geowissenschaften und Rohstoffe.

BICC 2007. *In Control of Natural Wealth? Governing the resource-conflict dynamic*, December 2007. Bonn: Bonn International Centre for Conversion.

Collier, P. & Hoeffler, A. 2005. Resource Rents, Governance, and Conflict. *Journal of Conflict Resolution* 49(4): 625–33.

Damania, R. & Bulte, E. 2003. *Resources for sale: Corruption, Democracy and the Natural Resource Curse*. St. Paul: Department of Applied Economics, University of Minnesota.

Das, U.S., Lu, Y., Mulder, C. & Sy, A. 2009. *Setting up a Sovereign Wealth Fund: Some Policy and Operational Considerations*. IMF Working Paper WP/09/179. Washington D.C.: IMF.

EITI 2003. *The EITI Principles and Criteria*. Oslo: Extractive Industries Tranparency Initiative.

EITI 2010. *The EITI Multi Donor Trust Fund*. Oslo: Extractive Industries Tranparency Initiative.

El Bakri, Z. 2009. *Natural Resources for Development*. Remarks given at a Plenary Session, 18 February 2009. Doha: African Development Bank Group.

Heinrich Böll Stiftung 2005. *To Have and Have Not. Resource Governance in the 21st Century*. Berlin: Heinrich Böll Stiftung.

ICMM 2010. *Sustainable Development Framework*. Boston: International Council on Mining and Metals.

Isham, J., Pritchett, L., Woolcook, M. & Busby, G. 2006. *The Varieties of Resource Experience: Natural Resource Export Structures and the Political Economy of Economic Growth*. Cambridge: Harvard University.

Kimberley Process 2002. *Kimberley Process Certification Scheme*. Kinshasa: Kimberley Process.

Lal Das, B. 2009. *Strengthening Developing Countries in the WTO*. Penang: Third World Network.

Land, B. 2009. Capturing a fair share of fiscal benefits in the extractive industry. *Transnational Corporations* 18(19): 157–74.

Medico International 2007. *Aus Konfliktdiamanten werden Diamantenkonflikte*. Frankfurt am Main: Medico International.

Müller, M. 2009. *Sachstand zur Extractive Industries Transparency Initiative (EITI) und zur Publish What You Pay-Kampagne (PWYP)*. Eschborn: Bereichsökonomie Afrika der GTZ.

NRC 2010. *Natural Resource Charter*. Oxford: Natural Resource Charter.

OECD 2000. *Die OECD-Leitsätze für multinationale Unternehmen, Neufassung 2000. Paris: OECD.*

PAC 2010. *Diamonds and Clubs: The Militarized Control of Diamonds and Power in Zimbabwe*, June 2010. Ottawa: Partnership Africa Canada.

PWYP 2010. *Transparency of company payments and government revenues*. London: Publish What You Pay Campaign.

Schnell, M. & Theissen, A.H. 2010. *Rohstoffgovernance im Afrika-Bereich der GTZ—Eine Bestandsaufnahme*. Eschborn: Bereichsökonomie Afrika der GTZ.

Task Force Rohstoffgovernance des Fachverbunds GGA 2010, *Resource Governance in Central and West Africa: Applying Early Lessons from the field towards an integrated Approach*. Eschborn:GTZ.

UNGC 2010. Overview of the UN Global Compact. New York: United Nations Global Compact.

Unmüßig, B. & Fuhr, L. 2007. *Haben und Nichthaben—Verantwortungsvolle Ressourcenpolitik im 21. Jahrhundert*. Berlin: Heinrich-Böll Stiftung.

Voluntary Principles 2000. *The Voluntary Principles on Security and Human Rights*. Boston.

Warner, M. & Kyle, A. 2005. *Does the sustained Global Demand for Oil, Gas and Minerals mean that Africa can now fund its own MDG Financing Gap?* London: Overseas Development Institute.

Williams, A. 2007. *Ethical Investment, Energy, Climate Change, and Security: New Research from Norway's PRIO*. Washington D.C.: Woodrow Wilson International Centre for Scholars.

Case studies

Geological Resources and Good Governance in Sub-Saharan Africa – Runge & Shikwati (eds)
© 2011 Taylor & Francis Group, London, ISBN 978-0-415-58267-4

Uganda's oil boom: Potential and risks

H.G. Babies & B. Pfeiffer
Federal Institute for Geosciences and Natural Resources, Hannover, Germany

ABSTRACT: Over the past decades, technological progress has enabled economies to get access to more plentiful energy resources than ever before. Ever since the Industrial Revolution, the global demand for energy resources has grown significantly. Today, the escalating volume of traffic and the economic growth of emerging nations have led to a massive increase in crude oil demand and thus in oil prices. Therefore, crude oil has become a crucial income source for many developing countries. In the year 2008, the oil producing Sub-Saharan African countries exported crude oil worth US$160 billion. As oil prices increased until mid 2008, a number of oil companies were attracted to invest in oil exploration and production in both traditional oil producing countries that already have a long history of oil production and non-traditional oil producing countries. As a result, Uganda gained a lot of attention from oil companies, which confirm that the Albert Basin has a multi-billion barrel reserve potential of crude oil. The present case-study analyses the development of the Ugandan petroleum sector, describes the geological potential, gives an evaluation of expectations relating to the benefits and characterises challenges in consequence of oil production.

1 UGANDA'S OIL SUPPLY AND SHORTAGES

Domestic oil consumption in Uganda is approximately 13,000–15,000 barrels oil per day (b/d) and has risen remarkably over the past decade from around 8000 b/d in 1998 to more than 13,000 b/d in 2008 (EIA 2009; Uganda Bureau of Statistics 2009). The most significant share in Uganda's oil consumption has diesel (Fig. 1). This is due to the fact that diesel is used in the transport sector as well as in generators to overcome electricity shortages.

Up to now, every single drop of Uganda's fuel respectively petroleum product consumption has to be imported. Since Uganda is not in possession of refinery capacities, the country has to import exclusively refined petroleum products such as diesel, gasoline and kerosene. Import expenditures on petroleum, petroleum products and related materials tripled between 2004 and 2007 from US$217.8 million to US$645.6 million (UN Comtrade 2009). This amount is a financial burden on the country's economy by representing 5.4% of Uganda's gross domestic product (GDP) and 18.5% of its total import spending in 2007 (International Monetary Fund 2009; World Bank 2009). Uganda's former energy minister, Daudi Migereko, sharply condemned Uganda's dependence on imported petroleum products as "expensive, unreliable, and not controllable." (Watkins 2009a). Apart from fluctuating international oil prices and refining costs, what makes Uganda's dependence from petroleum product imports expensive, uncontrollable and unreliable?

As a landlocked country, Uganda has to deal with ocean freight and port handling pipeline fees. Moreover, delays of oil deliveries cause serious petroleum supply bottlenecks, which is why Uganda's security of oil supply is particularly vulnerable. An aggravating factor for delivery failures and resulting market consequences are attacks by pirates. For this reason, insurances for ships transiting the piracy area of the Gulf of Aden increased tremendously. For kidnap and ransom cover, to cite an example, shippers have to pay up to US$30,000 in premium for a US$3 million in coverage for one trip through the Gulf of Aden (Phillips 2009).

So far, most of Uganda's petroleum product imports are routed through Kenya. Therefore, Uganda is extremely vulnerable to delivery failures from the Kenyan route. When post-election violence in Kenya affected the route of transports for fuel from Mombasa in January 2008, Uganda suffered from fuel shortages and pump prices rose to an all-time high (Schwarte 2008b). Additionally, the new regulation of the three-axle weight limit in Kenya has led to a reduction of oil truck loads from 42,000–44,000 litres per trip down to 30,000–35,000 litres in late 2008 and therewith limits the quantity of fuels which can be transported to Uganda by road (Mugerwa & Bonyo 2008). Moreover, the temporary closure of the Mombasa-Nairobi Pipeline due to capacity expanding activities has caused short-term

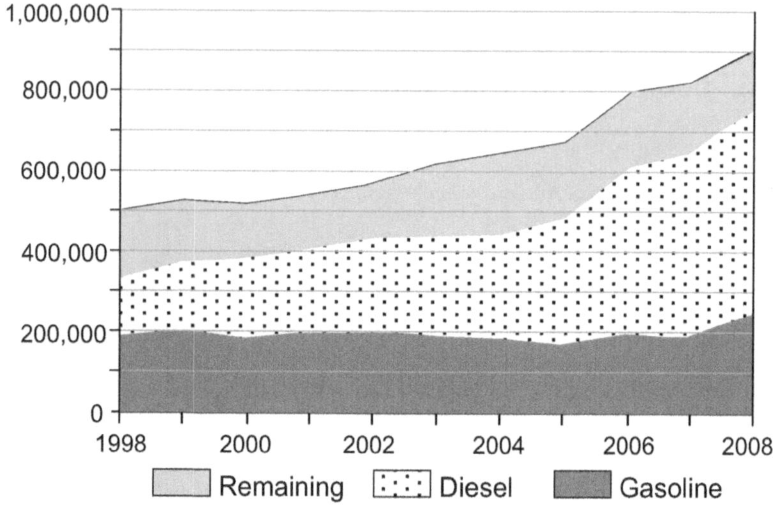

Figure 1. Fuel consumption in Uganda, 1998–2008 in m³ (Ministry of Energy and Mineral Resource 2009a and Uganda Bureau of Statistics 2009). Remaining = Kerosene, aviation fuel, fuel oil, liquefied petroleum gas.

shortages in petroleum product deliveries since 2007 (Anonymous, unpubl. 2008b). In early and late 2008 as well as in early 2009, there were serious shortages of petroleum products when the Kenya pipeline could not deliver the necessary quantities of fuel. The Tanzanian route, however, had not been in use for a long time and oil companies were reluctant to use it due to its limited infrastructure facilities and also because it was more expensive (Ministry of Energy and Mineral Development 2009b).

Which options does Uganda have to enhance its security of oil supply in the short-term? Because delivery delays have often been unpredictable, Uganda has to ensure that whenever uncontrollable circumstances occur, it has measures to react at short notice. Petroleum storage for the case of supply deficits can be a good precautionary measure. In 2009, Uganda's government commanded the reconstruction of national strategic fuel reserves which shall guarantee security of fuel supply for three months in case of delivery delays. Uganda's strategic fuel reserves were first established in the 1970s in Jinja (Kasasira 2009) and occasionally released to the market to moderate shortages whenever there were supply constraints from Kenya and Tanzania. The Jinja Storage Tanks, however, have been closed down in May 2008 due to budgetary constraints.

A new big storage facility is planned at the pipeline terminal at Nakilembe. The Ugandan government also intends to build four new regional storage facilities in Nakasongola, Gulu, Mbale and Kasese. According to a policy statement, the planned storage facilities will also be connected to the prolific Albertine Graben oil fields (Ministry of Energy and Mineral Development 2009b).

2 EXPLORATION HISTORY

The main prospective area for finding hydrocarbons in Uganda is along its western border. The area belongs to the northern extension of the western branch of the East African Rift Valley, stretching some 500 km from Lake Edward in the south to Sudan in the north. The most prolific area is the Albert Graben, mainly occupied by the Lake Albert, which is approximately 45 km wide and belongs partly to the Democratic Republic of Congo (DR Congo).

Uganda (Fig. 2) is one of the countries in East Africa where exploration for hydrocarbons began first. Already back in 1913, exploration activities began, when W. Brittlebank acquired a licence to explore for crude oil. Further licences were awarded in the early 1920s after E.J. Wayland described a series of natural oil and gas seeps at various locations along the basin margin. A number of reconnaissance field surveys, acquisition of gravity data, the drilling of shallow stratigraphic wells, and the drilling of two deeper exploratory tests were undertaken until 1940. The first exploration well in the Albert Graben, Butiaba Waki-1, drilled in 1938 by African & European Investment Company near an oil seep on the eastern side of the Lake

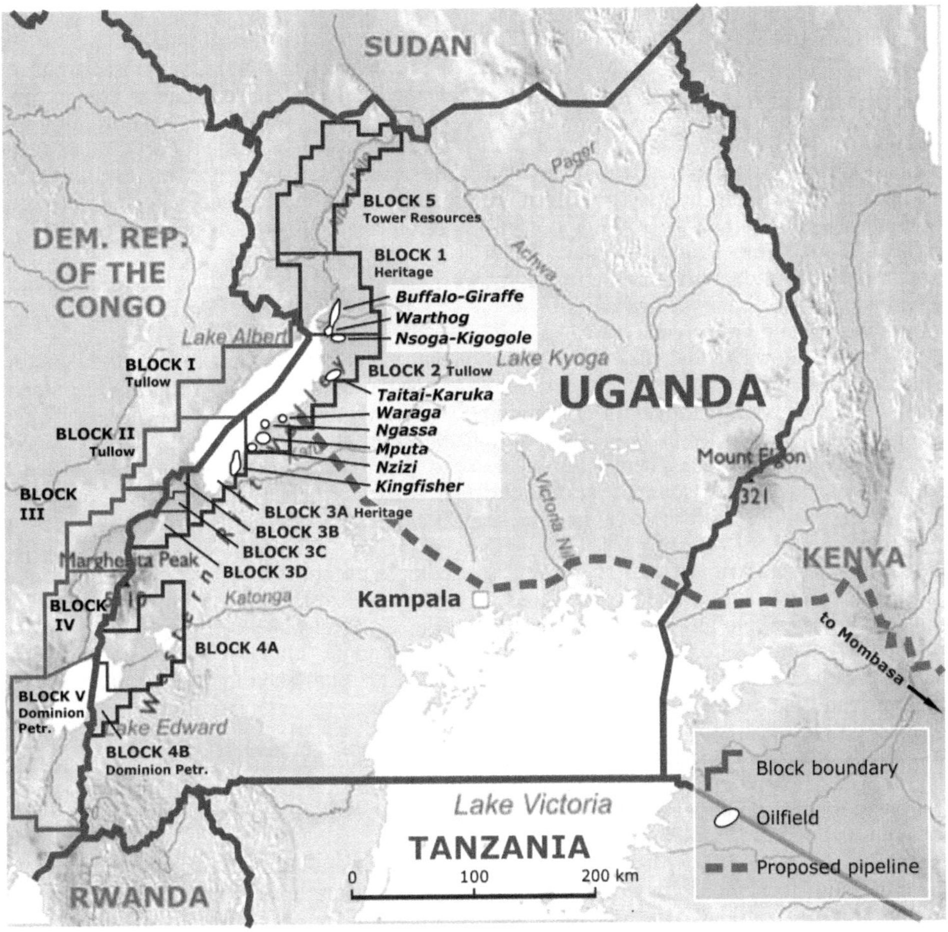

Figure 2. Uganda with oil discoveries along Lake Albert.

Albert with a total depth of 1237 m in basement encountered several oil shows.

Until the early 1950s, a number of shallow wells were drilled on behalf of the Ugandan government with no hydrocarbon indications. Between 1983 and 1984, an aeromagnetic survey over the Albert Graben area identified three major sub-basins of sufficient depth for oil generation. In 1991, the Petroleum Exploration and Production Department (PEPD) was formed by the Ugandan government; in the same year Petrofina of Belgium was awarded the entire Albert Graben area for petroleum exploration under a production sharing agreement. However, Petrofina prematurely withdrew from the licence as the deteriorating security situation in neighbouring DR Congo (the former Zaire) now discouraged further activity.

Since 1992, Universities of Columbia, Leeds and Lubumbashi carried out different gravity surveys on Lake Albert. Later, the prospective area along the Albert Graben was subdivided into five exploration zones by the PEPD. Heritage Oil Plc was awarded Block 3, which comprised the southern end of Lake Albert onshore and offshore in 1997. Hardman Resources Ltd. took Block 2 under licence in 2001. Exploration well Turaco-1 was drilled 2003 by Heritage Oil Plc in the Semliki plains of Block 3 to 2487 meters and encountered good oil and gas shows. The results of the well demonstrated the presence of a working hydrocarbon system. It took another three years until 2006 before well Waraga-1 drilled in Block 2 by Hardman discovered the first flowing oil well in East Africa south of Sudan with an initial test rate of 1500 b/d and 33.8° API (American Petroleum Institute, unit for the density). Heritage Oil Plc followed in the same year with the discovery well Kingfisher-1 in Block 3 with combined flow rates of 13,892 b/d oil (Anonymous, unpubl. 2006, Anonymous, unpubl. 2008a).

3 GEOLOGICAL SETTING

The Albert Graben in Uganda is bound by high angle dip and oblique-slip fault systems forming an asymmetrical graben (half-graben) that deepens to the west. Throws of the graben-boundary faults vary from 600 m to 1500 m. According to gravity, magnetic and seismic data the graben is divided into a number of sub-basins with estimated sediment thicknesses down to the pre-rift basement in the order of 6000 meters. The structural style is dominated by extensional rifting with indications of compression in the lower parts of some sub-basins resulting into thrusting. There is noticeable volcanic activity only in the southern part of the Albert Graben between the Lake Edward and Lake George area.

The geological setup of the Albert Graben is characterised by the pre-rift and syn-rift phases that make up its tectonic history. Basin development dates to about Miocene age. The pre-rift phase consists of Precambrian gneisses, quartzites and granites and of possible Permian to Jurassic unmetamorphosed clastics including tillites, mudstones, sandstones and conglomerates. Syn-rift rocks comprise Miocene to Quaternary sediments. They consist of conglomerates, sandstones and shales deposited mainly in a fluvio-lacustrine environment. The Upper Tertiary/Miocene to recent Kisegi and Kaiso formations are well exposed in the graben area and considered the principle reservoir rocks. Sandstone thicknesses are in the order of 150 m to 200 m (see Watkins, 2009e; Tullow Oil 2008).

4 WHY SO LATE?

Why has the oil industry neglected such a prolific petroleum basin for such a long time? Various challenges had to be met to convince the industry that it could be worthwhile to take a closer look into the Albert Graben region: political and security considerations, operational and logistical hurdles, and commercial considerations. After many years of unrest, Uganda started to be politically more stable at the end of the 1980s and represented a substantially more attractive environment for investment than in the years before. At the same time, the Ugandan government has made significant efforts to get security problems under control and, in particular, stabilised its western border with DR Congo. Another challenge for an operating company was the absence of any oil field supporting industry in the landlocked country. This necessitated seismic and drilling equipment and other materials to be mobilised from outside Uganda. Therefore, companies sought for cost effective solutions like slim hole drilling rigs. Due to the remoteness of the Albert Graben (1500 km from the coast, 150 km to the capital Kampala), commercial concerns made difficult both oil export by pipeline or processing in a refinery.

5 HYDROCARBON POTENTIAL

Since 2006, the amount of proven oil reserves in Uganda grew from zero to over 800 million barrels currently, with an estimated reserve potential of about 2 billion barrels of oil, which according to the Ugandan government could exceed 6 billion barrels. Of all wells drilled, about 95% have been successful. The most important discoveries so far are: Kingfisher and Buffalo-Giraffe field complexes followed by Kigogole, Mputa, Ngassa, Nsoga, Nzizi, Taitai, Turaco, Waraga and Warthog (Fig. 2). Meanwhile, even modest field sizes could be economic after the necessary infrastructure has been established for further processing of the crude oil.

6 EFFECTS ON NEIGHBOURING COUNTRIES

The discoveries in the west of landlocked Uganda are awakening exploration interest all along the Albert Graben and further on. Recently, Heritage Oil Plc together with Tullow Oil Plc acquired two blocks (Block I and II) on the western side of the Lake Albert in the DR Congo as well as Dominion Petroleum took Block V which covers the main part of Lake Edward in DR Congo opposite of its acreage in Uganda. In addition to the oil producing country Sudan with its production of 462,000 b/d in 2008 in the north of Uganda, some exploration activities started in the south. In the southern extension of the East African Rift in the Kivu Graben area of Rwanda, exploration efforts started with aeromagnetic and detailed gravity surveys. Burundi awarded already two blocks at the north-eastern end of Lake Tanganyika to an international oil company. Even Zambia plans a licence bidding round in the northeast of the country.

7 OIL COMPANIES

Meanwhile, a number of international Exploration & Production (E&P) companies are active in Uganda; some are highly interested to have a foot in the door. The pioneers Heritage Oil Plc and Hardman Resources (acquired by Tullow Oil Plc in 2006) acquired the first modern exploration data and discovered the first oil in the Albert Graben. Encouraged by the efforts, Tower Recources and Dominion Petroleum signed licence contracts for Blocks 5 and 4B (Fig. 2) recently, respectively. Other companies are keen to get involved in Uganda's oil exploration, mainly 'western' companies but also major Chinese and Indian companies.

Oil companies promised to implement social programmes in the districts and villages where they operate. Heritage Oil Plc, for example, intends to support road construction, development of clean water, schools and hospitals with US$2 million (Watkins 2009b).

8 EXPORT VERSUS LOCAL USE

Sufficient recoverable oil has been discovered in Uganda's Albert Graben to exceed the commercial threshold for development. The Buffalo-Giraffe complex (Block 1) has estimated recoverable oil reserves of at least 400 million barrels. Numerous other yet undrilled prospects and leads are mapped with substantial oil potential in the order of possible multibillion barrels of oil. Generally, the oil in the Albert Graben is a 30°–34° API low sulphur crude with a low gas-oil ratio and some associated wax. Planning of early commercialisation of the discovered oil reserves, with potential first production in 2011, are in progress. Heritage Oil Plc will focus initially on developing the Kingfisher field and utilising the existing railway network in East Africa. The discovered oil reserves will enable the country to produce up to 200,000 b/d for the next two decades (Anonymous, unpubl. 2009a). This is more than 13 times of the country's current petroleum consumption and therewith offers great opportunities for both local supply and export. Based on import expenditures on petroleum products in 2007, this could mean up to US$646 million of import expenditure savings per year. Up to 185,000 b/d or 67.5 million barrels per annum could be exported from 2011 to 2031. Based on an assumed constant international oil price of US$70, oil exports could reach a value of US$4.7 billion per annum.

The Ugandan government favours the option that crude oil is refined in the country and only products are exported. The government has commissioned a feasibility study to develop a refinery with a capacity of 100,000–150,000 b/d (Ministry of Energy and Mineral Development 2009b). First and foremost, petroleum products shall be supplied on the local markets. Refinery capacities do not exist yet, but the government is strongly wooing oil companies to invest in processing capacities. A refinery could contribute to the country's and the region's growing energy needs, guarantee security of fuel supply in the region, decrease the costs of fuel supply, generate export revenues and contribute to poverty reduction and development. Moreover, refinery capacities could attract other investors, create employment and support the industrialisation process through development of a petrochemical industry. As a long term goal, the development of petrochemical and energy based industries are planned.

Another option is to build a refinery near Lake Albert to supply the west of the country and to export the surplus of refined petroleum products via a 1300 km-pipeline through the port of Mombasa in Kenya, on the Indian Ocean. International oil companies like Tullow Oil Plc and its partner Heritage Oil Plc have raised concerns over Uganda's refining goals because of financial reasons. They prefer a construction of an oil pipeline from their fields in Uganda to the port of Mombasa to transport crude oil for global export. The Kenya's Mombasa-Eldoret pipeline could be extended via Kampala to Uganda's Albertine rift basin for this purpose (Watkins 2009c, d).

9 LEGAL FRAMEWORK

The hydrocarbon activities exploration and extraction are implemented under the Petroleum Act of 1985 (Exploration and Production). The Petroleum Supply Act of 2003 additionally addresses transportation, storage, distribution and marketing of petroleum products. However, refining of oil and utilisation of gas and oil revenue issues were not a part of the legal framework. For that reason, those issues have been addressed by the new National Oil and Gas Policy which was launched in 2008. However, the new policy already met with criticism for remaining "... vague on the anticipated involvement of local communities and civil society as to future benefit sharing structures and related decision making processes." (Schwarte 2008a, p. 47). The new Petroleum Legislations on Resource Management and Revenue Management should be enacted in April 2010 (Ministry of Energy and Mineral Development 2009b).

10 TRANSPARENCY

Generally speaking, transparency in the oil sector is comparatively lower than in other sectors; production sharing agreements are kept confidential. These contracts depend to a large extent on negotiations between state and companies. The Ugandan government claims that non-disclosure is necessary to protect the commercial interest of investors and to negotiate on an individual basis increasingly beneficial terms and conditions (Schwarte 2008b). However, civil society as well as the political opposition is pushing the government to make the agreements public before oil production starts. According to the Africa Institute for Energy Governance (AFIEGO), the failure of the government to reveal contracts and to subscribe to the Extractive Industries Transparency Initiative (EITI) could create trouble when production begins (Anonymous, unpubl. 2009b). The EITI supports transparency of company payments and government revenues from the oil, gas and mining sector. Nonetheless, AFIEGO believes that although Uganda is not yet a member of EITI, it has taken several steps towards greater transparency (AFIEGO, 2008). Furthermore, the new National Oil and Gas Policy's future action points include the participation in the EITI process (Schwarte 2008b).

11 DISPUTS

The distribution of the oil wealth will probably be defined according to the Mining Act. The largest share of revenues, 80%, from minerals mined in the country goes to the central government. The remaining 20% are split between local governments (17%) and landowners (3%). However, the discovery of oil in the region has deepened the conflict between tribes about land ownership because some tribes do not accept other tribes as equally legitimate (Oketch 2009). Other tribes, like the Bunyoro Kingdom, whose traditional territory covers the Lake Albert region, have demanded a much greater share of the oil revenues (Schwarte 2008b).

Border disputes between the governments of Uganda and the DR Congo especially about Rukwanzi Island in the south of Lake Albert have occurred but led to a jointly financed study that should help to define the border and thereby avoid future conflicts between both countries (Afrikaverein 2008a).

12 ENVIRONMENTAL ISSUES

Drilling for oil is carried out in the Murchison Falls Park, Uganda's largest national park (Afrikaverein 2009). This is problematic, since oil production can have adverse effects on wildlife and ecosystem. Local communities around Lake Albert have already voiced concerns about the likely impacts of oil exploitation in the area. They fear pollution and access restrictions to their fishing grounds (Schwarte 2008b).

The National Environment Act of 1995 is the legal framework for the sustainable management of the environment in Uganda. The Environmental Impact Assessment Guidelines by the National Environment Agency (NEMA) also apply to oil exploration projects. NEMA and other government agencies determine whether a project has a significant impact on the environment and if an Environmental Impact Assessment (EIA) is required. Then, the project developer has to submit an environmental impact statement including mitigation measures of potential impacts. The affected population is then allowed to submit comments on the project. Finally, the NEMA can approve, reject or modify a proposed project. In case of controversial issues, public hearings have to be organised in addition.

Uganda might have a comprehensive framework, however, public authorities dealing with the environment and natural resources are "understaffed and under-funded" (Schwarte 2008a, p. 50) The EIA therefore depends heavily on the expertise of private consultants (Schwarte 2008b).

13 CONCLUSION

Uganda has a promising geological potential which could meet the country's petroleum requirements in the near future and generate additional export revenues. Assuming that Uganda's petroleum product consume remains constant, production starts as planned and the oil price averages US$70 per barrel, it could additionally export oil worth US$94.5 billion in two decades or US$4.7 billion per annum respectively.

However, it is still too early to evaluate both beneficial and adverse effects of Uganda's oil production on the country's economic and social development. Once commercial oil production has started, export revenues could be used to fund poverty eradication and accelerate industrialisation. First and foremost, Uganda could increase its security of oil supply by gaining independence from imports and thereby contribute to a more attractive investment climate. However, unless local tribe conflicts and border disputes with the DR Congo can be settled as well as transparency issues can be solved and institutional weaknesses can be fixed, Uganda might face serious social and cross-border conflicts in the future. If Uganda manages to successfully solve disputes and use its potential for industrialisation, it could make big steps towards a well-off future.

REFERENCES

Africa Institute for Energy Governance (AFIEGO) 2008. Promoting the Extractive Industries Transparency Initiative (EITI) in Uganda's oil sector. *AFIEGO Research Series No. 3* 2008 (ed.), Kampala.

Afrikaverein, 2008a. Energieerzeugung/Erdöl: Uganda und Kongo bestellen neue geografische Studie vom Grenzgebiet, 16 June 2008, Hamburg.

Afrikaverein, 2008b. Energieerzeugung: Ab September werden alle Dieselgeneratoren im Land vom Netz genommen, 01 September 2008, Hamburg.

Afrikaverein, 2009. Bergbau/Erdöl: Heritage/Kanada bohrt im Murchison Falls National Park, 15 October 2009, Hamburg.

Anonymous, unpubl. 2006. Uganda discoveries turn heads, but production may be distant. *Oil & Gas Journal*, 04 September 2006, no place given.

Anonymous, unpubl. 2008a. Tullow, Heritage press exploration in Uganda. *Oil & Gas Journal*, 11 February 2008, no place given.

Anonymous, unpubl. 2008b. Fuel prices will remain high, warns govt. *Daily Monitor*, 29 November 2008. Kampala.

Anonymous, unpubl. 2009a. Tullow to Start Uganda Oil Devt Study Next Year. *Dow Jones Newswires*, 22 September 2009, no place given.

Anonymous, unpubl. 2009b. Oil Contract Rows Rock Uganda Ahead of Production. *Dow Jones Newswires*, 02 July 2009, no place given.

Bonyo, J. & Mugerwa, Y. 2008. Shell cuts Kenya fuel price by UShs375. *Daily Monitor*, 3 December 2008, Kampala/Nairobi.

Energy Information Administration (EIA) 2009. Uganda Energy Profile 2009, Washington.

International Monetary Fund 2009. World Economic Outlook Database, Washington.

Kasasira, R. 2009. Fuel crisis; Museveni orders construction of fuel reservoirs. *Daily Monitor*, 07 January 2009, Mukono.

Ministry of Energy and Mineral Development 2009a. *Uganda Energy Balance–Consumption of Petroleum Products 2009*, Kampala.

Ministry of Energy and Mineral Development 2009b. FY2009/10 Ministerial Policy Statement, Kampala.

Oketch, B. 2009. Uganda's oil bonanza overshadowed by tribal land dispute. All Africa, 08 September 2009, no place given.

Phillips, Z. 2009. Piracy driving up kidnap and ransom rates: Aon. *Business Insurance,* 09 April 2009, no place given.

Schwarte, C. 2008a. *Access to environmental information in Uganda.* Foundation for International Environmental Law and Development (FIELD) & International Institute for Environment and Development (eds.), London: FIELD.

Schwarte, C. 2008b. Public Participation and Oil Exploitation in Uganda. *International Institute for Environment and Development* (IIED, ed.): Gatekeeper Series, No. 138, December 2008, London.

Smith, B. & Rose, J. 2002. Uganda's Albert graben due first serious exploration test. *Oil & Gas Journal*, part 1, 10 June 2002; part 2, 17 June 2006, no place given.

Tullow Oil 2008. Tullow, Heritage press exploration in Uganda. *Oil & Gas Journal.*

Uganda Bureau of Statistics 2009. Sales of petroleum products by type (cubic metres) 2004–2008, Kampala.

UN Comtrade 2009. Uganda–Imports by principal commodities 2009, SITC Rev. 3 334, New York.

Watkins, E. 2009a. Uganda wants all of its oil refined domestically. *Oil & Gas Journal*, 16 March 2009, Los Angeles.

Watkins, E. 2009b. Heritage plans $2 million for Uganda 'social' works. *Oil & Gas Journal*, 20 March 2009, Los Angeles.

Watkins, E. 2009c. Kenya pipeline link to Uganda oil fields proposed. *Oil & Gas Journal*, 30 March 2009, Los Angeles.

Watkins, E. 2009d. Iran to help Uganda construct refinery. *Oil & Gas Journal*, 26 May 2009, Los Angeles.

Watkins, E. 2009e. Tullow finds more Uganda oil, nears award of Congo blocks. *Oil & Gas Journal.*

World Bank 2009. World Development Indicators 2009. Washington: Green Press Initiative.

Geological Resources and Good Governance in Sub-Saharan Africa – Runge & Shikwati (eds)

World Bank's failure in Chad-Cameroon oil project

G. Ngarsandje
Concordia University, Montreal, Quebec, Canada

ABSTRACT: When the World Bank decided to be involved in the exploitation of Chad petroleum project, its intention was to contribute largely to the population's well-being. Thus Chadians expected a better future. Cameroon got involved too because the project runs through its territory till reach the Pacific Ocean coast in Kribi. All the hope would turn to disarray when the Chadian government purchased arms to fight rebels at the East borders. The ill continued and forced the World Bank to suspend its relation with the government. Later Chad renewed it. In perspective of IPE and under constructivism approach, this paper analyses how the Chadian government carried out all his partners in the project and made people lose their benefits because of nepotism, clientelism and incapacity of the World Bank to coerce the government to respect its engagement.

1 INTRODUCTION

International organisations play an important role in domestic politics, economic growth and relations among states. This role goes from increasing pacific relationship to sustaining an internal economic development. They intervene politically as well as economically in domestic affairs; especially in developing countries that need their support. The Breton Woods Institutions; namely the World Bank and the International Monetary Fund.

The World Bank and the International Monetary Fund are sometimes used to pressurise countries to change their internal policies. The supposed goal is to help these countries to grow economically and administer good governance. However, the real purpose is to push those countries towards a democratisation of their national politics and transparency in governance that can generate development. It was one of these purposes that guided the World Bank to lend money to Chad in order to help it exploit oil.

Officially, Chad started to exploit oil in October 10, 2003 to develop further. However, rather than improving the population's situation, it is getting worse year after year. Life expenses became a harsh battle for households because of the dilemma that the government is facing of rebellions at East.

The paper aims to highlight the role played by the World Bank as an international institution to help making the Chad-Cameroon project a reality. The World Bank accepted to lend Chad money for this project to help it develop. Unfortunately, the political regime in place is not a perfect one for this kind of project aiming at improving the welfare of the population. As a matter of fact, negotiations for the project had been conducted as one of the familial enterprise. Only two persons close to the president, had been delegated to negotiate while the Cameroon neighbour had sent a bunch of experts from various fields. In addition, the key positions in the oil Ministry had occupied by relatives of the president. Most of those relatives who played key roles in this project are rebels at East of Chad and supposedly sustained by Sudan.

This oil exploitation was seen as an unprecedented project because of the mechanisms set up to give it better outcome. Unfortunately, after 6 years later, the goal of this project is to be reached. The revenues of this project are not invested in the sectors they were designated for, the mechanisms tools have been modified or completely changed; and the revenues buy arms and encourage fratricidal conflicts in Chad.

This study discusses the implication of the World Bank in the project and its incapability to coerce the Chadian government to respect the tools and agreements that have been put in place to make the project sustainable. This paper will first explore the process of Chadian oil from the exploration to the exploitation. An insight of the failure of the World Bank to constrain Chad to obey the agreements that led to the loan of the money that helped make concrete the project while the country uses the oil revenue to increase internal conflicts and built an authoritarian regime will be discussed. The conclusion of this paper will then situate the responsibility of each of the components of the project.

2 THE CHAD OIL

Chad is a land locked country situated in Central Africa. It was colonised by France in1908 and gained independence on August 11, 1960. At that moment, the French had told the first President François Ngarta Tombalbaye that the country had no natural or mineral resources.

In effect, the French ORSTOM started investing Chad's resources (1962–1965) in order to find natural resources but found nothing[1]. In 1965, Tombalbaye had visited the United States and was told that the country is full of mineral resources that can help Chad develop. He then appealed a US company named CONOCO which explored and discovered oil in 1969. Unfortunately, as Gee and Cadot point out, this oil was heavy and of poor quality (Gee & Cadot 1999) and could not be exploited with the technology at that time. However, in 1988, CONOCO sold the discovery rights to ExxonMobil and other allies, leadind to a struggle over the Chadian oil.

2.1 *Oil for power*

Since accepting that US companies exploit the oil in Chad was against France interest, France sustained a rebellion led by those who helped Hissein Habré[2] rule the country for over 7 years, namely Hassan Djamous and Idriss Deby to chase out Hissein Habré. In less than two years—April 1, 1999 to December 1, 1990—the rebellion overthrew the dictator with the intelligence of French attachés that influenced him to run through Cameroon and he found asylum in Senegal where he is today and has been sued for war crime. Beside this political battle, there was a feasibility of the project.

Even though this oil was heavy and of poor quality, new technologies available in 1988 helped to make it better. However, because of the location of the country and the high cost, "*the only viable solution identified was to build a pipeline from the oilfields in Chad to somewhere on or near the coast of Cameroon*" (World Bank 2006). This made it necessary for another partner to come in. Chad was a low-income country and could not finance the construction of the pipeline. Thus, ExxonMobil convinced the Chadian government to ask the World Bank to bail it out. The argument of reduction of poverty was raised and the process of financing the project began in 1994.

While the process was going on, an election took place in 1996 that was won by Idriss Deby against his last challenger Abdelkader Wadal Kamougué by fraud (European Parliament 2001). Later on, it was alleged that Kamougué had sold his victory to Deby in compensation of money promised by Elf Aquitaine, a French oil company. In 1998 because of a trial by Kamougué, the truth came out and Elf tried walk away from the project arguing it was not profitable (Jaillard 2005). This led to protests from the population that burnt France flags in many places in N'Djamena. However, this violent manifestation facilitated by the party in power, MPS[3], was not reprehended by the police as usual in this country and did not bring Elf back to the project. But, by 1999, the consortium—ExxonMobil, Chevron and Petronas—convinced the World Bank to bail out and make the project a reality.

2.2 *Actors and their contributions to the project*

Many scholars, Civil Society Organisations, NGOs, Human Rights associations/organisations and other types of structures were sceptical about this project. In many countries, it was proved that natural and mineral resources had brought only misery, wars, conflicts and other kinds of troubles to the countries that exploit them; especially the under developed or developing countries. However, the government has made poverty reduction an argument that the World Bank could barely refuse to agree on. Another argument and the most important one that the World Bank was truly concerned about—the tons of document on the subject confirm[4], that the project was keen on environmental sustainability. To be successfully run, the project needed several key actors.

The first actor was Chad, the beneficiary of the project. The location of 300 oil wells is in the southern part about 310 miles from N'Djamena. Chad was known a country whose economy was based on agriculture and livestock until the oil exploitation in 2003. The base of this agricultural economy was the "white oil", namely cotton, followed by peanuts. Agriculture was followed by the livestock of dromedaries and beef. As mineral resources, Chad focussed only on the exploitation of sodium carbonate that was never well developed to make it an international commodity. Later on, there were discoveries of bauxite, gold, uranium, tin and tungsten, but their exploitation could not occur because of civil war, fratricidal and ethnical conflicts and other instabilities. Then came a natural resource—oil that failed to help the country because of the political situation.

Chad is ruled by a regime that is supposedly democratic. The actual President, Idriss Deby Itno, came in power after a rebellious force in December 1990. After two years of power experimentation and in order to bring peace and "real" democracy in the country, Deby accepted to have a National Sovereign Conference that reunited all forces of the country. This lasted 4 months from January to April 1993. Instruments and guidelines had been set up to reach a real democracy and reconciliation but all these failed because of the stubborn will of Deby to stay in power. Local rumours and press in the country say he told the Prime Minister elected by the National Sovereign Conference, Doctor Fidel Moungar, that he came in power by guns and will leave only under fire guns and that this one came in the country by Air Afrique[5] and he could not stand before him. In 1996, simultaneously a referendum of a constitution, modified in 2005 to let Deby run again in 2006 for presidency and as long as he wants to, legislative and presidential elections were held. The "Oui[6]" won for the referendum, while the MPS won the majority in the National Assembly and Deby won the presidential election after two turns. The elections that followed were also won by the MPS and Deby. All those elections were highly contested. Despite lack of transparency, Deby wanted and convinced his partners to exploit the oil fields in the southern part of the country.

To exploit this oil the country needed to finance about US$1.8 billion. The project was supposed to last over 25 years from the starting date, peak a daily extraction of 225,000 barrels and generate, over the 25 years, US$2 billion (an annual revenue of US$80 million). All this could not be done by Chad itself. It made an appeal to partners for financing, extraction and shipping. The first partners were Cameroon, the World Bank, the consortium led by ExxonMobil; and, behind the World Bank were the European Union and the United States that lend to the Chadian government the necessary amount of money through those international/financial institutions.

Cameroon is the closest country to Chad that has access to the Atlantic Ocean. Thus, the two countries accepted to build a pipeline that runs from Doba to Kribi Coast about 7 miles. The total length of the pipeline is 1070 km (665 miles). Cameroon is supposed to benefit from this project about US$500 million, a quarter compared to Chad revenue. This is because of the degradation that the construction of the pipeline would cause to the environment there. Due to a well skilled and multidisciplinary team of negotiation, Cameroon had earned, if not the best, a better deal than Chad who had sent relatives to power without knowledge in oil industry and other related areas. This cooperation between the two states was made possible by the World Bank, despite their apparatus to a regional organisation; the Economic Community of Central Africa.

The World Bank, since 1994, had admitted that the project was a good deal for Chad, though risky. Cautiously they negotiated, built instruments and raised safeguards to make it a real instrument of reduction of poverty in both countries. Thus, the approval of US$3.72 billions to these governments had been given by the board executives in June 2000. This loan came from the IBRD (Investment Bank of Reconstruction and Development) and EIB (European Investment Bank) each giving US$39.5 million and €20.3 million to Chad and US$53.4 million and €35.7 million to Cameroon, respectively World Bank's financing was less than 3% of the total cost oriented towards capacity building. It maintained good environmental and revenue management plans in order to make this project an example model that would be adapted anywhere else if it succeeds. "*[t]he Chad model was important to the United States, if only because its success would have meant that similar programs might have a chance of working in Kazakhstan, Sao Tomé or even Iraq—places that might become friendly oil producers*", Margonelli (2007) speaking about United States implication in the project. Apparently, built on failure in precedent experiences, this project "*aims to maximize the development benefits of Chadian oil while avoiding the pitfalls that have plagued energy development elsewhere,*" the United States has invested US$3.7 billion which was "*the largest US investment in Africa*" as stated Honorable Edward R. Royce, presiding the hearing before the subcommittee of Africa on the Committee on International Relations (Congress April 18, 2002). This is not surprising though the United States is a larger contributor to the World Bank. But, what brought the World Bank to that affair is the consortium led by an American oil company.

The consortium that bought the owner rights from CONOCO is composed of ExxonMobil, Chevron and Petronas. While the first two are American companies, the third is Malaysian. Those companies shares are 40%, 35% and 25%, respectively ExxonMobil, the leader of the consortium, knew the real value of the project. According to Exxon, they needed to make this project to clean out the bad reputation oil companies have when exploiting natural resources. Another reason was also given by Royce during the April 18, 2002 hearing when he stated "*our country [the US] has a growing interest in Africa's energy development. By 2015, up to 25% of US oil imports are expected to come from Africa.*" Since the most share of oil in Africa belongs to other companies like Elf and Total, it was clear that they had to take up this project so as to meet their goal. Thus, besides washing out the bad reputation of oil companies and making profit from oil, it was to preserve American interests in Africa. This explains why the US had agreed that the World Bank approves the project and lends that money to the countries that are exploiting it. But the role played by the World Bank and the consortium had raised concerns from other actors in the social and economic domains.

Thanks to the loan that the two partner countries developed semi-private enterprises intervened in the oil industry. Those enterprises are COTCO (Cameroon Oil Transportation Company) and

TOTCO (Tchad Oil Transportation Company) in which Chad has 5% and 11% share respectively, while Cameroon has 10% in the first one. However, the boards leading those enterprises at their turn contributed to a rise of a stronger civil movement in both countries.

In Chad, a rise of commissions, associations, NGOs and other partners took place to protest against the process. They advocated for more transparency in the management of oil revenues, and an increase in the amount of compensation for the local peasants that lost their fields or other belongings that the construction of the pipeline would cause. The negotiation of the oil by the Chadian counterpart was not well conducted. The representatives did not know much about that kind of industry and were few (many of the structures quote two persons Abderamane Dadi, General Secretary for the President, and Ali Ahmed Lamine, Minister of Economic Promotion and Development both dead in a plane crash on February 2001), while Cameroon, though not the main beneficiary, sent experts from various fields that could be affected by the oil exploitation. There was also a private cabinet that integrated the CPPs (Commission Permanente Pétrole[7]), which contributed to many studies and expertise on the subject. It was the GRAMP/TC (Groupe de Recherches d'Alternatives de Monitoring du Projet Tchad–Cameroon[8]). These groups bound together and went to Washington in September 1999 to ask for a moratorium to mean for two to three years in order to make this project fight poverty in Chad. For them, the political climate at that time was not appropriate to run the project due to lack of respect of human rights, frauds during elections, rebellions at South, North and East and the high level of corruption in the public administration. It was clear that one day the project would be diverted from its first goal. Unfortunately another delegation went to Washington and to meet the World Bank President. This delegation was called the "other civil society" meaning a fabrication of Deby and the government. This delegation led by Ladjal Calixte, a journalist at the National Television channel in Chad, took over the one led by Djiraïbé Kemneloum Delphine following this three hour discussion they had with the former World Bank President, James D. Wolfensohn. This was decisive according to Serge Michailof, a former director of Central African countries at World Bank because, for Wolfensohn, there was no reason to wait two to three years to train people even though he doubted the full success of the project. There were people starving who needed their living conditions to be improved[9]. But those cons of the project did not act, and were not acting alone. Their allies in Cameroon did the same thing.

In Cameroon, protests rose to claim better compensations for peasants and transparency. They met regularly with the CPPs in Chad and in Cameroon. They sought support from other countries like Congo Brazzaville, Gabon and South Africa for strategies and training in any related field: environmental preservations, activism, budget monitoring etc. But reality shows a mere failure of the project. Everything started in 2004 when the country received the first royalties of the oil.

2.3 *Diversion and security matters for Chadian government*

Before the official launch of the oil exploitation on October 10, 2003, there was a defection within the army. A couple of generals and commandants close to Deby defected and took arms to settle a rebellion at the East of Chad. And when the royalties came, it was a matter of security question. While the population, especially government workers were expecting a substantial rise of their salaries and other forms of compensation, the government, meaning Deby in this case, decided to buy arms and heavy weapons to counterattack the rebels. Lisa argued "*[b]ut the project had barely gotten started when word got out that Chad's president Idriss Deby had spent $4.5 million in bonus money paid by the oil companies on weapons to fight rebels within his own borders*".

This was the first false note in the agreements among the three partners—Chad, World Bank and the consortium—of the project. After a while fighting this brotherhood rebellions—as matter of fact those rebels were from Deby's tribe—the rebels decided to reunite with the government in exchange of financial compensations. Other rumours from the presidency said it was a regiment sent by Deby to destabilise Sudanese government that had been stopped thanks to a defection from his cabinet that alerted the Sudanese President. But this is another case to be analysed concerning the Darfur crisis in which the two governments have specific roles.

A year later, the provision of future generations contained in a special law, part of resources management strategies, was amended and abolished the Future Generations Fund to be returned to the National Treasure in order to compensate the budget deficit generated by the high expending on arms and infrastructures building.

Since development means good infrastructures, key posts have been given to close relatives of Deby. His brother, Bichara Daoussa Deby, moved quickly from the Ministry of Finance to the Ministry of Infrastructures that comprises public buildings, roads and other related areas. At the same time the privatization of OFNAR (Office National des Routes[10]), the state enterprise that was responsible of building roads, was coming up. It was then sold to Bichara Daoussa Deby who left the Ministry of Infrastructure and became the Executive Director of SNER (Société Nouvelle d'Etudes et de Réalisations[11]). Along with SOGEA–SATOM a French company working on roads, SNER has the biggest part of any works market on roads and buildings. The public market of infrastructures is managed by the Ministry of Infrastructures through its Public Markets Office. While priority sectors had been determined for the project mainly education and health, the Ministry Infrastructures—whose current person in charge, Adoum Younousmi, is also a close relative to Deby—has the highest expenditures in the budget.

In 2005, the constitution had been modified and the prime concern was on the article about running for presidency and the number of terms. It stipulated two terms of 5 years but was modified to as long as the candidates want to run for the presidency[12]. This was to allow Deby to run for a third term in the 2006 elections. In fact, it was a clear stipulation for him to keep the power for life. This modification of the constitution, changes in the Ministry of oil and the coordination of the oil project led to another rebellion. They disagreed that Deby had to be in power for life and oppress the Chadians. The rebellions leaders were relatives to Deby and fellow tribesmen—the Zaghawa. Thus, a question of security rose because "a development needs to be protected" according to Deby and a quest of buying war weapons began.

In order to face all those rebellions that reunite or separate according to leader's mood, Deby had to expend on weapons of all kind. Other events and arguments that were helpful to Deby were the attack of September 11, 2001 and the fight against terrorism. The suspicion by the US of Sudan as a base of terrorism training that would lead to a bombing of a pharmaceutical industry was what he wanted. The argument was that the rebellions were mercenaries from Sudan and supported by Al Qaeda and other terrorist or islamist groups. Under this argument in benefit of the US, Chad received agreement from the US to train and support the Chadian army and the question of human rights, democracy and development was forgotten or sent to a Greek Calends.

The rebels had almost succeed to take over the power many times. The last one that was critical and harmful for the country was in February 2008. It had constrained Deby in the presidency for 3 days and he succeeded to fight with help from France. This affected the relationship between the World Bank and Chad. A big gap was found in the budget expenditures and was difficult to justify. Yet, after the modification of the law 001 provisioning the Future Generation Funds, the relationship had suffered tensions. The World Bank refused to finance many projects and threatened to leave the country and stop its support. Many financial agreements were simply frozen. But this last one made the World Bank shut representation in the country for awhile. After many rounds of negotiation between the country, the US and other interested countries, the World Bank announced on September 2008 that it would no longer support the project. Did the World Bank fail to succeed and is it responsible of this unsuccessful project?

3 SAME LANGUAGE BUT DIFFERENT PERSPECTIVES

All actors—the consortium, the World Bank, Cameroon and Chad—spoke the same language: the oil will improve the economy of the two countries that suffer poverty. Chad is the fifth poorest country in the world. During that time, 64% of the population lived below the level of poverty (Martin n.d.). Unfortunately, the outcomes show that the perspective of the actors are completely different and fit different personal interests. Theoretically, we can apply the Prisoner Dilemma, the two levels game, the Domestic Policy approach and the security dilemma to be associated to the institutional liberalism point of view.

Barnett and Finnemore's (1999) study on whether International Organisations are doing always what they are created for. The main concern is that often, they overpass their objectives and escape their creators' control. They become autonomous and more powerful than the states that created them. International organisations are created to further their common interests; that is, they are created to become autonomous and powerful. These characteristics derive from the fact that international organisations use bureaucracy and knowledge. They develop rules that creator states cannot control.

The World Bank was negotiating with a beneficiary that is a member but not an influential one. The World Bank did not expect Chad to fail to meet its promises—make the project a tool of reduction of poverty—and take the money to buy arms. In effect, at that time, the shadow of future was playing a great role and constrained Chad to cooperate. The main actors were four and as Axelrod and Keohane (1993) said (in *Achieving Cooperation under Anarchy: Strategies and Institutions*), cooperation is a difficult aim in an anarchic world. The higher the number of actors the harder the cooperation is likely to be. Thus, while the Consortium and the World Bank were aiming for the project to bring development, Chad was preparing to defect. The government of Chad cooperated at that moment because it weighted the benefits. Another expectation was to oppress the population and install an everlasting power, keeping Deby and his allies in power for life. This population is the same one that was used in the domestic policy approach since the project became public and the government was using it to help leaders profit of the oil revenues. It was difficult to know the intentions of negotiators from Chad.

Considering the two levels game (Putnam 1999), this case is difficult to analyse. The World Bank and the consortium had to deal with a supposed democratic regime though many other international and national NGOs gave warnings on the project. However, if the outcomes for Chad were not better than Exxon, it was because the exploitation contract is based on an old system of fifty-fifty share of the benefits. The World Bank and the consortium were negotiating with people that did not have support from the country. The support to the government within the company came when Elf refused to exploit—though for unclear reasons—and the political party in the power pushed its militants to tear down elf ownerships and France flag. This support won, and was the one against the internal civil society organisations that required a two to three year moratorium. Thus, the Chadian government negotiators had less power or maneuver margins to ask for better outcomes. Their own arguments had been used against them: they saw it as an opportunity to improve the development of Chad. But, they were just constructing a path to defect from the agreements with the World Bank.

Deby had planned all what was happening in the country, if the rumours (though a scientific thought cannot be built on rumours, it holds here because of the special characteristics of the country) are true. The rebellions are just building tools to divert the domestic audience, and even the international audience, towards national security. Apparently, the regime in power carefully built enemies and troubles to divert the population's attention: the country instability by mercenaries paid by Sudan. Despite Debys gesticulation to accuse Sudan to set those rebellious, it is odd that most of those rebellions leaders had worked in his cabinet even for more than two to three years. Tom Erdimi and his brother Timan Erdimi were the key managers of the oil revenues; but, now, Tom Erdimi is in exile in Texas while Timan Erdimi is leading the last coalition of rebellious factions in the East under the name of Union des Forces pour la Resistance[13] (UFR). Buying fighter planes like helicopters and other flying fighters of high capability of destruction and hiring mercenaries to master the tools acquired for the battle and overcome the rebellious forces, has become the focus of the government in Chad.

Nowadays, many measures are hardening the population living conditions. In the area where the oil is exploited, the cost of life had risen to ten times and the environment is polluted. The infrastructures that the consortium is supposed to build in completion to the government are of worse quality—concerning hospitals, schools and other social amenities—and those that the government has to build, are yet to built. The population does not know who to speak to. An address towards the consortium is sent back to be addressed to the government which never keeps its promises since the beginning of the project and vice versa. Everywhere in the country, the inflation is higher and not mastered by the structures put in place by the government because they are inefficient. Lately, cutting trees to produce charcoal, a combustible on which most part of the population rely on was forbidden. This increased the price of butane bottles.

4 CONCLUSION

A project that reunited many actors took place in Central Africa's Chad and Cameroon so as to improve social and economic life. However, this was not achieved. The shadow of feature had presided to the negotiation and made the parties cooperate because they were focused on common objectives.

Unfortunately, it clearly looks like thanks to globalisation, the multinationals in the consortium aimed at making profit from cheap labor,[14] and commercialisation to support their countries policies. Thus, they make more profit than they should in other countries where domestic policies are more consistent and with population support. Since the country is divided, unstable, and the area they exploit the oil is free of disruption, they are better off in their business.

The Chadian government used and is using constructivism strategies to build enemies and divert the domestic audience from the oil revenues management toward a national security that is more threatening. Argument about Islamism and terrorism is also used to have support from the United States and France. Moreover, France is convinced there is no one else capable of ruling the country and is supporting an illegitimate regime that oppresses the people it governs. Its last intervention before the United Nations to support the President and chase out or condemn the rebels attack confirms it.

The World Bank that cautioned and advocated for this project cannot coerce the country because of the sovereignty argument. Unable to constrain the country to respect the agreements rules, it withdrew from the project leaving those for whom—the poor in Chad—it supported the project. However, it is not too late to arrest the situation. Critics of institutional liberalism argue that international institutions are created to do what the powerful countries that created them cannot do by themselves. We clearly see how France and the United States had intervened directly by lending money or sending advisors to assist the country. Thus, they can use those advisors and other pressure measures to constrain the country to obey its agreements.

Finally, if the project has failed, it is not because of the World Bank, but the states' fault. Chad is not a democratic country unless democracy has different meaning—though the definition is the same anywhere—according to actors. Those powerful countries that have invested in this project do not really care about social life of the population and/or political situation in the country. A careful look at the investments helps us see that Europe has invested the higher amount of money compared to the United States, but the latter is compensated through its companies that are exploiting the oil. Thus, they can act consequently and force the elite in Chad to set back to the first goal of the project. France can act as the prime advisor and colonial power, while the United States can have the hegemonic power. A concerted effort by both to settle conflicts in Chad can help save a part of the goal of the project. If not, the model is lost and we will assist to remake Nigeria, Congo Brazzaville, Gabon and other countries where oil became a curse rather than blessings by making the elites richer. However, far to blame powerful countries, another work needs to be done on psychological profile of such leaders in order to advance the idea that failure of World Bank strategies and intervention are just a result of thirst of power displayed by these powerful countries.

ENDNOTES

[1] Chadians allegedly think French had done that on purpose in order to keep them too long so they can exploit it one day.

[2] Hissein Habré was President of Chad from June 7, 1982 to December 1, 1990 when he was chased out from the power by the actual president Idriss Deby Itno.

[3] Mouvement Patriotique du Salut–Patriotic Movement of Salute.

[4] Visit the website of the project to have a look: http://www-wds.worldbank.org/

[5] A former air company that bankrupted in 1999.

[6] As referendum election choices are "Oui" (Yes) and "Non" (No).

[7] Permanent Oil Commission.

[8] Alternatives Research and Monitoring Group of Cameroon-Chad Oil Project.

[9] In *Tchad: Main basse sur l'or Noir au Tchad*, op. cit.

[10] National Office of Roads.

[11] New Society of Studies and Realizations.

[12] Constitution of Chad, article 61.

[13] Union of Forces of Resistance.

[14] Local workers are paid under national convention rated from 1968.

REFERENCES

Axelrod, R. & Keohane, R.O. 1993. Achieving Cooperation Under Anarchy: Strategies and Institutions. In *Neorealism and Neoliberalism: The Contemporary Debate*, by David A. Baldwin, 85–115. New York: Colombia University Press.

Barnett, M. & Finnemore, M. 1999. The Politics, Power, and Pathologies of International Organisations, 53. *International Organisations*: 699–732.

Congress, National 2002. *The Chad-Cameroon Pipeline: A New Model for Natural Resource Development*. 107 Congress-Second Session, Washington D.C, April 18, 2002.

Ericksson, H. & Hangströmer, B. 2005. *Chad-Towards Democratisation or Petro-Dictatorship*? Nordiska Afrika-institutet, Upspsala, Sweden.

European Parliament 2001. *European parliament website.* June 14, 2001. http://www.europarl.europa.eu/omk/omnsapir.so/pv2?PRG=CALDOC&FILE=010614&LANGUE=EN&TPV=DEF&LASTCHAP=28&SDOCTA=19&TXTLST=1&Type_Doc=FIRST&POS=1 (accessed April 10, 2009).

Gary I. & Reisch, N. 2004. *Le pétrole tchadien: miracle ou mirage? Suivre l'argent au dernier né des petro-Etats d'Afrique*, Catholic Relief Services & Bank Information Center.

Gee, F. & Olivier, C. 1999. HEC Lausanne. *A HEC Lausanne website.* www.hec.unil.ch/ocadot/CASES/chad.doc (accessed March 17, 2009).

Gelb, A. & associates 1988. *Oil Windfalls: Blessings or Curse?* World Bank research Publication.

Jaillard, N. 2005. *Tchad: Main Basse sur l'or Noir*.

Margonelli, L. 2007. New York Times. *New York Times's blog.* February 12, 2007. http://pipeline.blogs.nytimes.com/2007/02/12/the-short-sad-history-of-chads-model-oil-project (accessed March 20, 2009).

Martin, J.P. *Columbia University.* www.columbia.edu/itc/sipa/martin/chad-cam/index.html (accessed April 12, 2009).

Petry, M. & Bambe, N. 2005. *Le pétrole du Tchad: rêve ou cauchemar pour les populations*?, Karthala, Paris.

Putnam, R.D. 1999. Diplomacy and Domestic Politics: the Logic of two-level games. In *Theory and Structure in International Political Economy: an International Organisation Reader*, by Charles Lipson and Benjamin J. Cohen, 347–381. Cambridge: International Organisation Readers: MIT Press.

World Bank 1999. Finance, Private Sector, and Infrastructure Network, *Natural gas: private sector participation andmarket development*, Washington, D.C.: International Bank for Reconstruction and Development.

World Bank 2006. http://web.worldbank.org/external/projects/main?menuPK=51447259&pagePK=51351007&piPK=64675967&theSitePK=40941&menuPK=64187510&searchMenuPK=51351213&theSitePK=40941&entityID=000020953_20070919092837&searchMenuPK=51351213&theSitePK=40941 (accessed February 23, 2009).

Geological Resources and Good Governance in Sub-Saharan Africa – Runge & Shikwati (eds)
© 2011 Taylor & Francis Group, London, ISBN 978-0-415-58267-4

Turning the curse into a blessing: A convenient Illusion. Lessons from the Nigerian EITI process

M. Müller
Bonn International Center for Conversion (BICC), Bonn, Germany

ABSTRACT: The Extractive Industries Transparency Initiative (EITI) is often cited by policymakers and advisors as an important means to overcome the 'resource curse'. In specific, its Nigerian format, the Nigeria Extractive Industries Transparency Initiative (NEITI), was hailed as a success by the international community. At the same time, Nigeria has been riddled by violent conflict in the Niger Delta since a decade. The current amnesty process presents an opportunity to investigate the potential of NEITI to contribute to greater democratic accountability and effective governance in the Niger Delta. The analysis of the conflict in the Niger Delta shows that revenue transparency is partially relevant to solving the conflict, although its origins involve issues that go far beyond transparency. This paper delineates clear limitations to NEITI's potential, however, arising from its design and its being embedded in the political context of a 'petro-state'.

1 INTRODUCTION, AIMS AND METHODS

In public discussions on the 'resource curse', the international Extractive Industries Transparency Initiative (EITI) is often cited as an avenue to better resource governance, which should help overcome the resource curse—that is: underdevelopment, authoritarianism, political unrest and violent conflicts. Policy-makers seem to hold on to this policy instrument as if it was the only anchor, besides the Kimberley Process, to hold onto in the 'good governance' boat. Rarely is it attempted to substantiate the link between transparency in revenue management and good resource governance, nor to find evidence for the assumption that the resource curse can be overcome by better governance.

This paper investigates to what extent the Nigeria Extractive Industries Transparency Initiative (NEITI) can contribute to better resource governance and to overcoming some of the woes associated with the resource curse. Nigeria is a pertinent case for this exercise because it is plagued with all the features of the curse—underdevelopment, authoritarianism, political unrest and violent conflict—but at the same time, the Nigeria Extractive Industries Transparency Initiative started off as a success and thus created high expectations among the international community.

As the resource curse is particularly felt in the oil-producing Niger Delta, the present study lays a focus on the violent conflict in that region. It is mainly confined to the potential effects that NEITI may have on developments in the Niger Delta, as it is still too early to measure any concrete impact of NEITI, it having only completed the first audit reports.

It is based on interviews conducted in Nigeria and a desk study covering academic literature, NGO and newspaper reports and official documents. The interviews were held in Abuja and Lagos over a period of two weeks in October/November 2009. The author conducted about 25 semi-structured interviews with representatives from government, NGOs, research institutes, oil companies and international agencies.

The paper starts with a short background on the Nigerian Extractive Industry Transparency Initiative (NEITI), its scope and the problems it revealed in the management of the Nigerian oil industry. This is complemented by a restatement of what constitutes the resource curse, using Terry Linn Karl's concept of the "paradox of plenty". In the second part, the paper analyses the general relevance that transparency has for overcoming the resource curse in Nigeria, by analysing violent conflict in the Niger Delta that is obviously the result of accumulated grievances related to the resource curse: the oil wealth produced underdevelopment in the oil region and political fissures on a national level. The general relevance of increased transparency in oil revenues is deducted from an analysis of the drivers of the conflict.

Finally, based on the entry points for transparency discerned in the second part, the concrete potential NEITI has to alleviate some of the drivers of conflict in the Niger Delta is assessed.

In specific, its contribution to more effective governance in terms of service delivery and to popular participation in decision-making is put under scrutiny. The author delineates the limitations to NEITI's impact that emanate from its design (scope and organisational structure) as well as its being embedded within Nigeria's political system.

As other countries in Central Africa are similarly dependent on oil exports as Nigeria, this analysis can help assessing the potential of EITI in these countries. Although those countries that mainly depend on off-shore oil exploitation will not experience the same production-site conflicts as Nigeria, countries with major mining operations may face similar problems. Moreover, they will still encounter similar challenges in terms of government oversight over company operations and accountability mechanisms for the citizenry.

2 BACKGROUND ON NEITI AND THE 'RESOURCE CURSE'

2.1 *NEITI*

2.1.1 *The scope of NEITI*

NEITI was constituted under former President Obasanjo in 2004, legally backed by the NEITI Act of 2007. It became a flagship of the international Extractive Industries Transparency Initiative (EITI) because its scope went beyond the requirements of this global initiative that had been launched in 2002 by UK Prime Minister, Tony Blair (Shaxson 2009, pp. 1–8).[1] EITI is sought to be a tripartite process involving mineral companies, 'civil society' representing the citizens, and governments. Hence, in principle, it represents the triangle of major actors in the Niger Delta conflict. The power relations within this triangle being skewed towards the Federal Government and the oil companies, an important contribution of NEITI would thus be to help the Niger Delta peoples empower themselves. NEITI can only do so if 'civil society' has a strong voice within the process and if it is able to represent the voice of the Niger Delta people.

The Extractive Industries Transparency Initiative seeks to increase transparency of government revenues from oil and minerals extraction, by comparing what mineral companies claim to pay to governments and what governments have recorded as revenues from companies. Information about government revenues is seen as a means of increasing accountability of governments towards their citizens, making the government-citizen relationship the core of the EITI set-up (Shaxson 2009, p. 8).

NEITI goes beyond the global EITI requirements in that it is backed by law[2] and that oil company payments are published on a company-by-company basis and not in an aggregated format.[3] From the beginning, it aimed at a value-for-money (VFM) audit that would put the costs being claimed by oil companies under scrutiny.[4] This would go much further than global EITI that only compares cash payments of companies without investigating the basis on which these payments were set in the first place (Shaxson 2009, p. 33). Moreover, NEITI is often hailed for including transparency in public expenditure and in federal allocations to state and local governments (3.j NEITI Act 2007). However, the publication of federal allocations to state and local constituencies was not initiated by NEITI, but by the Finance Minister Ngozi Okonjo-Iweala who published the allocations to state and local governments on the Ministry's website from 2004 on (Shaxson 2009, p. 14). With regard to expenditure, the NEITI law names "transparency and accountability by governments in the application of resources" (2.d NEITI Act 2007) merely as an objective of NEITI, without conferring upon NEITI any related precise function. The interviewees involved in NEITI confirmed that it was not concerned with expenditure and does not plan to do so in the near future. Thus, in practice, NEITI does not play any role in neither area.

Regarding the three tiers of government in Nigeria—federal, state and local—NEITI is only concerned with the federal level. The explanatory note at the beginning of the NEITI Act confines its scope to the "accountability in the reporting and disclosure by all extractive industry companies of revenue due to or paid to the Federal Government" (NEITI Act 2007).[5]

2.1.2 *Results of the audit reports*

In order to be accepted as compliant under EITI standards, EITI member countries need to produce audit reports by an independent consultancy

Table 1. Government revenues from the Nigerian oil industry (1999–2004).

Revenues	US$ billion
Sales of government's equity crude	62.8
Petroleum Profit Tax (PPT)	19
Royalties	10.5
Signature bonuses	0.485
Total	95
Government flows to the joint ventures	18.2
Net flows	77

Source: NEITI, Shaxson 2009, Vines *et al.* 2009.

firm and then go through a validation process. All candidate countries had to be validated by March 2010 so that EITI was touring national EITI chapters to ensure that they move to the validation stage. NEITI has so far commissioned two audit reports done by the British Hart Group: the first one covering the period 1999–2004 was published in November 2006, the other one covering the financial year 2005 was released in November 2009.

The 1999–2004 audit report documented that the Central Bank of Nigeria (CBN), that hosts the revenues on behalf of government, was unable to find receipts in its accounts for nine payments declared by companies (US$8.8 million). Two receipts of CBN, on the other hand, were not found in companies' records (US$1.8 million) (Hart Group 2006b, p. 8). A similar pattern can be observed in the 2005 audit report (Hart Group 2008, pp. 5–11). As the deviations listed in the audit reports are minimal, they are much less telling than the processes establishing these numbers. This concerns both the way the payments due by companies were calculated and the way NEITI reconciled the numbers (payments and revenues).

Regarding the payments due by the oil companies, both audit reports (1999–2004 and 2005) documented a glaring lack of government oversight over companies' declarations of payment. It noted a lack of cohesion among the various regulatory agencies: The Department of Petroleum Resources (DPR) has the overall responsibility for regulating and supervising the industry. It monitors and collects royalty liabilities, and compiles production data. The Federal Internal Revenue Service (FIRS) is responsible for assessing and collecting the Petroleum Profit Tax (PPT) and other direct taxes. Both DPR and NNPC act on instructions from the Presidency (Vines *et al.* 2009). While its main regulatory powers were outsourced to the DPR, the Nigerian National Petroleum Corporation (NNPC) is still implicated in contract negotiations, approvals of pre-qualifications for licenses, and consulting the Ministry of Petroleum on industry legislation. The oil companies self-assess their royalties and petroleum profit taxes (PPT), the DPR and FIRS not having the capacity to check whether the computed amounts are correct. Consequently, the taxes are possibly underpaid (Hart Group 2006b, p. 6). The FIRS does in fact ignore the exact allowances conceded to companies in the production contracts that can be deducted from the profit tax, because it does not even have access to the contracts. The only regulatory institution that knows these contract provisions is the NNPC, and the President's circle.[6] That is why both audit reports stated that "the regime amounted to unregulated self-assessment" (Hart Group 2008, p. 18).

Moreover, it has so far been impossible for the government oversight institutions to control the physical flows, i.e. the production volumes, on whose basis the companies make their declarations of payment. The question of who meters the oil and at what point of the oil flows is highly relevant at that point. The government itself has no metering capacity, but it has to rely on the oil companies for that. The World Bank consultant on NEITI confirmed: "Basically, what you can say is that before NEITI, there was no way to determine the accuracy of production figures because DPR didn't have that capability nor was it autonomous enough to carry out its mandate."[7] Both NEITI audits showed that the information on crude oil was unreliable, comparing production statements by DPR and producers (NEITI n.d., p. 6). What is taken as the basis for calculating company payments is actually the volume of exported oil, not the volume of produced oil.[8] The crude that is lost on its way from the oil well to the export terminals is not factored in at all. This is crucial, for the diverted volumes of crude fund the war economy in the Niger Delta. The question is why the companies had never been asked before by the DPR to provide a gross liquids mass balance that would capture all the flows in the different parts of the stream from wellhead to export terminal, quantifying the leakages, shrinkages and theft (NEITI n.d., p. 5, Shaxson 2009). NEITI has recently contracted a consultant to do a metering study that is looking into various measurement systems.[9]

Apart from the processes around establishing company payments, the reconciliation of the deviating numbers by NEITI is an interesting point to note. The deviation between company payments and government revenues finally mentioned in the audit report (US$16 million) is much less than an amount identified earlier by the Hart Group (US$232 million). The US$16 million represent merely 0.02% of the total US$95.5 billion cash flows to government over the six-year period (Shaxson 2009, p. 31, NEITI n.d., p. 12). Omiyi describes this informal process of reconciliation: "Initially we struggled with the numbers because data was still sort of different in different government departments, creating the impression that there was something wrong, but in the end, after the reconciliation, I think the figures match very, very well. (…) I'm really proud of those achievements."[10] The reconciliation exercise is problematic for it is not done in a transparent way: "When there was a gap established in the audit report, these companies go back to the FIRS and then work together. So the FIRS brings the data, the companies bring the data, and then they reconcile the figures."[11] As this seems to be done according to habitual procedures behind closed doors, on the initiative of the President (Shaxson 2009, p. 31), the collusion between oil companies and government institutions can continue uninterrupted.

The audit reports thus document the power of the oil companies and the lack of government capacity to regulate and control their actions—one reason for the Niger Delta conflict. This does not mean, however, that the Nigerian government is a helpless victim, but it may also mean that it is not willing to acquire the necessary capacities to regulate the oil industry. NEITI can be credited for having documented these well known problems: "Even before NEITI everybody already knew what the issues were. A committee (the Oil and Gas Sector Implementation Committee OGIC) that produced the (Petroleum Industry) Bill addressed those issues. It is such a good thing that this is what NEITI's audits revealed and that it brought evidence".[12]

2.2 *The paradox of plenty*

As the paper investigates to what extent the Extractive Industries Transparency Initiative can help overcome such a situation of 'resource curse', it is worthwhile to recall what the 'resource curse' is all about. The idea of a 'resource curse' was introduced by Richard Auty in 1993, referring to a situation in which wealth from natural resources hinders economic growth and social development instead of promoting it. The dimension of violent conflict was added by Collier and Hoeffler (Collier 1998, 2004) with a quantitative study that found that countries dependent on primary commodities experienced more violent conflict than others (Kaldor *et al.* 2007, p. 11). While there is evidence that oil-dependence and violent conflict are correlated (Kaldor *et al.*, 2007, p. 12), criticisms were raised against arguments suggesting a direct causal link between the two. It is argued here that the relationship between the two is not that straightforward, and that the context of oil exploitation matters, including the way the resource exploitation is governed (Basedau 2005, Basedau & Lacher 2006, Guesnet *et al.* 2009b).

However, the structural features of petro-states brought forward by Terry Lynn Karl in her 1997 book "The Paradox of plenty" can be a useful tool for understanding the situation that Nigeria finds itself in today. With regard to oil-dependent countries,[13] Karl (1997) argued that there are some common conditions that make very different oil-exporting countries suffer the same fate of economic deterioration and political decay. Her propositions are also of great interest to the various (African) countries such as Côte d'Ivoire, Chad, and Ghana that are only of late exploiting oil on a bigger scale or beginning to do so, but might become proper 'petro-states' in the future. Like them, Nigeria used to be an agricultural economy, deriving its fiscal revenues mainly from agriculture and mining.

The "paradox of plenty" starts from the idea that it matters which economic sector a state relies on most, i.e. whether a state relies on taxes from extractives, agriculture or foreign aid, because different sources of revenue have an impact on a state's institutional development (Karl 1997, p. 13). Roughly spoken, as developing countries derive their revenues more from external (foreign trade) than internal (their own citizens) sources, they have not developed the kind of highly institutionalised bureaucracies that exist in Europe. This reliance of many developing countries on external revenues is related to the fact that these states were not born out of some internally generated necessity, but were imposed by European colonial powers (Karl, 1997, pp. 58–64).[14] It negatively affects the state's overall capacity that can be defined as the "sum total of a state's material ability to control, extract, and allocate resources as well as its symbolic ability to create, implement and enforce collective decisions" (Karl 1997, p. 45). The reliance of 'petro-states' on oil exports has certain common features (Kaldor 2007, Karl 1997, p. 47):

- Oil exporters are much more dependent on oil revenues than other primary-commodity exporters are on, for example, cash crops;
- Oil-extraction generates very little employment: only between 1 and 2% of the workforce are usually employed in the oil sector; this produces a 'labor aristocracy' and underemployment among the unskilled workforce;
- Exploitation of oil generates very high rents due to the organisation of the international petroleum market and the special status of oil as a strategic resource;
- Oil extraction is highly capital-intensive and therefore often foreign-dominated.

'Petro-states' are confronted with some common conditions that further weaken their capacity (Karl 1997, p. 47; Soares de Oliveira 2007, pp. 33–36):

- A difficult economic environment: Internationally, policy-makers are confronted with sharply rising and falling petroleum prices. Domestically, the ability of the state to tap resources from oil tends to hinder production because the foreign exchange and the appreciated national currency allow them to import instead. In addition, the few oil workers do not represent large enough a consumer market to create a strong consumption effect.

- Private vested interests: Those people that stand to gain from the oil revenues—e.g. public officials and oil companies—become barriers to change.
- Rentier and distributive state: Political authority rests on the capacity of the state to extract rents from the international arena, less so from internal taxes, and to distribute revenues internally via political patronage to ensure consent. Patronage is similar to corruption—misuse of public offices for private enrichment—but distinguished from it by its political nature.

This rentier character of the state creates an "inextricable link between power and plenty" (Karl, 1997, p. 15): The high petroleum rents that accrue to the state tend to expand the state's jurisdiction. The state, not the private sector, becomes the center of oil-related accumulation. Through the state, elites have access to foreign exchange, import licenses, state contracts, privatised state property and the like (De Oliveira 2007, p. 129).[15] However, this refers to domestic rent-seekers only—foreign private companies are actually able to gain high rents from private oil business. Combined with the low government capacity, the expansionist tendency of the state works to weaken its authority. If Nigeria shows many signs of a 'petro-state', this will set clear limits to the impact of policy initiatives such as the NEITI. The following section will look into how Nigeria's dependency on oil exports has engendered conflict in the Niger Delta.

3 THE RELEVANCE OF TRANSPARENCY FOR CONFLICT IN THE NIGER DELTA

In order to assess to what extent the issue of revenue transparency is relevant to the Niger Delta conflict, this chapter analyses the drivers of the conflict in the Niger Delta. The major fault lines of this oil-related conflict are not easily apparent. In the 1990s, the protests of the Ogoni people of the Niger Delta seemed to have been primarily turned against the operations of the major oil company in their region, the Shell Petroleum Development Company (SPDC). The Niger Delta communities held the oil companies responsible for the destruction of their bases of livelihood and the general underdevelopment of the region, saying that "we don't see the government, but we do see you" (Ibeanu and Luckham 2007, p. 72). In the meantime, inter-ethnic clashes and fighting among various armed groups in the Niger Delta created the image of inter-communal cleavages. Today, the major militant groups in the Niger Delta openly challenge the central government, demanding "resource control", i.e. the management of a major part of the oil revenues directly by themselves instead of a centralised management by the federal government. This begs the question of where to draw the major fault lines of this oil-related conflict.

It is therefore helpful to consider the general relationship of oil exploitation and violent conflict. Generally, it is possible to distinguish three interrelated levels of resource-related violent conflict (Guesnet *et al.* 2009, p. 5):

- Revenue distribution. This dimension addresses the question of how revenues from the extraction of natural resources are being spent by the state. Violence is often triggered by conflict between different groups over the distribution of revenues between political constituencies.
- Production-site conflict dynamics. This includes conflicts between private companies and local communities often triggered by issues such as the compensation for expropriated land and environmental damage, hiring practices and other socio-economic changes brought by the onset of extraction activities.
- Resource-financed violent conflict. This includes violent conflicts where rebel groups are able to control natural resources. Where this is the case, political power might still be the ultimate aim of an armed group, but resource-exploitation during an ongoing conflict provides the necessary financial means to import goods such as arms, ammunition, food and fuel, while also providing financial incentives for the fighters.

This does not mean that grievances engendered by resource extraction on the level of revenue distribution or the production site necessarily lead to violent conflict. These levels merely denote avenues through which natural resource exploitation may foster conflicts, which can then degenerate into violence.

The sketchy picture of the Niger Delta conflict above already revealed the major actors in the conflict: the Nigerian government, oil companies, and Niger Delta communities. When trying to understand the various dimensions of the conflict, it is helpful to keep in mind the triangle of oil companies, government and Niger Delta communities. Both government and oil companies gain a lot from the current status quo, being the primary recipients of the oil revenues. Similarly, they are both a party to the conflict. The 2003 confidential WAC report, an internal study commissioned by Shell, confirmed this without any doubt when it stated that Shell "has become an integral part of the Niger Delta conflict system" (WAC Global Services 2003, p. 8).

Using the above-mentioned distinction between different levels of resource-related conflicts, the following analysis will explore how oil exploitation has contributed to conflict, and how transparency may alleviate it.

3.1 *Conflicts around revenue distribution*

The Nigerian central government plays an important role in the Niger Delta, because revenues from oil exploitation in the Niger Delta region are centrally managed by the federal government. This is the reason why many people in the Niger Delta are furious: they are confronted with abject poverty and destruction in the midst of plenty that goes to (company and) government coffers without producing development in the region.

3.1.1 *Centralised federalism*

It is easily apparent from the laws that govern petroleum production in Nigeria that the state has actually acquired a dominant role in the ownership and management of the oil industry. The 1969 Petroleum Act places the complete ownership of oil resources in the federal government, and so does the Nigeria Constitution under its section 44(3) (Omeje 2005, p. 326, Ebeku 2001). Hence, the government gives out licences for oil exploration, leases the land to oil companies and receives all taxes and royalties that are due from oil operations. There are three related features that make the Nigerian state a major player in the Nigerian 'oil complex': the national petroleum company NNPC (1), the security apparatuses of the state (2) and fiscal centralisation (3) (Watts 2004, p. 60).

1. Due to the indigenisation policy of the 1970s, the state's equity stake in the total oil production increased to over 50%. The nationalised Nigeria National Petroleum Company (NNPC) operates on behalf of the state and cooperates with the oil majors who are granted concessions. Following the 2000 Memorandum of Understanding (MoU) between government and oil companies, the Nigerian state takes over 70% of the total oil revenues from the joint venture operations (equity stake plus taxes, rents and royalties) (Omeje, 2005, p. 326):
2. The security apparatuses of the state protect costly investments and are to ensure that the oil operations continue uninterrupted. Especially during the period of military rule, they have come to be known for the crushing of any kind of protests of the local population against the negative impacts of oil operations (Human Rights Watch 1999, p. 20–23; Luckham 2007, p. 62; Okonta 2007, p. 4). Unfortunately, while the repressive character of the political regime has receded with the transition to civilian rule in 1999, security forces sometimes still serve to intimidate protesters who demand, for example, the cleaning-up of oil spills. This happens often in collaboration with oil companies (Social Action, 2009a). Not only in the Niger Delta but also in other regions extrajudicial killings by the police force or the military continued, e.g. in the central Nigerian city of Jos to crush the Islamist group Boko Haram in November 2008 (HRW 2002, 2009).
3. Nigeria also developed a very pronounced fiscal centralism—despite being a federation consisting of, by now, 36 states. The federal character mostly ensures that every state is represented in the federal government, but it does not imply a regional autonomy (Heinemann-Grüder 2009, pp. 40–46; Guichaoua 2009, pp. 26–29). Between 1980 and 2002, the financial contributions of the federal government to state budgets averaged 67%, reaching 86% at a point in time. Hence, the states are highly dependent on central government fiscal allocations. Not only that: the federal government, and by implication all non-oil producing states, rely to a large extent on revenues from oil. Between 2000 and 2004, oil and gas resources exported from the Delta region produced 79.5% of total government revenues and 97% of foreign exchange income (Technical Committee 2008, p. 102). All the oil revenues are collected in the Federation Account and then allocated to the different states according to certain criteria.[16] While the fiscal principle of 'derivation' ensured that the federation returned to each state the revenues that it itself generated, successive military regimes have eroded this federal character of fiscal allocation. The percentage of derivation was cut back from 50% of the Federation Account allocation to 30% in 1970 at the height of the Biafran civil war,[17] and then to a mere 1.5% in 1984, to be increased to the current 13% by former President Obasanjo (Guichaoua 2009, p. 28; Ibeanu & Robin Luckham 2007, p. 60).

This expansion of the state's jurisdiction was mainly felt in the period of military dictatorship (1966–1999). At the same time, a proliferation of sub-national administrative structures set in that continues until today. These dynamics can only be understood in the context of a multi-ethnic federal state: oil production was "inserted into an already deeply ethnic policy" (Watts 2004, p. 73).

Nigeria is regionally and ethnically divided—a legacy of pre-colonial and colonial political divisions. The (still British) Lyttleton Constitution of 1954 grouped the northern, eastern and southern regions together in a federal structure, each with considerable autonomy (Guichaoua 2009, pp. 21–25). This institutional structure that favored the three majority groups (Hausa-Fulani, Yoruba, Igbo) characterised Nigerian politics even after independence. Its tripartite character ignored the multiplicity and fluidity of ethnicity in the Nigerian territory and created a "federal imbalance", with the Northern region comprising three quarters of its territory (Suberu 2001, p. 4).

The proliferation of sub-national administrative units set in towards the end of the Biafra War, when Colonel Gowon replaced the former four regions with 12 states to undermine further secessionist efforts (Ibeanu & Luckham 2007, p. 60).[18] Nigerian oil politics became a "state making machine" to claim oil revenues, for which ethnicity was a political tool. This sometimes led to violent confrontations between ethnic groups over supremacy in new Local Government Areas (LGAs) (Anugwom 2005, p. 94; Watts 2004, pp. 72–73).[19] Suberu goes as far as saying that "these communities are not primordial but have been shaped by the evolution and reconfiguration of the federal state itself" (Suberu 2001, p. 1).

The current debate over 'resource control' advanced by Ijaw militants is thus the product of indigenous claims-making against the Federal State. However, it is also a reaction to the centralised management of the oil revenues that are generated in the Niger Delta whose 'minority groups' are marginalised at the national center of power (Anugwom 2005, p. 92). There is a strong rhetoric among some Niger Delta activists against the so-called 'tripod' of the three majority groups and 'Northern' domination more general (Oporum-Briggs 2009).[20] The militant struggle of the Ijaw is therefore not only one for greater control over 'their' resources but also one for a greater political weight on the national level.

3.1.2 *Weak state authority*

Thus, the oil revenues account for the paradox of centralised federalism described above, that made federalism in Nigeria "simply a guarantee of ethnic and religious group representation in the institutions of government, no matter how centralised" (Adamolekun & Kincaid 1991, quoted in: Suberu 2001, p. 16). The multi-ethnic federal structure, on the other hand, holds the key for understanding how the expansion of the state's jurisdiction concomitantly weakened the state's authority, as suggested by Karl (1997). The continuous fragmentation and recreation of sub-national units reinforced the concentration of economic and political power in the centre (Suberu 2001, p. 15). Being financially dependent on the Federal Government and constantly recreated, the states and LGAs did not develop financial responsibility and institutional effectiveness. Moreover, regional representativeness partly undermines meritocratic placement in the bureaucracy (Suberu 2001, p. 127).

This also explains the lack of administrative institutionalisation, one of the features of 'petro-states' identified by Karl: As there is little routinisation, state actors constantly redefine the way in which the system would operate (Karl 1997, p. 63). The outcome of it is known as 'the Nigerian factor'—"a euphemism for bungling every good policy", as a local newspaper describes it (This Day, 8 December 2009).[21] That is why one development plan or special fund for the Niger Delta follows the other, but none has been properly implemented so far. The Technical Committee that was set up in 2008 to come up with recommendations on how to solve the Niger Delta crisis, commented on this issue: "The terrain is littered with the output of several committees (…) all of which have been barely implemented. Frustration with this cyclical situation led stakeholders from the Region (…) to reject the idea of another summit on the Region" (Technical Committee, 2008, p. 2, p. 130).

There are some indications that the fiscal centralism of the Nigerian state also helped perpetuating "traditional concepts of authority as the personal patrimony of rulers", as identified by Karl (Karl, 1997, p. 62). The statement of an expert in the Nigerian oil industry that the author asked who signed the contracts for oil exploitation, is telling in that respect. He first laughed out loud, to finally answer:

"The way in which that happens, that is very, very political, you know. The Ministry [of Petroleum] is—in terms of who is establishing it—given the power of giving these licences, but of course we know that you don't just give out licences if the President or some of those big men doesn't say so. That is a very, very political decision, and it is never transparent."[22]

This serves to illustrate how all decision-making power relating to oil business is concentrated in the President and his advisors. Until the former President Ya'ardua named Rilwamu Lukman, a former secretary general of the oil cartel Opec, Minister of Petroleum in December 2008, the President of Nigeria used to perform this task, as did Ya'ardua's predecessor Olusegun Obasanjo. The NNPC and DPR, the main regulator, are both subordinated to the president. During the course of privatisation in the energy sector and ostensibly competitive new bidding rounds for oil fields, former President Obasanjo tried to sell out refineries and oil fields to his cronies (Vines 2009).

As the management of the oil sector is concentrated in the federal government, there is no popular participation in decision-making processes.

People in the Niger Delta feel excluded from the process of deciding if, how and by whom the oil is extracted. In addition, military rule and relative independence from domestic tax payers have undermined transparency in fiscal affairs. The rampant corruption of political elites increased the fury of Niger Delta people who saw the oil funds generated from their region being channeled to foreign bank accounts by corrupt leaders (Oporum-Briggs 2009). Some of the public funds embezzled by former military ruler Sani Abacha have been returned by Swiss Banks in 2005 (US$458 million), but many more foreign bank accounts still harbor Nigerian oil money. Apart from the official oil revenues, bribes paid by international companies involve also enormous sums that can be very attractive to state officials. The Halliburton Case revealed that the American company Halliburton and its former subsidiary Kellogg Brown & Root (KBR) had paid US$180 million to officials to secure a construction contract for the liquefied natural gas plant in Bonny Island in Niger Delta (Garuba 2009).

3.1.3 *Governance problems at sub-national level*

The transition to democratic rule in 1999 has not fundamentally improved mechanisms of accountability and popular participation in decision-making. The enormous oil revenues concentrated in the state still turn the wheel of "high stake politics" among domestic stakeholders that struggle to have their stake in oil-related accumulation (Omeje 2006b, pp. 5–8). Both the governors and the LGAs are now elected. These dynamics have created an attitude to political office in Nigeria, which a Nigerian researcher describes very drastically: "We had 49 candidates of the same party [the ruling PDP] at the governorship elections in Anambra State. Why this rush to political offices—to serve selflessly? Because the state is the richest point in society, it is endowed with enormous resources and the governance of these resources is not transparently done (…). So electoral positions, office becomes like an investment, it's not a service to the people."

The states and their governors have become an important political entity to deal with, being provided with the 13% of derivation funds that involve huge sums of money. Problems at the national level are reproduced on the state as well as the local level in the Niger Delta (Heinemann-Grüder 2009, p. 42). Infrastructure development is as much a problem of state and local governments, for they are charged with such functions at their respective administrative level.[23] Bayelsa State alone was allocated N 435.9 billion (€ 1.9 billion) between 1999 and 2007, which is more revenue than four northern states—Bauchi, Gombe, Ademawa and Taiaba—received together (Pöyry 2008, p. 37). Still, there are states with comparatively less revenues that have a comparatively better road network, health facilities, and public schools than Bayelsa State: "It is a problem of governance."[24] Therefore not all the woes can be blamed on the central government.

3.1.4 *Relevance of transparency (I)*

Hence, there is a great lack of effectiveness in governance in terms of its developmental outcomes. This is combined with a lack of transparency in the governance of the oil sector, for the state officials and their dependants in the 'private' economy hold no interest to reveal how much revenues would be due to the state and how much they actually take to the side. Thus the high expectations of the Niger Delta people to benefit from the oil wealth, are constantly frustrated. Centralisation of revenue management, politicisation of the multi-ethnic constitution of the state, lack of institutionalisation and a tendency to personalised rule excluded popular participation in decision-making processes on the oil industry. Opacity in oil revenue management that favored corruption was certainly a factor contributing to the grievances so that the revenue transparency initiative seems pertinent.

On the one hand, transparency that exposes who gets what could ease inter-ethnic mutual suspicion, e.g. the suspicion of Niger Delta people that they are exploited by the north could be countered by the fact that some northern states are worse off than Niger Delta states in terms of revenues and human development indicators. However, suspicion between communities about who benefits from oil company payments is not addressed by NEITI. Knowing the exact numbers of company payments may enable the Niger Delta people to hold the Federal Government accountable. At least, they are informed about the extent of the total wealth that is created on what they regard as their soil. On the other hand, mere transparency does not alter the fact that very little trickles down to the minorities in the Niger Delta.

The strong centralising tendency has been reversed in recent years with the turn to a more decentralised management of oil revenues by the states. This presents an opportunity for the Niger Delta people to be more involved in the decision-making processes on oil exploitation. Revenue transparency is therefore imperative not only on the federal but also on the state and local level.

3.2 *Production-site grievances*

This section is concerned with conflicts generated at the level of the production site. In many oil-producing countries these are minor issues, for they mainly host off-shore facilities, but in Nigeria, on-shore oil extraction is significant. The 2003 WAC Report commissioned by Shell delineates clearly how the operations of international oil companies are engendering conflicts among the Niger Delta communities (WAC Report 2003). While the oil companies negotiate territorial concessions and royalty payments with the Federal Government, they still have to acquire what they call a 'social license to operate' from the communities who live in these areas and who often consider themselves to be the rightful owners of the land. Hence, issues at stake are the ownership of land and water, compensation, and the distribution of payments from oil companies among community members.

3.2.1 *Meagre right to compensation*

From the outset, oil-bearing communities are in a disadvantaged position what regards legal entitlements to compensation for their land. The Land Use Act (LUA) of 1978 has removed the ownership of land from individual Nigerians to the state: "All land ... is vested in the military governor of the state and such land shall be in trust and administered for the benefit of all Nigerians." (section 1). Before 1978, the communities concerned were paid annual land rents. Communities are only entitled to compensation as 'holders' and 'occupiers' of land, under the Minerals Act or the Petroleum Act. Section 77 of the Minerals Act sets as compensation rates "such sums as my be a fair and reasonable compensation for ... disturbance of surface rights ... and for damage done to the surface of land ..., any crops, economic trees, buildings ... removed or destroyed." (Ebeku 2001).

Moreover the disputes over the quantum of compensation are to be settled administratively and not by courts (section 47(2) of LUA). In practice, compensation for land acquisition and damages caused by oil spills is set by the Oil Producer's Trade Section (OPTS), the association of oil producing companies operating in Nigeria, based on government compensation rates (Amnesty International 2009a, p. 71). This amounts to a self-regulation of the oil companies, as the government rates mainly follow the recommendations of OPTS.[25] The fatal consequences from this are depicted by the World Bank in its familiar language: "The compensation rates create a market failure because the opportunity cost of lost indigenous production is not included in the operational costs, such that oil companies consume excessive land and cause excessive environmental damage" (World Bank 1995, quoted in: Amnesty International 2009a). While agreeing with the outcome, this paper disagrees with the World Bank's diagnosis: if perpetrators set themselves the fines they have to pay, this is a failure of government. The Nigerian government blatantly fails to fulfil its regulatory and oversight functions.

As a result of all this, communities receive very little in compensation. Oil companies offer communities minor amounts of compensation money for their houses destroyed by pipelines that criss-cross through villages above the ground, one-off payments for trees and crops that they can no longer harvest or for oil spillages that destroy farmland or fishing grounds:

Compensation for environmental damages arising out of oil operations is even harder to get. A recent law suit of some Ijaw people against Shell ruled that adequate compensation should be paid to the victims, but there is no evidence they ever did (Ebeku 2001).[26] The more than 6000 oil spills between 1976 and 1996 totaled more than 4 million barrels (Pöyry 2008). SPDC blames a large percentage of the oil spills on sabotage. While intentional sabotage to get money for the clean-up in collaboration with Shell employees, is certainly a problem (WAC Report 2003), it cannot be an excuse for its own failures. Many of the pipelines have not been renewed since decades. Moreover, contractors of oil companies are under pressure to deliver services at minimal costs, which negatively affects the maintenance of oil installations (Frynas 2000, p. 179).

Another enormous problem is that more than 80% of the gas associated to oil extraction in Nigeria continues to be flared, contrary to practices of the oil industry in other parts of the world (Pöyry 2008, p. 40). The resulting huge flames are of great nuisance to the local population (24 hours daylight, acid rains) but also a major world-wide source of CO_2 emissions (about 70 million metric tons per year) (Social Action 2009b).

Another issue that social activists from the Niger Delta, including the militants, often come up with is the insufficient employment of local people by the oil companies that is perceived as unfair. This corresponds very closely with one characteristic of 'petro-states': the oil industry creates very, very few jobs. The desperate attempt to create more jobs for Nigerians in the oil industry leads to an extreme economic nationalism, such as enshrined in the draft for a new local content law of the Nigerian

Table 2. Oil industry compensation rates (for selected crops).

Crop	Maximum amount per hectare of crop (US$)	Maximum amount per crop/stand (US$)
Maize	58.84	–
Yam	369.23	0.31
Cassava	136	–
Mango	–	7.69
Banana	–	2.36

Source: Oil Products Trade Section (OPTS), Lagos Chamber of Commerce and Industry (1997) (in: Akpan n.d., p. 6).

House of Representatives: Besides job placement issues, the new law stipulates that Nigerians "shall be given first consideration in the award of oil blocs, oil field licenses, oil lifting licenses and shipping services; and all projects for which contracts are to be awarded in the Nigerian oil and gas industry." (This Day, 8 December 2009).

3.2.2 *Opacity—the price for the social license to operate*

The pronounced dissatisfaction among communities conjured up by such practices poses a continual threat to oil production. In consequence, oil companies developed various pacification strategies to achieve the social license to operate nevertheless. Among these are community development projects, scholarship programs and informal payouts to community leaders or to armed youths. The WAC report on Shell practices shows that these payouts are given only in response to real or perceived threats to oil extraction facilities, and not to legitimate and peaceful complaints by communities (WAC Report 2003). Due to the informal nature of these payments, the process is not transparent. It is difficult to discern whom the oil companies pay money to and how much:

My quarrel is: if you [oil companies] are giving money to the communities—why do you keep quiet? Every now and then the people are fighting, are killing, and all we here is: silence. We really need to know: Who is giving what? How much money are you giving to the people in the Niger Delta? Whom are you giving the money?[27]

The payments used to be made to traditional community leaders on behalf of their communities. This creates several problems: as the oil companies finally determine whom they accept as the rightful leader, they are often accused of 'divide and rule' tactics—as favoring one family on the expense of other families in the village (Social Action 2009a). As there is no transparency in the payment process, the chiefs on the other hand are not accountable to their communities: "The oil company gave food items for Christmas: one cup of rice per family! (…) The communities don't know how much is due to them, the total amount is unknown, so everyone takes its share. The leaders sell scholarships to other communities; the communities don't know about the scholarships".[28] Another fundamental problem is the definition of the oil-bearing community that benefits from the company payments.

This often led to inter- and intra- communal quarrels and fighting between allegedly 'indigenous' and 'settler' groups (Ibeanu & Luckham 2007, p. 63). Accordingly, inter-ethnic fighting broke out if community development projects seemed to favor one ethnic group about the other. Another factor came in when oil companies started to arm local thugs for the protection of their pipelines, distancing themselves from the Nigerian military after the transition to constitutional rule in 1999 (WAC Report 2003, Okonta 2007).

Community members developed various strategies to appropriate oil-based benefits from IOCs (Ikelegbe 2005, pp. 216–220). Traditional rulership structures used youth protests to blackmail oil companies, sometimes helped by oil company officers. In the Nembe kingdom in Bayelsa State, e.g., chiefs collaborated with youth and cultural organisations to ask for compensation for false ritual sites. The youth groups increasingly used violence, such as the occupation and shutting down of flow stations, the seizure of equipment and sabotage. When they turned too violent, they were provided 'stand by' payments by the companies to keep calm, and hence were turned into a "protection company" for Shell (Watts 2004, pp. 62–64). As the youth groups grew into local security forces, they became a challenge to customary authorities. This pattern can be observed in many communities in the Niger Delta.

In recent years, there have been some efforts by oil companies, together with NGOs, to improve the way the community development projects are implemented. The latest developments are the "General Memorandi of Understanding" (GmoU) introduced by Chevron that group several villages into one cluster per region. In each region there are elected councils who decide upon the projects. This new process has created some level of transparency because people from one village or ethnic group can see what is done in other villages, possibly decreasing the level of mutual suspicion.[29] However, there is still a wide-spread feeling that the process is top-down (GMOU Participatory Stakeholder Evaluation 2008).

3.2.3 *Relevance of transparency (II)*

In conclusion, the oil companies operated in the Niger Delta with as much opacity as did the Federal Government in the management of its oil revenues. Similarly, their approach to community relations was as much top-down as that of the government. Despite some improvements in the preceding years, communities still raise similar complaints. There is therefore ample space for improvement in the oil company-community relations. Enhanced transparency in the oil companies' operations, including the management of payouts to communities, would represent a possible avenue for improvement. This is clearly beyond NEITI's scope that is merely concerned with state revenues from oil. However, the management of official oil revenues by communities could soon become an issue for NEITI if former President Ya'ardua's promise becomes true that oil-bearing communities will have a 10% equity stake in the oil Joint Ventures.

3.3 *Resource-financed wars*

The struggle among ethnic and community leaderships, elites, businessman, politicians and youths, in the last decade, led to many deaths, large-scale population displacements and the destruction of whole villages. Since 1999, the estimated annual fatalities in the Niger Delta have continued to exceed 500 and can be considered a "low intensity" armed conflict (Ibeanu & Luckham 2007, p. 63).[30] The high number of deaths not only stems from inter-communal violence, but also from the proliferation of a high number of sometimes ethnicity-based armed militias, vigilante groups and so-called 'cults' (political syndicates), and rivalry among those (Ibeanu & Luckham 2007, p. 63, 85; Florquin & Berman 2005, pp. 19–26).

The grievances that the current militants under the lead of MEND denounce were described in the previous sections. It is not realistic to assume any significant role of transparency initiatives in preventing the spiraling into violence of these grievances. Some triggering factors were already mentioned in the preceding section: mobilisation of youth groups as protection agencies; reluctant response to legitimate complaints by both the Nigerian government and oil companies; suppression by force or payouts to violent groups. In addition, politicians in the Niger Delta armed youth groups to help them win elections, promising them a material compensation in case they won the seat (Ibeanu & Luckham 2007, pp. 81–85, Florquin & Berman 2005; HRW 2008). A Nigerian researcher goes even so far as stating that "the basis for the development of militant tendency is in the process of the electioneering campaigns."[31]

3.3.1 *"In the confusion in the entire delta there is wealth"*[32]

The militants turned to the profitable illegal theft and trading of oil ('bunkering'), when they were dumped by politicians after elections. In the first place, they were used for oil bunkering, not organising it themselves. The bunkering of refined and crude oil had started much earlier, and was dominated by people from outside the Niger Delta.[33] By now, crude oil is bunkered on a large scale and with a sophisticated industry which uses advanced technology to tap crude. The stolen oil is sold very cheaply to Africa, Europe, Asia and North America. The amount of stolen crude is difficult to assess, but the estimates exceed the entire production of oil in Côte d'Ivoire (about 50,000 bpd) (Ikelegbe 2005, pp. 221–225).

Being equipped with arms provided for by politicians, oil companies, and increasingly bought by themselves, militants started to control the oil bunkering routes. The attacks on oil installations and kidnappings of oil workers created a seemingly ungovernable situation and general confusion. This worked to the advantage of many actors in the Niger Delta besides the militants.

For one, it widened avenues of diversion of state funds because a great part of the public expenses was earmarked for security, "but only a tiny fraction really goes to security".[34] Those politicians who do not gain access to public office can benefit from the confusion by illegally selling oil. It was also convenient for the governors of the Niger Delta because the state of emergency increases their political weight on a national level.[35] In addition, the oil companies are not under government scrutiny neither. The official production volumes rapidly and drastically declined from a peak in April 2009 at 2.2 million bpd to 2 million bpd by July, fluctuating between 800,000 and 1 million in August, to inch towards 2.1 bpd million already by the end of October 2009, after the acceptance of the amnesty (Agbo 2009, p. 20). This implied a concurrent reduction in royalties that the oil companies had to pay, while avenues of illegal sale of oil remained open- as long as operations continued.[36] And lastly, the international buyers of bunkered oil mentioned above have benefited from the cheaply available petroleum.

Therefore it is plausible that the large-scale oil theft has happened in collaboration with parts of the military, political and economic elites, as this statement suggests:

Oil companies are part of it. The army generals are part of it. Government functionaries are part of if. They are all involved. It is not a business for common people, because the vessels that they use, they are not bicycles, they are huge equipment. … The bunkerers, the generals, the thieves … they reach an agreement with oil company officials who say: 'Ok, so and so quantity will be attributed to you, this is your quota'.[37]

Since 2004, there has been a crackdown by the Joint Task Force on the Niger Delta armed militants and illegal oil and arms trading. The militias, since 2005 united under the roof of MEND, started building solid military bases, such as Camp 5 located in Gbaramatu Kingdom in Warri South in Delta State, which was under the central command of Tompolo (Smith 2009). In May 2009, violence escalated when the JTF stormed Camp 5 and raised down whole villages where they suspected militants (Amnesty International 2009c). MEND, on the other hand, assaulted and exploded the pipelines at a major oil distribution point in Lagos, the major economic city of Nigeria (calling itself the "Centre of excellence"), on 13 July 2009 (McGregor 2009). Only then, when both sides were decidedly hurt and the support for MEND began to decline among the population of the Niger Delta due to the immense human suffering, did the government offer an amnesty program, that MEND leaders finally accepted in October 2009.

3.3.2 *Relevance of transparency (III)*

As is apparent from the factors triggering violence mentioned above, transparency initiatives can do little to prevent the spiraling of grievances into outright violence. What this section highlights is situations of generalised violence and confusion undermine transparency—transparency in company statements of production volumes, transparency in public expenses. One result of the Niger Delta crisis was the decrease in public attentiveness to issues of transparency in the management of oil revenues: "The discussion on the amnesty [makes] that little is heard on what is being done on NEITI. So there's a shift. At a point in time, it was more like: 'Let's talk about the extractive industry, about transparency issues around', but now it's more: 'Let's focus on the people causing the trouble, let's bring the people out of the issue'."[38] However, increased transparency on the volumes of oil production in the Niger Delta, as attempted by NEITI, is imperative to stop the large-scale oil theft on which the war economy thrives.

4 NEITI'S IMPACT ON GOVERNANCE AND POPULAR PARTICIPATION

This chapter examines the potential that the Nigeria Extractive Industries Transparency Initiative (NEITI) has to contribute to more effective governance and popular participation in the Niger Delta. It depends on its own political clout, whether the potential can be realised. It delineates the opportunities and limits arising from NEITI's design and the political context of a 'petro-state'.

4.1 *Limitations arising from NEITI's design*

Relevance 1: NEITI is certainly relevant on the first level of resource-related conflict. Via the audit reports, it has collected and published detailed information on oil revenues for the first time, and stirred some public discussion on the topic. The NEITI secretariat even esteems that it has instilled some trust and openness among the major actors in the oil industry—including the non-governmental organisations (NGOs).[39] Therefore, NEITI has the potential to contribute to greater accountability of the government.

However, there are clear limitations to NEITI's impact that arise from its design. As NEITI does not publish the production volumes per wellhead, the Niger Delta communities still do not know how much oil was pumped from their community area. They can only approximately deduct these amounts from the distribution by the Federal government of the 13% derivation money among the oil-producing states.

NEITI's organisational structure also raises some concerns, for civil society groups are only weakly represented. The civil society representatives in the National Stakeholder Working Group (NSWG) that functions as the board of NEITI are appointed by the President and do not have the necessary expertise of the extractive industry. There is hardly any contact between these civil society representatives and the Publish What You Pay (PWYP) Coalition in Nigeria[40] that is specialised on transparency issues in the extractive industry.[41] Out of the 15 members of the NSWG, only three come from civil society groups. The quorum for decision-making is eight (NEITI Act 10). Hence, apart from the representative of the South-South Zone who happens to be a civil society activist, the Niger Delta communities are only weakly represented within NEITI.

The NEITI Secretariat clearly sees itself as a government institution.[42] Its staff are civil servants, appointed by the government. The dominance of the government seems to hamper NEITI's drive of action, as is seen in a quarrel between the Executive Secretary, a civil servant, and the Chair, a professor and civil activist (former chair of Transparency International Nigeria).

Moreover, the discussion on oil revenue transparency is largely confined to elite circles. Apart from the publications on its website, NEITI has organised road shows throughout the country during which the audit reports are disseminated and discussed at public meetings. Yet, other informants noticed that NEITI did not disseminate the audit results widely enough and in an appropriate format that could reach the wider population, e.g. via radio or in a local vernacular. A lot more effort in this direction would therefore be needed to reach ordinary people in the Niger Delta: "Ordinary people hardly bother about revenue, how much goes to the state, how much is spent etc. This is a very sophisticated question for rural people living in the creeks. What they are asking for is: daily livelihoods."[43]

An important area for improving governance is actually the state and local governments that receive increasing volumes of oil revenues. As NEITI is confined to the federal state level, its impact is minimal in that area. The discussions on transparency opened up by NEITI on the national level, have nevertheless had some repercussions on the state and local level. At the NEITI road shows, for example, representatives of state and local governments participated in the debate on revenue management. Moreover, the financial support from international donors for NEITI is extended to other projects and actors that are concerned with budget transparency. The NEITI Secretariat is supported by a Multi-Donor-Trustfund; World Bank and DFID provide consultancy services. The governor of Bayelsa State who has launched his own Bayelsa Expenditure and Income Transparency Initiative (BEITI) is supported by the Revenue Watch Institute.[44] As civil society is supposed to play a vital role in the EITI process, there is a whole range of Nigerian NGOs that receive support from international agencies, such as the Revenue Watch Institute, Oxfam, USAID, ActionAid, Cordaid, German political foundations etc.

Several of these NGOs have created a Niger Delta Citizens and Budget Platform to monitor budgets of the state and local governments (Niger Delta Citizens and Budget Platform 2009). The difficulties they met in accessing information on state budgets underline the importance of their efforts. Another related initiative, the "Public Eye Project", has managed to install expenditure monitoring commissions in the three oil-rich states Ondo, Edo, and Delta backed by a law that prescribes that 50% of the state oil revenues are spent on infrastructure projects only (ANEEJ 2008).[45] Some interviewees were sceptic about the political strength of most of these NGOs, arguing that they were mainly interested to "eat and survive". Nevertheless, the various initiatives are a sign that some Niger Delta populations are already mobilising themselves and are becoming increasingly conscious of their rights: "The level of awareness of Nigerians to fight corruption now is far higher than it used to."[46]

The PWYP coalition had acquired some political strength, due to, among other things, an engaged national coordinator, Reverend David Ugolor. However, the coalition has got in some disrepute, and lost most of its donor funding. The new leadership is currently regaining its strength by reassembling the existing members.

Relevance 2: The above analysis has shown that enhanced transparency in the oil companies' operations, including the management of payouts to communities, would represent a possible avenue for improvement. This is clearly beyond NEITI's scope that is merely concerned with company payments to the state, not to the communities. However, if NEITI finally goes ahead with a Value-for-Money-Audit and helps addressing the issue of production measurement, this will go a long way into a greater accountability of companies towards the Niger Delta people.

Relevance 3: The current amnesty process represents an opportunity to address the drivers of the war economy in the Niger Delta—the illegal oil theft. NEITI's audit reports have laid open the lack of government oversight over actual production volumes and have thus laid some basis for reaching transparency in companies' production volumes.

4.2 *Limitations due to NEITI's political context*

NEITI being in fact a government institution, it is clearly embedded in a political context that puts further limits on its impact. It is subject to the same dynamics of a 'petro-state' that were already introduced—the lack of institutionalisation, the tendency to personalised rule and the politics of patronage: "Right now, NEITI is seen as an extended bureaucracy. (...) It simply has a National Stakeholder Working Group and the Secretariat that is made up of civil servants that are being deployed from various ministries, and mostly the deployment is actually based on politics, nepotism, favoritism and all of that".[47]

NEITI was instituted in a period of reform in Nigeria aiming at transparency and accountability in the public sector and at privatisation of the economy, especially in the years 2003–06. It was underwritten by a team of officials around Obasanjo and the International Financial Institutions (IFI).[48] Part of the reforms was a new Economic and Financial Crimes Commission (EFCC) that pursued financial misappropriations. It is this context that provided the necessary political drive that pushed NEITI that far (Shaxson 2009, pp. 9–20).

Former president Obasanjo's backing for these reforms was crucial. One of his main motivations seems to have been his wish to be an international donor darling; he wanted Nigeria to be seen as the economic powerhouse in Africa (Shaxson 2009, p. 15). Hence, NEITI was born out of international pressures radiating from the EITI and a group of Nigerian reformers. The reform initiative was driven from above and outside—much less from internal pressures from below. NEITI was probably also just easier to implement because it was less threatening to interests of the ruling party PDP and the NNPC than another reform of the Nigerian oil sector that was more encompassing and is now under way, the Petroleum Industry Bill (Shaxson 2009, pp. 15–18).

As these reform efforts have lost considerable drive following the change of government in 2007, so has the pace of NEITI's work. This is a sign of the lacking institutionalisation identified by Karl (1997) . The achievements of reform in Nigeria were difficult to sustain after the few persons sustaining them had left, such as the head of EFCC, Ribadu. He was ousted just weeks after the former Delta State governor James Ibori was arrested in December 2007 on charges of corruption and money laundering by the EFCC (VOA News, 12 December 2007 and AFP, 29 December 2007). Major cases accusing incumbent or former high-level politicians of corruption and fraud are now blocked as files somewhere in the corridors of EFCC or have failed in court. 31 Nigerian state governors and former president Obasanjo were exonerated of corruption charges in late 2008, after a Senate Committee had indicted Obasanjo and his Vice-president Abubakar for corruption over an oil fund, the Petroleum Technology Development Fund (PTDF), in 2007 (CISLAC 2008).

Patronage politics related with a personalised rule may also account for the slow-down of NEITI's and EFCC's political clout. As political leaders try to fill administrative and political positions with their entourage, new governments tend to mistrust the institutions created by their predecessors, and thus aim to create their own institutions with their own loyal people. In addition, the previous President Ya'ardua came to power without any political ambition, being physically weakened by some illness. He was therefore probably less able to confront established interests that had been under attack by institutions such as the Economic and Financial Crimes Commission (EFCC) and NEITI.

NEITI is as much subjected to the personalised rule as any other government institution in Nigeria. Even the implementation of NEITI is dependent on the President's will: some domestic oil companies would refuse to comply with NEITI's requirements, until a phone call at the President's office would reign them in.[49] NEITI will therefore hardly be able to overcome the political patronage that engulfs Nigeria's political system. Nigeria's rating on the global Corruption Perceptions Index by Transparency International (TI) worsened in 2009, dropping from 121th to 130th position out of 180 countries, after having moved up 27 places in 2008. The index measures the perceived levels of public-sector corruption (Akosile 2009, Idonor 2009).[50]

Similarly, the low effectiveness of state institutions in 'petro-states' obviously also diminishes NEITI's impact. Since the publication of the first audit report in 2006 that recommended an overhaul and harmonisation of the various regulatory institutions' procedures (so-called "remediation"), little has happened in that regard.[51] Moreover, the bureaucratic character of the NEITI Secretariat seems to have crippled its own capacity:

NEITI at the moment is understaffed. They almost completed the staff recruiting process. (…) For it to function effectively, it was established by the NEITI Act 2007. The present government didn't appoint the board until last year [2008]. So once the board and the executives of the secretariat were appointed, they then began to staff the secretariat. But first, they needed a salary structure, confirmed by the Office of the Secretary of the Federation.[52]

The political context also has repercussions on NEITI's ability to enforce compliance of oil companies with the legal requirements of information disclosure (NEITI Act, 2007). Several interviewees brought up the topic of compliance by themselves. Some NGOs felt that NEITI should do more to enforce compliance while other interviewees agreed that NEITI has no right to pursue offenders, but is dependent on the Nigerian judiciary or the EFCC to do so.

To conclude, NEITI itself is not the driver of change, but other political forces, such as the group of Nigerian reformers, the "international community" (EITI, the audit consultancy firms, international donors), and Nigerian NGOs under the umbrella of PWYP are agents of change. Within the confines of the Nigerian state, NEITI does not seem able to generally improve governance effectiveness as it is dependent on other Nigerian political and judicial institutions to have an impact. It does, however, have important side-effects. The NGO activities in the Niger Delta as well as the Bayelsa government transparency initiative represent one step towards greater accountability and effectiveness in government spending. However, it is questionable that transparency can actually achieve accountability given the patronage politics and violence surrounding elections in the Niger Delta.

5 CONCLUSION

As the Extractive Industries Transparency Initiative (EITI) seeks to improve governance, the paper investigated the potential of NEITI to contribute to greater popular participation and more effective governance to enhance living conditions in the Niger Delta, which are part of a solution to the Niger Delta crisis. The analysis of the oil-related conflict in the Niger Delta has shown that transparency—or rather the lack of it—is relevant on all three levels of conflict, but at the same time that the conflict involves issues that go far beyond transparency. On the first level, this included the distribution of revenues from oil in a multi-ethnic state, where production of oil in a so-called minority area meets fiscal centralisation. On the second level, the destruction of sources of livelihoods through oil extraction, questions of land ownership and compensation, and the very limited capacity of the oil industry to employ people, are of relevance.

The third part of the paper delineated limitations to NEITI's potential to contribute to overcoming the resource curse in the Niger Delta, arising from NEITI's design and the context of a 'petro-state'. For one, the transparency initiative does only weakly represent the Niger Delta people in its organisational structure. Moreover, the accountability of oil companies is only marginally increased. However, through the public debate it stirred and the publication of the oil revenues it may help to hold the government more accountable—although expenditure and local and state governments are not addressed. Also the financial support of international donors to NGOs in the Niger Delta may convey a greater political weight on the Niger Delta communities as a side-effect. The activities of NGOs such as the PWYP coalition may also indirectly enhance effective governance, if the increased accountability of governments leads to a better use of public budgets.

Karl's (1997) analytical framework of a 'petro-state' was used to analyse limitations to NEITI's effectiveness and political clout. NEITI's own capacity as well as its impact on general government effectiveness and accountability is limited, for it is a government institution and subject to the same political dynamics as other public institutions in 'petro-states'.

Although the limitations of NEITI were analysed in relation to the Niger Delta—being a worst-case scenario-, the production-site grievances relating to land ownership and environmental destruction can be found in many other onshore oil-producing or mining areas in developing countries. Countries with mainly offshore oil exploitation may not experience the production-site conflicts related to the onshore drilling of oil in the Niger Delta. They will still encounter similar challenges in terms of government oversight over company operations and accountability mechanisms for the citizenry.

While using the analytical framework of a 'petro-state' (Karl 1997) to analyse NEITI's limitations, the paper does not want to suggest that any attempt at reform in a 'petro-state' is futile from the beginning. However, there are some limitations, such as the weak institutionalisation, that can hardly be overcome by external intervention. If external ambitions such as EITI meet with a group of local reformers, this can represent a temporary opportunity. Moreover, continued donor support for local civil society organisations in the Niger Delta such as the PWYP coalition are essential to pressure for increased oversight over the oil industry.

The analysis of the war economy in the Niger Delta having shown how generalised violence and confusion undermines transparency in the management of the oil industry, the current post-amnesty process represents an opportunity to make better use of the transparency initiative. NEITI has made the first step towards transparently measuring oil production volumes in the Niger Delta. Efforts in this direction will be imperative to stop the large-scale oil theft on which the war economy thrives. It will also go a long way in holding oil companies accountably towards the Niger Delta citizenry.

Referring back to the—surely polemic—title of this article, the results of this analysis call into question policy priorities among the international community regarding problems of corruption, environmental destruction and social dislocation in oil-producing countries. Turning the curse into a blessing by revenue transparency is a convenient illusion for governments in extractive countries—if they thereby avoid addressing crucial issues of revenue distribution within their societies and of production standards. Turning the curse into a blessing by revenue transparency in producing countries is a convenient illusion for European governments—if they thereby avoid holding their extractive companies legally accountable for their actions abroad. Given the inherently weak state structures in terms of effectiveness, accountability and transparency in many oil-rich African countries, the EU should concentrate its efforts on its own countries and companies. If EU countries committed themselves to EITI, they would increase pressure on their mineral companies to comply with EITI requirements. More importantly, mandatory rules for EU companies operating abroad would help avoid some problems African governments are faced with now when trying to regulate international investors. The EU even has a responsibility to do so, as European investors in the extractive industries of developing countries are implicated in the political dynamics of these 'petro-states'. This being said, it does not mean that transparency is not an important means for avoiding the curse to appear in the first place—certain conditions being fulfilled. If politicians are not elected into their offices by the people but with the help of some political protectors and violence, the prospects are much worse for transparency in oil revenues and budgets actually leading to greater accountability towards the people.

ENDNOTES

[1] See www.eitransparency.org.

[2] Another EITI candidate country that has enacted an EITI law is Liberia.

[3] ExxonMobil accepted this as an exception, while usually not allowing for the disaggregation of data in its world-wide operations (Shaxson 2009, p. 13).

[4] The costs incurred by companies reduce the revenues paid to government by lowering company profits.

[5] According to the Nigerian constitution, NEITI is only able to monitor at the federal level, not at the state and local levels (Shaxson 2009, p. 38).

[6] Interview with Amanda Feese, World Bank, EITI Consultant, Abuja, 28 October 2009; Interview with Dayo Olaide, OSIWA, Abuja, 28 October 2009.

[7] Interview with Amanda Feese, World Bank, EITI Consultant, Abuja, 28 October 2009.

[8] Interview with NEITI Secretariat, Abuja, 29 October 2009.

[9] Interview with Amanda Feese, World Bank, EITI Consultant, Abuja, 28 October 2009.

[10] Interview with Basil Omiyi, Shell and NEITI, Abuja, 27 October 2009.

[11] Interview with Amanda Feese, World Bank, EITI Consultant, Abuja, 28 October 2009.

[12] Interview with Amanda Feese, World Bank, EITI Consultant, Abuja, 28 October 2009.

[13] Oil-dependent are those countries with a high share of oil production in GDP and of oil exports in total exports. The Guide on Resource Revenue Transparency of the IMF sets this share at 25% of total exports as a minimum.

[14] For further reading on the origin of European 'modern' nation-states see the discussion among scholars of Historical Sociology, namely between neo-Weberian theorists such as Theda Skocpol, Michael Mann (1984), Peter Evans, John M. Hobson (1998) and Charles Tilly (1985) and neo-Marxist scholars such as David Harvey, Hannes Lacher (2007), Benno Teschke (2003), Alex Callinicos (2007) and Ellen Meiksins Wood (1991, 2002).

[15] Again, this is not only true for petro-states, but for most African countries, as scholars such as Jean-Francois Bayart (1993) or Bruce Berman (1998) would contend. As African states lack a sizeable class of autonomous entrepreneurs, the state rather than the private economy is the primary object of rent-seeking. An indication of this are the close ties to the political elite of successful Nigerian entrepreneurs. Examples in Nigeria include: Wale Tinubu, the CEO of Oando, a major oil products distribution company, whose uncle was a former governor of Lagos state; Aliko Dangote, Chairman and CEO of Dangote Group that trades also petroleum products, who had a close relationship with former President O.Obasanjo (Africa Report 2009, p. 58).

[16] The revenues are divided in the following way: One part are the statutory allocations that are due to all states based on their respective land mass and population, one part is kept for the federation, and the last part is 'derivation' money allocated to the oil-producing states based on their oil production volumes. (Guichaoua 2009, p. 30).

[17] The Biafra War between 1967 and 1970 began after the southeast region, with a majority of Igbo, declared its secession from the Nigerian Federation. Control of oil resources found in the region was one of many aspects of this war. It left more than a million dead. (Guichaoua 2009, p. 20).

[18] Thereby, the oil-bearing communities of the Niger Delta (today's South-South) were separated from the Igbo-dominated southeast (today's South-East).

[19] The violent clashes between Ijaw and Itsekiri in Warri, Delta State, in 1997–1999 followed the creation of new Local Government Areas and led to the deaths of 100 people. Since then, there has not been a crisis of that magnitude (Institute for Peace and Conflict Resolution 2008, pp. 182–184).

[20] Interview with Mr. Bari Ara KPALAP, MOSOP, Lagos, 30 October 2009.

[21] Another expression of the limited capacity of policy implementation is the very low score of budget implementation in Nigeria. By December 2009, for example, less than 50% of the projects contained in the 2009 budget were implemented (Vanguard, 10 December 2009).

[22] Interview with Dayo Olaide, OSIWA, Abuja, 28 October 2009.

[23] Interview with Dr. John Emeka Akude, Cologne, Cologne, Germany, 9 September 2009.

[24] Interview with Dr. Etham Mijah, National Defense Academy Kaduna, Abuja, 28th October 2009.

[25] Interview with Reverend Kevin O'Hara, founder of CSCR, Lagos, 3rd November 2009.

[26] A promising sign is the recent court case, brought by four Nigerian victims of Shell oil spills, in conjunction with Friends of the Earth Netherlands, in December 2009 in the court at The Hague, Netherlands. The Dutch court will also address whether it has jurisdiction over Shell's Nigerian subsidiary (remembersarowiwa.org).

[27] Interview with Dickson Orji, President of WAANSA Nigeria (West African Small Arms Network), Abuja, 26th October 2009.

[28] Extract from interview with Bridget Osakwe, WANEP, Lagos, 4th November 2009.

[29] Interview Dennis Flemming, Chevron Nigeria, Community Engagement Advisor, Lagos, 2nd November 2009.

[30] The numbers listed by the Stockholm Peace Research Institute (535 battle-related deaths in 2004) do only account for fighting that involved the state military. As much of the violence in the Niger Delta is intercommunal, there is hardly any reliable data on annual fatalities and the numbers are contested (Mähler 2010, pp. 11–13).

[31] Interview with Dr. Etham Mijah, National Defense Academy Kaduna, Abuja, 28th October 2009.

[32] Interview with Dr. Etham Mijah, National Defense Academy Kaduna, Abuja, 28th October 2009.

[33] Informal conversation with Prince Joseph Ettela Harry, "Godfather" of some MEND militias, Abuja, 6 November 2009.

[34] Interview with Dr. Etham Mijah, National Defense Academy Kaduna, Abuja, 28th October 2009.

[35] Interview with Dickson Orji, WAANSA, Abuja, 26th October 2009.

[36] Shell declared 87 incidents of oil theft in 2008, and that it had to shut down all operations in the western delta from early 2006 until the end of 2007 (Shell 2009).

[37] Interview with Celestine Akpobari, Social Action and OSF, 5th November 2009, Lagos. This is not a singular statement, but one of several very similar ones of different interviewees, also referring to evidence seen by members of the Technical Committee on the Niger Delta set up in 2008.

[38] Interview WARDC (member of PWYP), Lagos, 3rd November 2009.

[39] Interview with NEITI Secretariat, Abuja, 29 October 2009.

[40] PWYP Nigeria forms part of the international campaign "Publish What You Pay" (PWYP). (*www.publishwhatyoupay.org*).

[41] Telephone Interview with Faith Nwadishi, National Coordinator of PWYP Nigeria, 7 November 2009.

[42] Interview with NEITI Secretariat, Abuja, 29 October 2009.

[43] Interview with Dr. Etham Mijah, National Defense Academy Kaduna, Abuja, 28 October 2009.

[44] The initiative is not concerned with the extractive industry but with transparency in all fiscal matters. It clearly alludes to NEITI in name and organisational structure, e.g. the Multi-Stakeholder Working Group (BSWG), but is not directly related to NEITI in reality. Therefore attention needs to be paid to what will be done in substance. See also: Bayelsa Expenditure and Income Transparency Initiative (BEITI). n.d. "Report of the Journey-so-Far".

[45] Interview with Agbojo Adewale Enoch, HBS (Heinrich Böll Stiftung) Nigeria, Program Manager Governance, Lagos, 3 November 2009. The Commissions are called Ondo State Oil Mineral Producing Areas Development Commission (OSOPADEC) and Edo State Oil and Gas Producing Areas Development Commission (EDOGPADEC). A law was passed in Edo State in May 2007, and EDOGPADEC was established in July 2007. The NGO ANEEJ critically monitors its work.

[46] Interview with Innocent Adjenughure, CAAT, Abuja, 5 November 2009; Interview with Agbojo Adewale Enoch, HBS (Heinrich Böll Stiftung) Nigeria, Program Manager Governance, Lagos, 3 November 2009.

[47] Interview with Dayo Olaide, OSIWA, Abuja, 28 October 2009.

[48] The team consisted of the Finance Minister Ngozi Okonjo-Iweala, Oby Ezekwesili, one of the founders of Transparency International Nigeria who was appointed coordinator of the NEITI Working Group, Charles Soludo as Central Bank Governor, Nuhu Ribadu, chairman of the new Economic and Financial Crimes Commission (EFCC), Bright Okogu, special advisor to the Finance Minister 2004–07, and Nasir El-Rufai, former director-general of the Nigerian privatisation agency, minister of the Federal Capital Territory (Shaxson, 2009, pp. 9–20).

[49] Interview with NEITI Secretariat, Abuja, 29 October 2009.

[50] It is important at the same time to remember whose perception is taken as the standard measure for this corruption index: it is mainly Western businessmen and bureaucrats who are interviewed in these surveys.

[51] Interview with Amanda Feese, World Bank, EITI Consultant, Abuja, 28 October 2009.

[52] Interview with Amanda Feese, World Bank, EITI Consultant, Abuja, 28 October 2009.

REFERENCES

Africa Network for Environment and Economic Justice (ANEEJ) 2006. *Oil of Poverty in Niger Delta*. 2.

Africa Network for Environment and Economic Justice (ANEEJ) 2008. An independent Civil Society analysis of Edo State Oil and Gas Development Commission (EDSOGPADEC): 2007–2008 Budget & Expenditure. November 2008.

Agbo, A. 2009. "The Gains of Amnesty", *Tell*, 44, Weekly, 2 November 2009.

Akosile, A. 2009. Corruption Perception—Nigeria's Rating Worsens. *This Day*, 18 November 2009, Available at: http://allafrica.com/stories/200911180368.html (Accessed 28 November 2009).

Akpan, W. n.d. *Oil, people and the environment: Understanding land-related controversies in Nigeria's oil region*. Available at: http://www.codesria.org/Links/conferences/general_assembly11/papers/akpan.pdf (Accessed 10 December 2009).

Amnesty International 2009a. *Nigeria: Petroleum, Pollution and Poverty in the Niger Delta*. June 2009.

Amnesty International 2009b. *Killing at Will. Extrajudicial Executions and Other Unlawful Killings by the Police in Nigeria*. December 2009.

Amnesty International 2009c. *Tens of thousands caught in crossfire in Niger Delta fighting*. 21 May 2009. Available at: http://www.amnesty.org/en/news-and-updates/news/tens-thousands-caught-crossfire-niger-delta-fighting-20090521 (Accessed 14 December 2009).

ANEEJ. See Africa Network for Environment and Economic Justice.

Anugwom, E.E. 2005. Oil minorities and the politics of resource control in Nigeria. *Africa Development* (Dakar), 30 (4) (2005): 87–120.

Baker, L. 2008. At issue: Facilitating whose power? IFI policy influence in Nigeria's energy sector. *Bretton Woods Project, At issue*, April 2008, London. Available via www.brettonswoodsproject.org

Basedau, M. 2005. Resource Politics in Sub-Saharan Africa beyond the Resource Curse: Towards a Future Research Agenda. In Matthias Basedau and Andreas Mehler, (eds.) *Resource politics in Sub-Saharan Africa*. Hamburg African Studies, 14, Hamburg: Institut für Afrikakunde.

Basedau, M. & Lacher, W. 2006. A Paradox of Plenty? Rent Distribution and Political Stability in Oil States, *GIGA Working Papers*, 21, April 2006.

CISLAC 2008. Anti-Corruption groups vow continued pursuit of tainted former Nigerian State Governors. 12 October 2008. Available via www.cislacnigeria.org (Accessed 23 October 2009).

De Oliveira, R.S. 2007. *Oil and Politics in the Gulf of Guinea*. London: Hurst and Company.

Dokubo, A. 2007. Me, Henry Okah, 'Jomo Gbomo', Judith Asuni and the Niger Delta. *Sahara Reporters*. 19 October 2007. Available at http://www.ocnus.net/artman2/publish/Africa_8/Me_Henry_Okah_Jomo_Gbomo_Judith_Asuni_and_the_Niger_Delta_Insurgency.shtml (Accessed 11 December 2009).

Ebeku, K.S.A. 2001. Oil and the Niger Delta People: The Injustice of the Land Use Act. *CEPMLP Journal*. November 18. Available at: http://www.dundee.ac.uk/cepmlp/journal/html/vol9/article9 - 14.html (Accessed 28 November 2009).

Fasan, R. 2009. Nigeria: This Boiling Niger-Delta. *The Vanguard*. 26 May 2009. Available at: http://allafrica.com/stories/200906020040.html (Accessed 13 December 2009).

Federal Republic of Nigeria. n.d. *Niger Delta Regional Development Master Plan.*

Florquin, N. & Berman, E.G. (eds.). 2005. Armed and Aimless: Armed Groups, Guns, and Human Security in the ECOWAS Region, *Small Arms Survey (SAS)*, May 2005, Geneva.

Frynas, J.G. 2000. *Oil in Nigeria. Conflict and litigation between oil companies and village communities. Politics and Economics in Africa*. Volume 1. Hamburg: Lit Verlag.

Garuba, D.S. 2003. Oil and the Politics of Natural Resources Governance in Nigeria. Paper presented at the XIV *Biennial Congress of the African Association of Political Science (AAPS)*. Durban, South Africa, 26–28 June.

Garuba, D.S. 2005. Gunning for the Barrel: The Oil Dimension to Inter-Communal Violence in the Niger Delta City of Warri. Paper prepared for presentation at the 15th *Biennial Congress of the African Association of Political Science (AAPS)*. Cairo, Egypt, 19–21 September.

Garuba, D.S. 2009. Nigeria: Halliburton, Bribes and the Deceit of "Zero-Tolerance" for Corruption. *Revenue Watch News*, 9 April 2009.

Guesnet, L., Schure J. & Müller, M. 2009b. Natural Resources in Côte d'Ivoire: Fostering Crisis or Peace? The cocoa, oil, diamond and gold sectors. *brief* 40, Bonn: BICC.

Guesnet, L., Schure J. & Paes, W.C. (eds). 2009a. Digging for Peace. Private Companies and Emerging Economies in Zones of Conflicts: Report of the Fatal Transactions Conference, Bonn, 21–22 November. *brief* 38, Bonn: BICC.

Guichaoua, Y. 2009. Oil and political violence in Nigeria. In Jacques Lesourne, (ed.) *Governance of Oil in Africa: Unfinished Business.* Les études, May 2009. Paris: Institut Francais des Relations Internationales (IFRI): 9–50.

Hart Group. 2006a. *Nigerian Extractive Industries Transparency Initiative.* Report on the Financial Audit 1999–2004. Presented to the National Stakeholder Working Group of the NEITI by Hart Nurse Ltd. in association with SS Afemikhe & Co, Final Submission: November 2006, Information as at 30th June 2006.

Hart Group 2006b. *Nigerian Extractive Industries Transparency Initiative. Final Report. Combined Executive Summary*. Presented to the National Stakeholder Working Group of the NEITI by Hart Group in association with CMA Limited and SS Afemikhe & Co, Submission: December 2006, Information as at 20th December 2006.

Hart Group 2008. *Nigerian Extractive Industries Transparency Initiative (NEITI). Financial, Physical and Process Audits, 2005.* Summary of Recommendations, Presented to the National Stakeholder Working Group of the NEITI by Hart Nurse Ltd, Chartered Accountants in association with SS Afemikhe & Co, Chartered Accountants, November 2008.

Hart Group 2009a. *NEITI. Physical and Financial Audits 2005*. Presentation to the Federal Executive Council, May 2009.

Hazen, J. & Horner, J. 2007. Small Arms, Armed Violence, and Insecurity in Nigeria: The Niger Delta in Perspective. *Small Arms Survey (SAS). Occasional Paper 20*, December 2007.

Heinemann-Grüder, A. 2009. Föderalismus als Konfliktregelung. *Forschung DSF 21*, Deutsche Stiftung Friedensforschung.

Heller, P. 2009. The Nigerian Petroleum Industry Bill: Key Upstream Questions for the National Assembly, *Revenue Watch Institute.*

HRW. See Human Rights Watch.

Human Rights Watch. 1999. *The Price of Oil. Corporate Responsibility and Human Rights Violations in Nigeria's Oil Producing Communities*. January 1999.

Human Rights Watch 2002. *Nigeria. The Niger Delta. No Democratic Dividend*. 14, 7, October 2002.

Human Rights Watch 2007. Chop Fine. *The Human Rights Impact of Local Government Corruption and Mismanagement in Rivers State, Nigeria. 19*, 2, January 2007.

Human Rights Watch 2008. *Politics at War. The Human Rights Impact and Causes of Post-Election Violence in Rivers State, Nigeria*. 30, 3, March 2008.

Human Rights Watch 2009. *Arbitrary Killings by Security Forces. Submission to the Investigative Bodies on the November 28–29, 2008 Violence in Jos, Plateau State, Nigeria*. July 20, 2009. Available at: http://www.hrw.org/en/node/84015 (Accessed 9 December 2009).

Ibeanu, Okey & Luckham, R. 2007. Nigeria: political violence, governance and corporate responsibility in a petro-state. In Mary Kaldor, Terry Lynn Karl and Yahia Said, (eds.) *Oil wars.* London: Pluto Press: 41–99.

Idonor, D. 2009. Nigeria: Corruption Rating -FG Blames Private Sector. *Vanguard*, 19 November 2009.

Ikelegbe, A. 2005. The economy of conflict in the oil rich Niger Delta region of Nigeria. *Nordic Journal of African Studies*, 14 2 (2005): 208–234.

Institute for Peace and Conflict Resolution. 2008. Strategic Conflict Assessment of Nigeria. Consolidated and Zonal Reports. Second Edition. Abuja, Nigeria: Institute for Peace and Conflict Resolution.

International Crisis Group. 2009. Nigeria: Seizing the Moment in the Niger Delta. *Policy Briefing. Africa Briefing 60*, 30 April 2009, Abuja/ Dakar/ Brussels.

Kaldor, M., Karl T.L. & Said, Y. 2007. "Introduction". In Mary Kaldor, Terry Lynn Karl and Yahia Said, (eds.) *Oil wars.* London: Pluto Press: 1–41.

Karl, T.L. 1997. *The paradox of plenty: oil booms and petro-states*. Berkeley, California: University of California Press.

Mähler, A. 2010. Nigeria: A Prime Example of the Resource Curse? Revisiting the Oil-Violence Link in the Niger Delta, *GIGA Working Paper Series*, 120, January 2010. Available via http://www.giga-hamburg.de.

Maier, K. 2001. *This house has fallen: Nigeria in crisis.* London: Penguin Press.

McGregor, A. 2009. Niger Delta Militants Mount First Ever Raid on Lagos Oil Facilities. *Terrorism Monitor.* 7, 22.

NEITI. N.d. *Nigerian Extractive Industries Transparency Initiative*. Audit of the Period 1999–2004 (Popular version). Available via www.neiti.org.

Niger Delta Citizens and Budget Platform. 2009. *'Carry Go'. Citizens Report on State and Local Government Budgets in the Niger Delta 2008*. Social Development Integrated Centre (Social Action).

Obi, C.I. 2004. Nigeria: Der Ölstaat und die Krise des Nation-Building in Afrika. In Hippler, Jochen (Hg.). *Nation-Building: ein Schlüsselkonzept für friedliche Konfliktbearbeitung?* EINE Welt—Texte der Stiftung Entwicklung und Frieden, 17. Bonn: Stiftung Entwicklung und Frieden: 159–175.

Oilwatch. 2008. *World Bank acknowledges serious flaws in West African Gas Pipeline. But questions on compensation for displaced, 'upstream' impacts, and gas flaring remain unanswered. 15 December 2008*. Available at: http://www.oilwatch.org/index.php?option=com_content&task=view&id=602&Itemid=224&lang=en (Accessed at 12 December 2009).

Oji, G. 2009. FG Revives Amnesty, Inaugurates Five Committees". *This Day*, 17 December 2009. Available at: http://allafrica.com/stories/200912170057.html (Accessed 19 December 2009).

Okonta, I. 2007. The Niger Delta Crisis and its Implications for Nigeria's 2007 elections. *Situation Report*, 5 April 2007. Institute for Security Studies.

Omeje, K. 2005. Oil Conflict in Nigeria: Contending Issues and Perspectives of the Local Niger Delta People. *New Political Economy*, 10, 3: 321–334.

Omeje, K. 2006a. The rentier state: oil-related legislation and conflict in the Niger Delta, Nigeria. *Conflict, Security and Development*, 6 (2006): 212–230.

Omeje, K. 2006b. *High stakes and stakeholders: oil conflict and security in Nigeria*. Aldershot, UK: Ashgate.

Opurum-Briggs, A. 2009. Niger Delta Amnesty: What Next? Options for Continuing Peace and Implementing Change. London: Chatham House Talk, 16 October 2010, Author's notes.

Oxfam. 2009. Lifting the Resource Curse. How poor people can and should benefit from the revenues of extractive industries. *Oxfam Briefing Paper*, 134, December 2009, Oxfam International. Available via www.oxfam.org.

Pöyry, Econ *et al.* 2008. Common cause, different approaches. Risks and mitigation strategies in Nigeria—Chinese, Nigerian and Norwegian perspectives. *Econ-Research Report 2008–014*, 23 January 2008.

RTI International, Search for Common Ground, Consensus Building Institute (cbi). 2008. *GMOU Participatory Stakeholder Evaluation. A Joint Evaluation of the Global Memoranda of Understanding between Chevron, Community Organisations and State Governments in the Niger Delta. Final.* October 2008.

Shaxson, N. 2009. *Nigeria's Extractive Industries Transparency Initiative. Just a Glorious Audit? November* 2009. London: Chatham House.

Shell. 2009. Shell in Nigeria. The operating environment. *Briefing Notes*, May 2009.

Smith, Ankar. 2009. Hornests' nest. Camp 5: Nigeria's strongest militant base. *Jane's Intelligence Review*, 21. 8, August 2009.

Social Action. 2009a. Fuelling Discord. Oil and Conflict in Three Niger Delta Communities. *Social Development Integrated Centre (Social Action)*, Port Harcourt, Nigeria.

Social Action 2009b. Flames of Hell. Gas Flaring in the Niger Delta. *Social Development Integrated Centre (Social Action)*, Port Harcourt, Nigeria.

Stratfor Global Intelligence. 2009. *Nigeria's MEND: Odili, Asari and the NDPVF*, 18 March 2009.

Stratfor Global Intelligence. 2009. *Nigeria's MEND: A Different Militant Movement*, 19 March 2009. Available at: http://web.stratfor.com/images/writers/NigeriaPartThree.pdf (Accessed 13 December 2009).

Suberu, R.T. 2001. *Federalism and Ethnic Conflict in Nigeria*. Washington D.C.: United States Institute for Peace Press.

The Nigerian-German Business Quarterly. 2009. The Nigerian-German Energy Partnership—The Story so far. *The Nigerian-German Business Quarterly. A Joint Publication of the Delegation of German Industry and Commerce in Nigeria and the Nigerian-German Business Association*, April–June 2009.

The Technical Committee on the Niger Delta. 2008. *Report of the Technical Committee on the Niger Delta*, November 2008.

Vines, A. *et al.* 2009. Thirst for African Oil. Asian National Oil Companies in Nigeria and Angol". *Chatham House Report*, August 2009. London: Chatham House.

WAC Global Services. 2003. *Peace and Security in the Niger Delta. Conflict Expert Group Baseline Report*. Working Paper for SPDC, December 2003.

Watts, M. 2004. Resource Curse? Governmentality, Oil and Power in the Niger Delta, Nigeria. *Geopolitics*, 9, 1: 50–80.

Young, M. 2009. Energy Development and EITI: Improving coherence of EU policies towards Nigeria. *Policy brief 4*, November 2009, FRIDE.

NEWS

AFP, 29 December, 2007. *Nigerian anti-corruption chief dismissed: police*. LAGOS. Available at: http://afp.google.com/article/ALeqM5jZ47 NfUaw1FLiwITaQY-2wwXbnn7A (Accessed 7 December 2009).

The Africa Report. 2009. *The Naira Republic*. 19 (October/November 2009). Paris: Group Jeune Afrique.

This Day, 8 December 2009. *Nigeria: The Local Content Bill*, http://allafrica.com/stories/200912090247.html (Accessed 9 December 2009).

Vanguard, 10 December 2009. Nigeria: Rep Decries Poor Budget Implementation. Available at: http://allafrica.com/stories/200912100019.html (Accessed 9 December 2009).

VOA News, 12 December 2007. *Nigerian Officials Arrest Former Niger Delta Governor*.

Geological Resources and Good Governance in Sub-Saharan Africa – Runge & Shikwati (eds)
© 2011 Taylor & Francis Group, London, ISBN 978-0-415-58267-4

Geological resources and transparency in the Central African commodities sector—examples from Equatorial Guinea and the Central African Republic

J. Runge
Department of Physical Geography, Goethe-University Frankfurt & CIRA, Centre for Interdisciplinary Research on Africa, Frankfurt am Main, Germany

ABSTRACT: Geological resources in Sub-Saharan Africa and the potential revenues from oil, gas and minerals can transform economies, reduce poverty and raise the living standards of entire populations in resource-rich countries. Central Africa has been known as an area of enormous geological richness ever since the colonial age, but there still seems to be a strong contradiction expressed by the term 'paradox of plenty' meaning that even an abundance of natural resources does not automatically lead to economic welfare for the entire population of a country. In view of this problem, the Extractive Industries Transparency Initiative (EITI) was launched in 2002 at the UN World Summit on Sustainable Development in Johannesburg, South Africa. It brought together a global coalition of governments, companies, civil society organisations and investors to promote greater transparency in the payment and receipt of natural resource revenues. If a country decides to implement the EITI, its government is making a commitment to strengthen the transparency of its natural resource revenues, and its citizens are making a commitment to hold the government to account for how it uses the revenue. In 2006, due to increasing awareness about transparency in the extractive sector and the fight against poverty in economically rich African countries, the Economic and Monetary Community of Central Africa (CEMAC) decided to ask Germany for technical support in implementing the EITI within its member states. The project REMAP (*Renforcement de la gouvernance dans le secteur des matières premières en Afrique Centrale*—Strengthening Governance in Central Africa's Extractive Sector) was set up in November 2007 on behalf of the German Federal Ministry of Economic Cooperation and Development (BMZ). During an initial three-year implementation period, the different activities of commercial and traditional mining in the sub-region were studied, and practical support was given to help implement the EITI by suggesting multi-stakeholder dialogues. Individual problems in the mining sector and the regulations that govern implementation of the EITI are examined, using the example of large-scale commercial offshore oil exploration and extraction in Equatorial Guinea and small-scale diamond mining in the Central African Republic.

1 COMMODITIES BOOM VERSUS DEVELOPMENT?

Mineral resources have a bad reputation in Africa, as a result of corruption, mismanagement and a lack of transparency, frequently under authoritarian governments, and they are seen as partly responsible for numerous crises in Sub-Saharan Africa (Basedau & Lay, 2005). In states such as these, control over resources is the most important factor in the competition for economic and political power. Associated with this is a heightened risk of violent conflict. Disputes over resources, conflicts financed by resources and the emergence of fragmented warlord territories striving towards secession are characteristic, e.g. the call for the creation of the République Logone-Chari in the northern part of the Central African Republic (CAR) (*Le Democrate*, 10 September 2008).

The expression 'paradox of plenty' (Karl, 1997) is frequently used here, describing a situation where there is often no relationship between a wealth of resources and general prosperity. The 'blood diamonds' and coltan (a rare mineral containing niobium and tantalum used for the condensers in cell phones) from the civil war-torn regions of Sierra Leone, Liberia and the Democratic Republic of Congo have created international awareness of the need for mechanisms to promote more efficient and equitable use of geological resources.

The Kimberley Process Certification Scheme (KCS) of 2003 (www.kimberleyprocess.com) and the Extractive Industries Transparency Initiative of 2002 (EITI—www.eitransparency.org) are two instruments designed to bring us closer to this goal.

Many countries in Sub-Saharan Africa have benefited in the past ten years from an increase in revenue, particularly from the sharp rise in oil prices and global growth in demand for commodities, above all from the booming Chinese economy. The Economic and Monetary Community of Central Africa (CEMAC—Chad, Cameroon, Equatorial Guinea, CAR, Gabon, Republic of Congo) is doing particularly well out of this economic boom. However, because of bad governance, mismanagement and corruption, it is virtually impossible to use the immense profits for infrastructural and social measures (poverty reduction—see Tab. 1).

Besides established oil producers such as Nigeria and Gabon, new actors have arrived: Equatorial Guinea, with offshore production in the Gulf of Guinea, and Chad with its Doba oil fields, linked to the ocean by the 1,051 km pipeline to Kribi, which went into operation in 2004. As a result of oil production, state revenues in Equatorial Guinea rose by over 1200% (BMZ 2006) between 1997 and 2000, a unique example in economic history. In the past ten years, the Central African region has seen oil production grow by 36% (compared to 16% for the rest of the world), making it the fastest growing oil-producing region in the world.

Most of the oil reserves in the Gulf of Guinea are in the Niger delta or close to the coast on the continental shelf (200 m isobath), and further large reserves are expected. The oil is high quality, with a low sulphur content. Geopolitically and for transport, the region is favourably placed for Europe, North America and—to a limited extent—for Asia as well. With the exception of the CAR, all the countries in CEMAC are oil-producing states, although none is an OPEC member (Tab. 2).

It is believed that there is also oil in the CAR in the border region with Chad. A US company recently announced that it intends to carry out seismic studies, followed in the medium term by exploration drilling, in this politically unstable region (rebels). The immense wealth and potential that the Gulf of Guinea offers is leading to major long-term investment in exploration and in production and processing plants. The market is dominated by numerous multinational groups such as Chevron, Exxon-Mobil, Shell, Total, BP, Petrobras and recently Chinese consortia as well (Sinodec). The criteria for government awards of oil, gas and other concessions are frequently not clear, and there are rarely documents providing a transparent picture of payment flows in the oil sector in this region. The NGOs Publish What You Pay (PWYP, www.publishwhatyoupay.org), Transparency International (www.transparency.org) and Global Witness (www.globalwitness.org) are examples of civil society groups that are working for greater transparency and better governance in

Table 1. Social indicators for CEMAC member states in 2003: HDI = Human Development Index; GDP = Gross Domestic Product; CAR = Central African Republic.

CEMAC state	HDI	HDI Ranking	Life expectancy, years	Literacy rate (>15 years)	Annual per capita GDP (US$)
Equatorial Guinea	0.497	121	43.3	84.2	5773.9
Gabon	0.635	123	54.5	71.0	4585.5
Cameroon	0.497	148	45.8	67.9	840.4
Congo	0.521	142	52.0	82.8	1107.9
Chad	0.341	178	43.6	25.5	352.0
CAR	0.355	171	39.3	48.6	306.2

Source: UNDP (2006), Bank of Central African States—BEAC (2005), cited from Tamba *et al.* (2007, p. 39).

Table 2. Macroeconomic importance of oil production in the CEMAC region in 2004 (exc. CAR); GDP = Gross Domestic Product.

Country	Oil production in million tonnes (2004)	Share (%) of African production	Share (%) of GDP from oil production	Share (%) of oil in exports
Equatorial Guinea	17.8	3.8	92.3	92.2
Gabon	13.5	2.9	44.8	81.9
Cameroon	4.5	1.0	5.6	40.9
Congo	11.2	2.4	53.7	86.3
Chad	8.7	1.9	47.4	80.8
CEMAC (total)	55.5	12.0	37.3	78.4

Source: Banque de France (2004), Rapport Zone Franc, modified from Tamba *et al.* (2007, p. 33).

the commodities sector and that have laid the basis for the EITI process.

Multinational involvement in the Gulf of Guinea is easy to understand if we look at the projections of the International Energy Agency (IEA, 2005) up to 2030. Starting from 2005, demand for oil will increase from its present 4 Gtoe to 6 Gtoe in 25 years, an increase of a further 50% (the same applies to gas and coal).

In the context of the commodity governance programme assisted by the Deutsche Gesellschaft für Technische Zusammenarbeit GmbH (GTZ) on behalf of the German Federal Ministry for Economic Cooperation and Development (BMZ), the present paper explains the mechanisms of the EITI transparency initiative and outlines the prospects for and problems in implementing such programmes, taking the examples of the commercial oil sector in Equatorial Guinea and informal small-scale diamond mining in the CAR.

2 THE EITI TRANSPARENCY INITIATIVE

In the Sub-Saharan, particularly the Central African region, there is enormous economic potential from oil, gas, diamonds and other commodities. However, despite these resources, there has so far been no success in implementing the Millennium Development Goal (MDG) agreed in 2000 on the Millenium Summit in New York of halving hunger and poverty in developing countries by 2015. The Food and Agriculture Organisation of the United Nations (FAO) thus admitted in late 2006 that there had been no progress on this goal (*FAZ*, 31 October 2006)—on the contrary, the overall situation in Sub-Saharan Africa had shown a marked deterioration.

At the UN Summit for Sustainable Development (Rio+10) in Johannesburg in September 2002, Tony Blair (the UK's Prime Minister) announced the Extractive Industries Transparency Initiative (EITI). The goal of this initiative is to reduce mismanagement and corruption in the natural resources sector. The assumption is that problems such as corruption and low economic growth which often characterise states rich in natural resources can only be solved if the monetary flows in countries with natural resources are made transparent. Transparency is to create the basis for forcing public administrations to be accountable for the use of revenue from commodity production. The initiative calls on participating nations to publish regular reports on government tax revenue and tax payments by oil and mining companies. The plan is to follow this with a comparison of the data by an independent party. In addition, all the companies operating in a country—whether public or private sector—together with representatives of local civil society are to be involved in implementing the initiative. National EITI processes are steered by stakeholder groups where representatives of the companies, civil society and government actors come together.

The narrow focus of the initiative and the multi-stakeholder approach are seen as the main reasons for the political success of the initiative. The EITI is supported by a large number of donor nations, including Germany, France, the UK and the USA, the World Bank, NGOs, companies in the extractive industries and institutional investors. A total of 27 developing countries are now (2009) implementing the initiative. In 2007, Norway also joined the EITI as a non-developing oil-producing country.

As a highly technocratic and standardised process, the EITI follows certain rules and procedures. The important thing is that the validation process is repeated regularly, establishing transparency and the associated basic democratic structures. The 'validator' plays an important role in this, as someone who may (but need not be) from the world of politics and must be nominated and accepted by a multi-stakeholder community (www.eitransparency.org).

The report on the commodity and tax data collected must be approved by the international EITI Board in Oslo, the government in question and the multi-stakeholder group (MSG) before it can be published and made generally available (e.g. through the internet or in print form).

The G8 Summit held in Evian, France, in June 2003 adopted a declaration on 'Fighting Corruption and Improving Transparency'. Finally, in September 2004 in Paris, the finance ministers of the CFA franc zone decided to implement the EITI in their member states.

3 THE AFRICAN SUPRAREGIONAL PROGRAMME REMAP

As shown above, the Central African nations have so far made virtually no use of their extensive natural resources to generate any significant stimulus for development. Firstly, it is difficult for the public to obtain any concrete and disaggregated data on the extraction of commodities and revenue from this, and secondly, the revenue is still not being used to the desirable extent to initiate sustainable economic development and reduce poverty. In this situation, the CEMAC Executive Secretariat applied to the Federal Republic of Germany in 2006 for assistance in implementing the EITI and other measures for transparency in the commodities sector in CEMAC nations. After reviewing the programme on 'Commodity governance in Central Africa', BMZ provided EUR 1 million for a two-year exploratory phase, which started on 1 November 2007. A GTZ expert has since been advising CEMAC in Bangui.

GTZ's programme Strengthening Governance in Central Africa's Extractive Sector (REMAP) is involved in the CEMAC region in implementing the international transparency initiative EITI for publicising payments flows in the commodities sector. The initiative also has strong political support from Germany. As a supplement to the synergies in implementing the EITI at national level, REMAP contributes to the pro-development use of commodity revenue at regional level and to establishing the regional policy consensus this requires.

The regional cooperation partner is the Central African economic and monetary community CEMAC, within which five of the six member states (all except Chad in 2009) have already declared their willingness to implement the EITI. The programme complements cooperation at regional level and cooperates with individual CEMAC member states. Besides the development of standards and benchmarks based on good practices, drawing on experience from other regions, support for EITI implementation in problem states (e.g. Equatorial Guniea and Chad)[1] in cooperation with other donors is receiving particular attention. Going beyond support for the creation of revenue transparency (EITI implementation), the programme aims to improve adding value in commodity management in the CEMAC region for the benefit of the population and to help boost sustainable economic development. This includes improving the environment for private investment and integrating the private sector in its role and function as a prospector for natural resources. The programme contributes towards tackling a complex problem with high potential for social and economic conflict, towards implementing international transparency and reform efforts in the commodities sector, and towards achieving Millennium Development Goals (MDGs) for poverty reduction. The German contribution supplements efforts by the World Bank, the EU, France and other relevant donors.

The main objectives and important indicators of REMAP are:

- Commodity management in the CEMAC region is increasingly used for poverty reduction and sustainable economic development.
- The national statutory framework conditions in the commodities sector in the CEMAC states are aligned with international regulations and standards (EITI commitments).
- Gender-specific socioeconomic effects of commodity management (integration of local firms in value chains, provision of social services, growth for local enterprises) can be demonstrated.
- International NGOs give a positive rating to the development of transparency in the signing of contracts and award of licences for new investment in the commodities sector in the Congo Basin (e.g. through hearings, correspondence, publications).
- Selected oil and mining companies in the six countries comply with their obligations under the EITI and international standards, and make an increased contribution to socially and environmentally responsible commodity management and regional development.
- There is an obvious increase in the share of commodity-induced revenue in the total budget of the CEMAC states.

The programme's target group is the population. It is ultimately the population that is to benefit from improved transparency and improved resource allocation through poverty reduction. The intermediaries are experts and managers with a key role in the sector. These include government institutions (ministries of economics, finance, energy, mining and justice), state and private oil and extractive companies, members of parliament, NGOs and church representatives.

4 CASE STUDIES

4.1 *The oil boom in Equatorial Guinea*

In the Gulf of Guinea, an offshore area of around 15,000 km² was allotted in 1990 for oil and gas exploration. Ten years later, the area allotted for concessions was already 237,000 km² (Ford 2002, Fig. 1). Of the CEMAC states, the smallest state—Equatorial Guinea, surface area: 28,051 km², population: 616, 459, under the dictatorship of the Obiang family—has the largest production at 320,000 b/d (GTZ 2008, see Tab. 2). As late as the 1980s, there was a proposal among Equatorial Guinea's politicians to merge the island nation with Cameroon. The argument was that it was unable to survive economically by itself, as it was an agrarian state and because of the adverse geographical constellation (main island with the city of Malabo and a small mainland region with the city of Bata between Cameroon and Gabon, Fig. 1).

Figure 1 shows the territory of Equatorial Guinea with the various oil exploration areas currently allotted to multinational groups. To date, oil production has been limited to the oil fields of Alba, Zafiro (northwest of Malabo) and Ceiba (southwest of Bata). There is great additional potential for this small island state from oil. Known reserves are estimated at 1.3 billion barrels (Rosellini 2006).

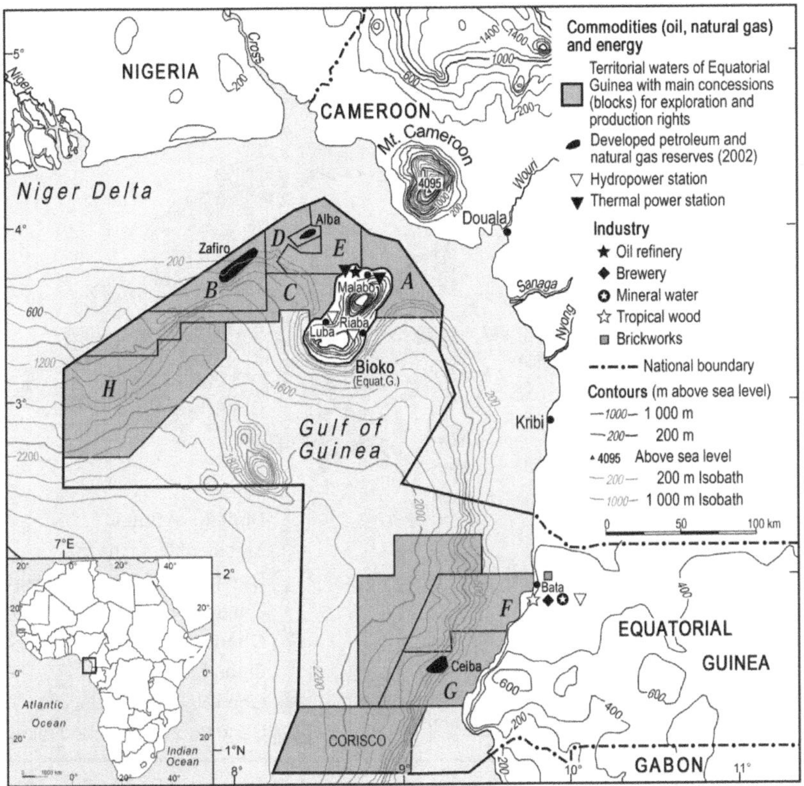

Design: J. Runge, Cartography: U. Olbrich

Block A: GEPETROL (Equatorial Guinea)
Block B: EXXON-MOBIL, TEXACO, CHEVRON (USA)
Block C: EXXON-MOBIL, TEXACO, CHEVRON (USA)
Block D: YUKONG (South Korea), GEPETROL (Equatorial Guinea)
Block E: NOMECO (USA)
Block F: TRITON (USA)
Block G: TRITON (USA)
Block H: SASOL (South Africa)

Figure 1. Oil and gas reserves in Equatorial Guinea, showing the oil exploration and production concessions (blocks) granted to various multinationals, and industrial locations. The relatively broad continental shelf and the Cameroon Volcanic Line (CVL) running SW-NE, with a submarine massif, the island of Bioko and Mount Cameroon are clearly shown. (Sources: Les éditions Jeune Afrique (2001), www.offshore-technology.com (last accessed: 10 September 2008; SRTM altitude and GEBCO bathymetric data).

At the current daily production of 320,000 b, the boom could last another decade.

According to rumour, the son of dictator Obiang regularly invests petrodollars in Hollywood productions. There are private accounts with Riggs Bank to which large sums have been transferred from oil production (Washington Post, 28 January 2005). Equatorial Guinea's opposition politician Severo Moto, alleged to have been linked to the failed coup in March 2004 (a venture by South African and UK mercenaries), was recently sentenced in absentia to 63 years' imprisonment for high treason. Moto is currently (2008) serving a prison term in Spain for arms smuggling (FAZ, 18 June 2008). These scandals highlight the depth of corruption and cronyism in this Central African nation.

The oil boom in Equatorial Guinea did not start until 1993, and it was followed in just a decade by an unparalleled economic upswing, which could well continue. A great deal has since been invested in the infrastructure (e.g. roads and a new airport at Malabo), but the general socioeconomic and civil rights situation of the population continues to be precarious.

4.2 *Informal diamond mining in the Central African Republic*

The Central African Republic (surface area: 622,984 km², population: 3,895,139) is rich in natural resources, although in recent years only just under 5% of government revenue has come from the extractive sector. Among other things, this

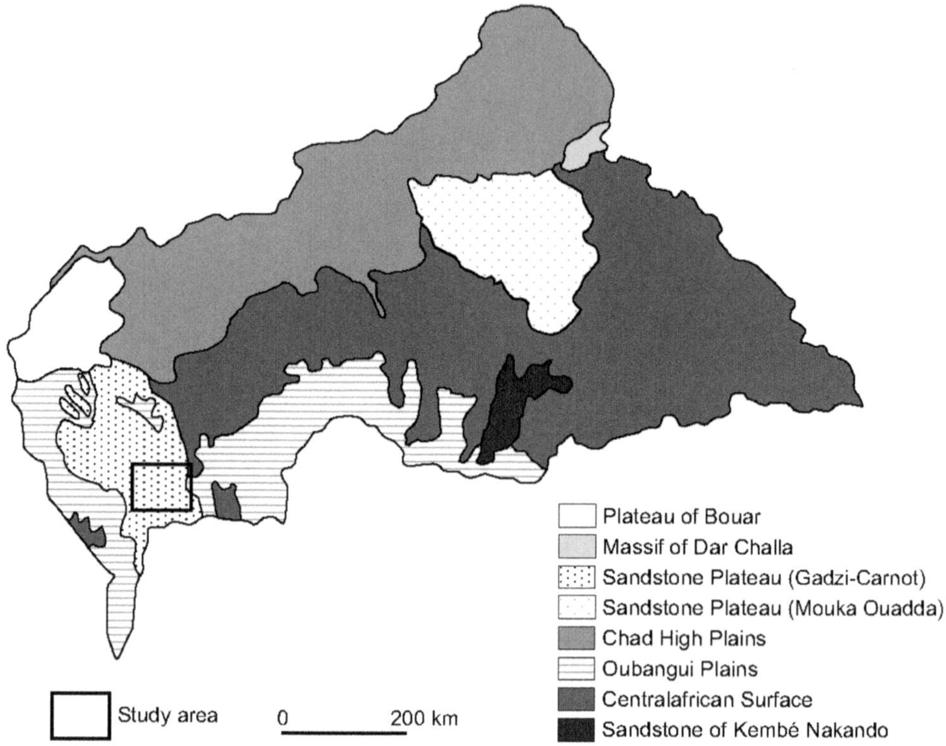

Figure 2. Geological and geomorphological overview of the Central African Republic. The fluvial plans around the Gadzi-Carnot and Mouka Ouadda Mesozoic sandstone plateaux are among the most important diamond areas. Three diamond-bearing alluvial profiles (Fig. 3) were recorded in the Mbaéré valley floor near Ngotto (see 'Study area' box) in the context of physio-geographical studies (Neumer *et al.* 2007, p. 124).

is because diamond and gold mining as a whole have been handled entirely by the informal sector (artisan minier). Commercial mining companies will probably start a number of large opencast mines from 2010. For uranium, the partner at Bakouma is the French state company AREVA, which took over the former South African Uramin Ltd. in 2006 and signed a contract with the CAR in August 2008 on uranium exploitation rights for the next 40 years. The British-Canadian AXMIN Inc. (Aurafrique) will open an opencast gold mine near Bambari (Passendro).

Diamonds found in the CAR are exclusively alluvial and are mined by informal miners (small-scale mining); the same goes for gold. A systematic search for Kimberlite pipes by the South African group De Beers was abandoned in 2005. Two Mesozoic sandstone plateaux (Fig. 2) and the river systems draining to the south are the main locations for diamonds. The 3 to 5-metre deep, heavily sandy and relatively loose alluvial flood plains (Fig. 3) are prospected using simple tools (spades, shovels, sieves, Fig. 4), particularly in the dry season, or even by diving from rafts in the river.

In places, rivers are dammed or diverted to give easier access to the sedimentary layers with the diamonds. Locally, this results in major ecological damage and conflict with those engaged in agriculture.

The quality of Central African diamonds is high, and around 80% of the stones found are jewellery-grade diamonds ('river' or D/E colour grades). The rest are suitable for industrial use. Taxation (currently 12%), export and certification to the Kimberley Process (see section 1) is performed by the Ministry of Mines in Bangui, and specifically the *Bureau d'évaluation et de coordination de diamants et d'or* (BECDOR—Office for Valuation and Coordination of Diamonds and Gold), which was set up in 1995. Of the 13 officially licensed purchasing agents (bureaux d'achat) in Bangui (in the CAR's Poverty Reduction Strategy Paper (PRSP) for 2008–2010 there are only eight, cf. Ministry of Economy, Planning and International Cooperation, 2007), only five agents regularly supply the statutory export statements for diamonds and gold. The others either supply nothing or, more probably, circumvent taxes and the Kimberley Process and

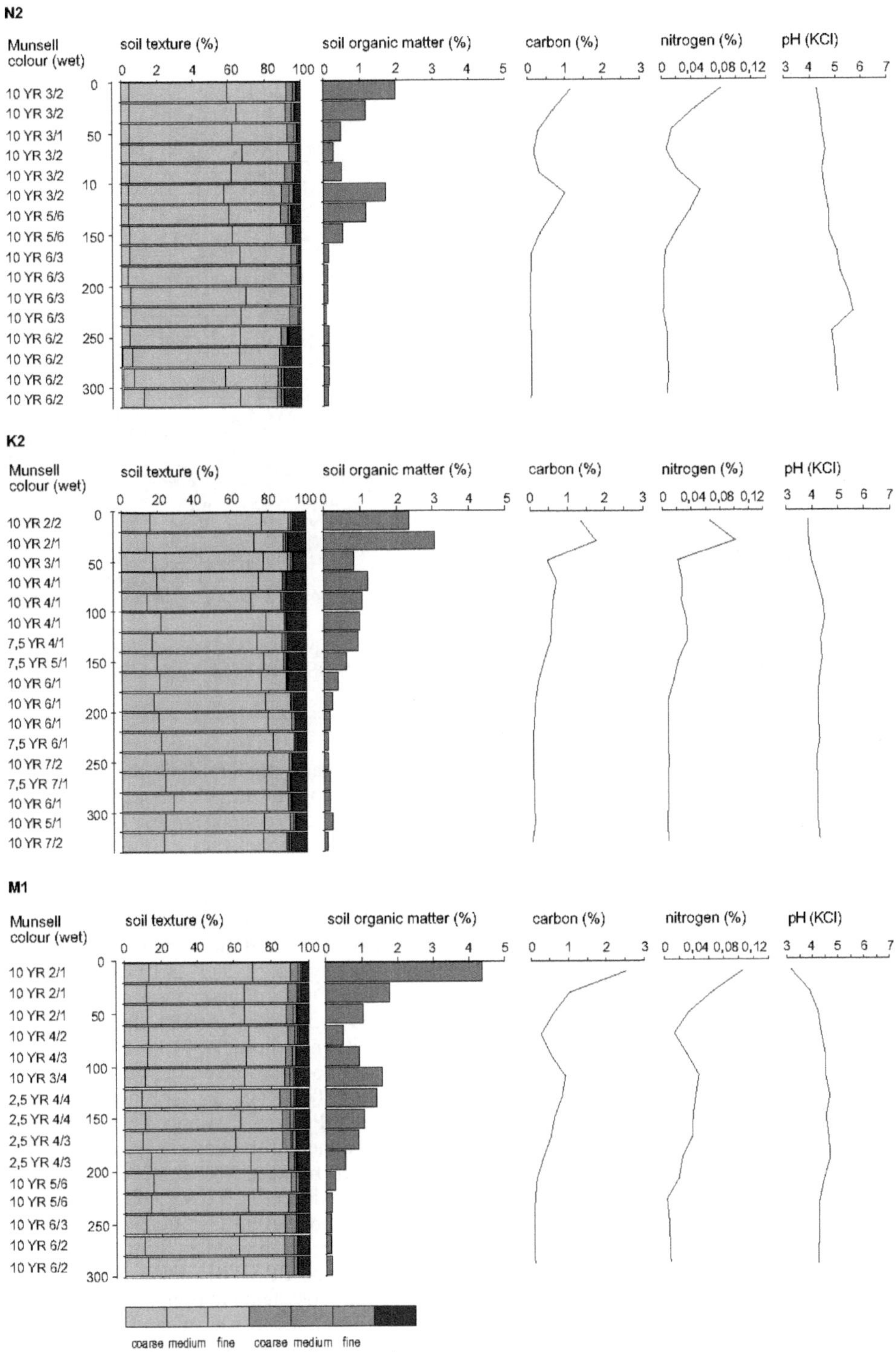

Figure 3. Soil profile records of three potentially diamond-bearing alluvial sediments on the Mbaéré alluvial floor. As a result of the Mesozoic Carnot sandstone in the catchment area, the alluvial deposits consist of over 90% medium and fine sand, which is relatively easy for informal miners to work at low water (Neumer *et al.* 2007, p. 130).

Figure 4. Traditional informal diamond prospecting in the alluvial flood plains of the Lobaye river by young people and children near the village of Bossoui (see also contribution on HIV/AIDS by Runge & Ngakola in this volume) near Ngotto in the Boda *sous-préfecture* (photo: J. Runge, March 2008).

Figure 5. Changes in diamond production in carats (ct) in the Central African Republic (CAR) since 1984, and export income in US$ from various sources (own figure).

export stones directly. According to GTZ REMAP studies, the share of legally exported diamonds is only 20–30%, with potential added value and tax revenue of 70–80% lost to a lack of transparency, the informal economy and corruption. The share of 'only' 30% of illegal diamond exports cited in 2003 by a Kimberley Process Review Mission seems much too low (Pampaloni *et al.* 2003). The mining police (brigade minière) lack the necessary staff and technology to provide efficient control.

Under these circumstances, it is difficult to generate reliable statistics on the Central African diamond sector. However, based on data from the Stockholm Raw Materials Group, the German Federal Institute for Geosciences and Natural Resources (BGR, Hannover), the Poverty Reduction Strategy Paper 2008–2010 (PRSP) and data from BECDOR, Figure 5 shows the trends in diamond production (carats) and export income (US$ millions—since 1999) for the period 1984–2007. Diamond production in the CAR peaked in 1994 at 650,000 ct; up until the introduction of the Kimberley Process, it dropped below 350,000 ct. Political upheaval and unrest approaching civil war in the CAR between 1996 and 2003 have decisively weakened this sector. It is very difficult to estimate export income, given the differences of up to US$ 50 million between the data from the Raw Materials Group and PRSP/BECDOR (Fig. 5). In 2007, production rose again to over 400,000 ct. However, as a result of the loss in value of the US$, CAR's income was again only around US$60 million (BECDOR). Indirectly, Figure 5 provides some idea of what government revenue from the diamond sector could be if 70–80% of the gems were not smuggled out of the country.

Figure 6. Illustration from the *Guide du Code Minier en République Centrafrique—à l'usage des artisans et ouvriers miniers* (Guide to the CAR Mining Code for small-scale miners and mineworkers) produced by the project *Développement du Diamant Artisanal en République Centrafricaine* ('Development of small-scale diamond mining in the CAR'—DPDDA). This booklet is used to explain the statutory basics in the CAR diamond sector to informal mineworkers. (Source: ARD Inc., Associates in Rural Development, USAID, Bangui 2008).

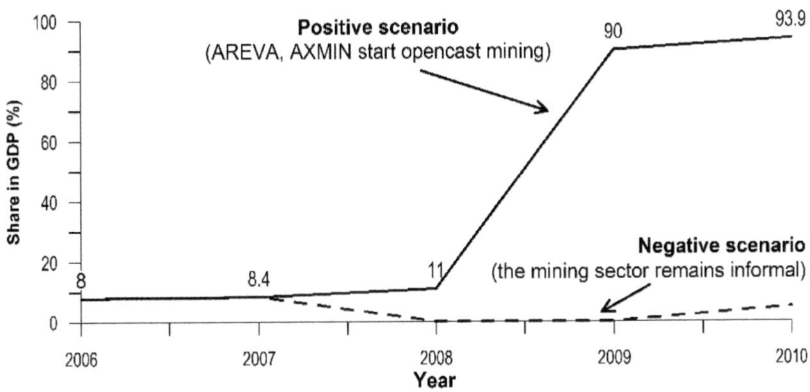

Figure 7. Projections of future developments in the CAR mining sector (2006–2010) according to the PRSP (Source: Ministry of Economy, Planning and International Cooperation 2007, own figure).

According to the statistics, the informal miners—estimated at over 100,000—continue to live on US$1 a day or less, although they could (and possibly do) earn considerably more (Fig. 4). The current report on the PRSP 2008–2010 (Ministry of Economy, Planning and International Cooperation 2007) defines the poverty level in the CAR as an annual income of XAF 156,079 (EUR 1 = XAF 655), equivalent to around EUR 238 (EUR 0.65/day). The national poverty index overall is 67.2%, with urban regions (Bangui) doing 'better' (49.7%) than rural areas, which suffer extreme poverty (71.7%) in monetary terms (Ministry of Economy, Planning and International Cooperation 2007).

The artisanal miners are frequently unaware of their rights and obligations, which are determined by the 2004 Mining Act (Code Minier). The GTZ REMAP programme is collaborating closely with the US company ARD Inc. (Associates in Rural Development), which has produced a *Guide du Code Minier* (Guide to the Mining Act) in comic format together with the United States Agency for International Development (USAID, Fig. 6) to help communicate the text of the Mining Act in simple words, with numerous illustrations. At the lowest social level, this makes an effective contribution towards increasing transparency in the informal mining sector.

The conditions for implementing the EITI transparency initiative have now been met at state level in the CAR as well. In a joint campaign by the World Bank and GTZ, in collaboration with the Central African EITI *comité de pilotage*, a national work plan 2008–2010 and a budget plan were adopted. The CAR was accordingly granted EITI candidate status at the end of 2008.

5 RESULTS AND CONCLUSIONS

The World Bank and the International Monetary Fund (IMF) have recognised the importance of the EITI process and made its implementation an issue in its own guidelines for action. The IMF published a Resource Revenue Transparency Guide in 2005, for example, while an internal evaluation report by the World Bank entitled Extractive Industries and Sustainable Development (2003, cited from BMZ 2006) called on the organisation 'to pursue country- and industry-wide disclosure of government revenues from extractive industries and related contractual arrangements'. This resulted in the creation of an Extractive Industries Advisory Group (EI-AG) at the World Bank. The Federal Republic of Germany is also represented on this body. It was also decided that the World Bank would only assist projects in the extractive sector which respect the EITI transparency criteria in the countries in question. In addition, implementation of the EITI by the relevant governments is to be made an integral element of the PRSPs (BMZ 2006, p. 11).

This puts in question the originally (2002) intended 'voluntary' nature of the national EITI process, which the affected governments, e.g. in CAR, criticise as 'Western' and donor-oriented 'meddling' in national policy (EITI as a form of economic and development policy pressure). This also affects the way Norway's 'exemplary' action in joining EITI as a non-developing oil-producing country should be judged.

As the EITI is a highly formal and 'technocratic' instrument, there is also criticism stressing that the EITI is not a genuine control instrument, and is more likely to contribute towards the creation of new and expensive functionary positions. This would unnecessarily increase an African administrative apparatus which frequently suffers from corruption (Centrafrique Matin, 29 August 2008). One example of this is the widespread demand for *jêtons de présence* (attendance money), where cash payments of EUR 500–1000 per participant are expected for mere attendance at EITI meetings in Cameroon and the CAR.

Even so, sustainable management of oil and diamond profits through advisory services to political decision makers in ministries of energy, economic affairs and mining will continue to be a goal of German development cooperation (DC). This also assists the governments in question with their goal of increasing their negotiating competence with oil and mining groups (cf. BMZ 2006).

In the case of Chad, only an estimated 12.5% of oil income went to the state (*SZ*, 28 August 2006), and was mainly used to by military equipment. This figure is well below international standards, and must be renegotiated. President Deby is reported to have described the contract as 'a fool's agreement' (BMZ 2006). As there has been no visible progress to date (2008/2009) towards improving transparency, the World Bank recently announced it was terminating cooperation with Chad[2].

In the CAR, only just under 5% of state revenue came from the commodities sector in 2007, which is incredibly low considering the existing geological resources. At the same time, the presence of major investors such as AREVA (uranium) and AXMIN Aurafrique (gold) and the start of implementation of the EITI with GTZ and World Bank assistance give grounds for hope that the value to the state of the Central African mining sector will increase in the coming years.

This positive scenario for the development of the mining sector is entirely dependent on the involvement of major groups (Fig. 7). The informal sector, by contrast, is—incomprehensibly—regarded as having virtually no potential for increasing state revenue from diamonds and gold through focused promotion and transparency. The CAR seems to have accepted the de facto impossibility of controlling the informal mining sector, and is relying entirely on commitments by foreign investors and help from the international donor community.

It should, however, be noted here that in addition to the involvement of German DC and other donors in the extractive sector, there are growing calls, e.g. from the Organisation for Economic Cooperation and Development (OECD), for a return to increased promotion in future of the neglected rural areas and agriculture, on which, despite the commodities boom, the majority of the African population are still dependent (*SZ*, 14 May 2008).

ENDNOTES

[1] Chad returned to the EITI community in 2010 whereas Equatorial Guinea was suspended for non-reporting on its extractive industries data.

[2] Chad rejoined the EITI in 2010.

REFERENCES

Basedau, M. & Lay, J. 2005. Conceptualising the 'Resourse Curse' in Sub-Saharan Africa: Affected Areas and Transmission Channels. *Hamburg African Studies*, 14: 9–24.

BMZ (German Federal Ministry for Economic Cooperation and Development) 2006. Entwickelt Öl? Möglichkeiten der entwicklungsorientierten Nutzung der Öleinnahmen in Subsahara Afrika ('Does oil mean development? Options for the development-oriented use of oil revenue in Sub-Saharan Africa'). *BMZ Diskurs* 8: 1–20.

Centrafrique Matin (29 August 2008). '*Initiative pour la transparence dans la gestion des industries extractives et minières: mythe ou réalité?*' ('Initiative for transparency in the management of the extractive and mining industries: myth or reality?').

GTZ—Deutsche Gesellschaft für Technische Zusammenarbeit 2008. Oil fact sheet. GTZ Good Financial Governance Programme Fact Sheet. *Ghana's Upstream Petroleum Sector: A Brief Overview by S. Bauchowitz and D. Nguyen-Thanh.*

Ford, N.A. 2002. *Oil and Gas in the Gulf of Guinea*, 1, West Africa, London.

FAZ—Frankfurter Allgemeine Zeitung (31 October 2006). 'UN: *Kampf gegen den Hunger erfolglos*' ('UN: Fight against hunger unsuccessful').

FAZ—Frankfurter Allgemeine Zeitung (18 June 2008). '*Anklage wegen Umsturzversuchs—Prozess gegen britischen Söldner in Äquatorialguinea*' ('Charge of attempted coup—trial against British mercenaries in Equatorial Guinea').

IEA (International Energy Agency) 2005. *Development of worldwide energy consumption and projections until 2030*. International Energy Agency, Paris, www.iea.org.

Karl. T.L. 1997. The Paradox of Plenty. Oil Booms and Petro-States; Berkeley: University of California Press.

Les éditions Jeune Afrique 2001. *Les Atlas de l'Afrique—Guinée Équatoriale* ('Atlas of Africa—Equatorial Guinea'), Paris: 1–64.

Le Democrate, Quotidien indépendant centrafricain (10 September 2008): '*Création de la République du Logone-Chari: que dire, que faire?*' ('Creation of the Republic of Logone-Chari: what should be said, what should be done?').

Ministry of Economy, Planning and International Cooperation 2007. PRSP 2008–2010, *Poverty Reduction Strategy Paper*, 1–113, Bangui, RCA www.minplan-rca.org.

Neumer, M., Becker, E. & Runge, J. 2007. Palaeoenvironmental studies in the Ngotto Forest: alluvial sediments as indicators of recent and Holocene landscape evolution in the Central African Republic. *Palaeoecology of Africa*, 28: 121–137.

Pampaloni, C., Villemain, J., Kourouma, D., Dandéka, M., Selekane, L., van Bockstael & Yearsley, A. 2003. *Report of the review mission of the Kimberley Process to the Central African Republic*, 8–15 June 2003: p. 11

Rosellini, C. 2006. La répartition de la rente pétrolière en Afrique ('Distribution of oil wealth in Africa'). *Problèmes économiques* 2902: 20–25.

SZ—Süddeutsche Zeitung (28 August 2006): '*Tschad weist zwei Ölkonzerne aus*' ('Chad expels two oil cooperations').

SZ—Süddeutsche Zeitung (14 August 2008): '*Afrika profitiert von hohen Wirtschaftspreisen*' ('Africa profits from high economic prices') and '*Gefährlicher Wohlstand—Das Ölgeschäft bringt Afrika hohe Einkommen, aber keinen Wohlstand*' ('Dangerous prosperity—The oil business brings high incomes but not prosperity to Africa').

Tamba, I., Tchatchouang, C. & Dou'A, R. 2007. *L'Afrique Centrale, le paradoxe de la richesse: industries extractives, gouvernance et développement social dans les pays de la CEMAC* ('Central Africa, the paradox of plenty: extractive industries, governance and social development in the CEMAC countries'). Friedrich Ebert Stiftung Yaoundé: p. 162.

UNDP 2006. *Human Development Report 2006*, UNDP, New York.

Washington Post 28.01.2005. '*Riggs Bank Agrees to Guilty Plea and Fine'—Civil, Criminal Penalties Would Total $41 Million.*

Geological Resources and Good Governance in Sub-Saharan Africa – Runge & Shikwati (eds)
© 2011 Taylor & Francis Group, London, ISBN 978-0-415-58267-4

Whither communities? Restorative justice in the Tiomin Kenya Ltd. titanium mining case

C.A. Khamala
Kenya School of Law and High Court of Kenya, Nairobi, Kenya

ABSTRACT: Kenya is bound by international environmental principles to establish public law regulations which prevent environmental destruction and protect people from harm to their health. Hence the High Court restrained a foreign-owned company, Tiomin Kenya Ltd., from mining titanium for failure to submit an Environment Impact Assessment (EAI) study. Yet excessive criminal sanctions may discourage mining processes altogether. The government therefore used its superior bargaining power to buy the relevant private land. Dissatisfied landowners omitted to pursue any public law remedies in appropriate forums. Instead, at successive proceedings not only were their collective sale contracts upheld, but also individual landowners were estopped from rescinding or invoking their constitutional rights against forced expropriation. Information scarcity deprives communities from participation. Environmental law therefore requires common law reinforcement to attain good governance of polycentric natural resource disputes. By linking policies to project management, restorative justice processes make law responsive to stakeholder needs for sustainable development.

1 POLYCENTRIC PROBLEMS AND IMPERFECT PROCEDURES

Policymakers, administrators and managers concerned with a goal of maximising the social benefits derived from natural resources are required to balance between environmental degradation and community participation. A polycentric adjudicative problem (Fuller 1978, p. 401) arises when different interest groups emphasise promotion of either of those valid goals at the expense of the other. At one extreme, liberal market proponents argue that government-ownership of minerals discourages entrepreneurship, private investment and production. However, environmentalists insist that private ownership rights must be subordinated to potential environmental harms by property. At another extreme, the mining industry may require use of private land in order to gain access to minerals. Yet, third generation human rights proponents increasingly recognise rights of local communities to participate in decision-making regarding development projects affecting them. While balancing between these extreme positions, the government, in exercise of its sovereign right of eminent domain conferring police powers over natural resources, is expected to stimulate investment in order to raise capital to facilitate sustainable economic development.

Liberty to perform acceptable actions, such as maximising profits, requires no rules (Mill 1859, ch II). However, the liberal response to harmful and undesirable activities, whether environmental crimes, trespass to land, or tax evasion, is to prescribe sanctions. While punitive responses may deter or discourage offensive behavior, they may simultaneously prove counter-productive in enhancing profitability of a mining enterprise. Instead, prosecution may precipitate capital flight thus denying communities an opportunity to share potential benefits of mining investments altogether. Rather than using the 'efficient' and 'effective' bivalent criminal justice system, an optimum resolution to complex environmental-cum-expropriation-cum-taxation conflicts, can be more fairly achieved through a restorative justice approach. Yet lawyers, analysts and decision-makers may fail to clearly distinguish between overlapping dimensions of natural resource governance. Regarding fragmentation between environmental and human rights norms at international level, environmental jurist Tim Stephens notes that: 'Potentially three main problematic scenarios can arise from jurisdictional competition, namely forum shopping, simultaneous proceedings in multiple fora, and successive proceedings in different tribunals' (Stephens 2009, p. 248, 274).

Conflicting interpretations emerge from mining companies, environmentalists, communities and the public. Competing interests of profitability, conservation, participation and development tend to generate tensions. Cultural theorists argue that different possible combinations of governance concepts can be measured by their various ideologies whether fatalism, individualism, hierarchism or egalitarianism. Correspondingly, restorative regarding environmental offenders dispute resolution may entail varying degrees of official responses ranging from neglectful, permissive, punitive.

The hypothesis is that failure by the Kenyan Environmental Management and Co-Coordination Act[1] to expressly require community involvement in the preparation of Environmental Impact Assessment (EIA) studies while simultaneously vesting the Director General of the National Management Environment Authority (NEMA) with near unfettered discretion to ignore applications and subsequently 'circumspect' over development creates uncertainty regarding determination of a project's Net Present Social Value (NPSV) (Mikesell 1993, pp. 23–25). This proposition shall be tested by considering what EMCA means by the 'environmental management' and 'sustainable development' criteria of a titanium mining project in Kenya.

To environmental law scholar Charles Okidi: 'Environmental law is the ensemble of norms ... statutes, treaties and administrative regulations to ensure or to facilitate the rational management of such resources for human development' (Okidi 1988 cited in Ojwang 1996, p. 47). In addition to laws that 'ensure,' constitutional lawyer J.B. Ojwang maintains a strict distinction between two meanings attributed to 'laws that facilitate.' My rationale is that the recent Kenyan environmental governance legislation fails to sufficiently appreciate Ojwang's distinction between environmental law's punitive and democratic dimensions. I argue that clumsy institutions possess certain advantages in resolving polycentric disputes. Yet unless decision-making is sufficiently sophisticated, administrators may not necessarily fully promote the liberal rights standards expected by mining companies in the economic dimension while facilitating the public interest of sustainable development. However, no detailed discussion shall be made of the vast emerging restorative justice literature, suffice to note that cultural theory argues that policy choices are influenced by four different myths of nature. Section 2 introduces perspectives regarding the constitutional basis for environmental rights and their contingent limitations on liberal property rights. Combining restorative justice and cultural theory, section 3 attempts to define environmental governance in relation to balancing polycentric interests associated with mining projects in the African context. Sections 4 and 5 present evidence from successive recent Kenyan High Court decisions concerning Tiom in Mining Ltd. Section 6 analyses our court's innovative use of common law principles and equitable doctrines to fill the gap in the environmental legislation. The outcome of successive proceedings is seen as determining the project NPSV by a triple restorative management response to resolve environmental, land acquisition and mining licensing issues. It is observed that punitive environmental governance under Kenyan policy and law inordinately confers environmentalists with bargaining power to resist access to minerals. In conclusion the government's 'fully restorative' response of facilitating purchase of community land on Tiomin's behalf, was justifiable in the public interest. However from participating in negotiations which yielded unegalitarian outcomes, the Tiomin cases were only 'partly restorative' from the perspective of landowners who felt 'excluded.' Hence a formal framework to facilitate community participation is recommended.

2 ENVIRONMENTALISM MEETS THE MARKET

2.1 *The constitutional basis of individual rights to a clean and healthy environment*

Ojwang (1996, p. 40) begins by quoting Goodland *et al.* (1991) proposition that 'in conflicts between biophysical realities and political realities, the latter must eventually give ground'. He argues that: 'The (Global Biodiversity) strategy calls for an integrated legal management and policy framework, as the best approach to the sustainable use of natural resources. It is obvious that the environmental crisis has to be accorded priority in the political and legal arrangements of any well managed state' (Ojwang 1996, p. 41). His assertion is justified since 'the current reality of national sovereignty, and the international power structure, place the most crucial environmental functions in the hands of individual states. The constitutional arrangements in these states, accordingly play a decisive role in the reality of environmental protection'. Alluding to the need for multiple responses to polycentric natural resource problems. Ojwang observes that: 'The constitutional basis for environmental conservation, in the Kenyan experience, comprises several factors: the state's public mandate to provide for the people's welfare; the crystallisation of specific rights, out of the state's public welfare initiatives; individual rights claims vis-à-vis public development programmes; and the emergence of new public institutions, and their constitutional characteristics' (Ojwang 1996, p. 51). Furthermore, as regards the state's mandate in public welfare: 'The constitution views the environment, a basic capital input for economic and social development, as a prime sphere of executive policy-making; In so far as conservation of the environment supports health and ensures the sustained production of products and resources essential to life it is a matter of concern to the governmental process'. Because the United Nations Declaration of Human Rights[2] 'is essentially a moral code for the conduct of states in their domestic affairs,' therefore Ojwang argues that the state's role in rights-creation is derived from universal rights to social security (art. 22), work (art. 23) and to a standard of living adequate for health and well-being (art. 25). Moreover the UNDHR's 'terms are substantially reproduced in international instruments of a binding nature (for parties to the instruments)'[3] (Ojwang 1996, p. 51 citing Starke 1978, pp. 113–131).

While 'international laws impose a duty to observe constitutionalism in domestic practices,' Ojwang notes that such duty requires a twin approach to governance. Duality is necessary because, for one thing: 'conditions, as regards Africa, are still in the ferment of socio economic dynamics which must be regulated and processed in development efforts' (Ojwang 1996, p. 53). To contain command governance implies a negative duty in restraint of public power. For another thing: 'Such developments must be pursued as they alone will enhance the material base for enjoyment of social and personal liberties ... the African context is ... a context of ... construction, of larger rights and liberties, through orderly and well-conceived management of national resources'. This 'positive duty is the creative one facilitating national development ...'. This African scheme of rights-creation, is contrasted with Western countries 'which have attempted to replace the top-down orientation in environmental protection, by a bottom-up approach, which specifically empowers private individuals and interest groups to employ the state's legal machinery in the control of actions which are likely to lead to environmental degradation.' Ojwang (1996, p. 54) regrets however that, 'its effectiveness is unlikely to be as great in developing countries as it would in developed countries, since litigation is an expensive means of redress which is only available to a limited degree in developing countries.' Nevertheless, his formulation of environmental rights is person-centered. To him 'The rights of the *self* are an abstraction from a secure social condition which in turn, rests upon a fully protective economy. The relationship between economic stability and the environment makes the latter a vital factor in the ideal social condition that would facilitate the full enjoyment of the rights of the *self*'. Such democratic participation would represent the Western neo-Kantian tradition (Kant 1996).

2.2 *Collective rights and new environmentalism*

Okoth-Ogendo's (1999, p. 46) formulation of environmental rights 'are clearly about how best to guarantee human survival, maintain the quality of nature, ensure transgenerational equity and institutionalise choice'. Thus 'an *environmental right* may be described as freedom to exploit an environment adequately but responsibly for long-term survival'. He categorically rejects: 'The dominant Western jurisprudence ... that rights only exist as properties of individuals. They alone are capable of enjoying them'. He reasons: 'That, is too narrow and dangerous a view. It is now widely accepted that communities, whether or not organised, and corporate entities, associations or groups, defined simply by social and cultural ties, can and do have and enjoy rights by reason of their collective character'. Therefore, Okoth-Ogendo (1999, p. 43) asserts that while: 'the common property regimes ... recognise individual claims over certain categories of resources, access to a great deal of the most important resources can only be had collectively ... certain environmental resource rights could accrue only to individuals or to individuals and groups, others can only be group or collective rights. And since environmental rights are about quality assurance, individual rights are necessarily subordinate to group rights'. He acknowledges that: 'The demands of community living must be credited with the evolution of simple (Anglo-Saxon) rules of reciprocity based on control and use of natural resources.' However, 'in other societies community living produced no rules of reciprocity but rather, resource access principles managed and controlled by various levels of community organisation. The very complex regimes pertaining to acquisition, use and stewardship of common property resources in Africa are evidence of this development.' Deploying a Hohfeldian analysis of jural postulates—correlative to environmental rights—Okoth-Ogendo (1999, p. 48) derives four environmental duties. They comprise 'action requirements intended to ensure that entitlements generated by a regime of rights are not only expected and protected but also, in fact, achievable'. First, 'to refrain from activities injurious to the environment or any component of it.' This duty 'involves both abstention ... "nature knows best,"' and positive action, 'to intervene only where the rules of nature appear to have broken down'. Second, 'to perform specific tasks on a regular basis to ensure environmental quality at all times' (Okoth-Ogendo 1999, p. 49). This duty requires 'either anticipatory action (the precautionary principle) or action founded on objective scientific criteria (the foreseeability principle).' Third, 'to guarantee a floor of quality enough to ensure that all survive, especially the human species.' That is, 'to prevent the loss of quality'. Fourth, 'to police, supervise, monitor and evaluate the performance of individuals and agencies on which the first three or any other duties are imposed.'

This author decries that spectacular 'technological advances have posed near irreversible destruction on the eco-climatic system. Thus new form(s) of environmentalism demands strict ecological accounting in policy formulation, audit in decision-making, the primacy of social policy—over proprietary freedom, and attention to transgenerational management' (Okoth-Ogendo 1999, p. 43). Yet he concedes that 'environmental issues are often incapable of adequate resolution without indirect intervention.' Hence Okoth-Ogendo 1999, p. 50) concludes that 'these demands are now firmly established in the national agenda'. The first base of national environmental law 'is the doctrine of police power ... that political sovereignty over land *per se*, implies a residual duty to ensure that its use does not harm the public welfare' (Okoth-Ogendo 1999, p. 52). It justifies 'suppression or limitation of using private property ... to protect the public against dangers arising or likely to arise from misuse of that property.' It 'is now widely used as a positive instrument of land use planning'. Second, written constitutions in Africa 'merely restate(s) the doctrine of police power and makes its exercise an express exception to the guarantee of fundamental rights and freedoms'. This is reflected in 'a characteristically wide language and permit intervention for a variety of reasons including public order, morality, environmental safety and governance'.

2.3 *Risks of environmental depletion*

Liberal economics' cost-benefit approach reflects Jeremy Bentham's (1781) belief that 'the road to individual liberty and societal wealth is paved with individual's freedom to convert land into a commercial good; for the individual to have rights to land, traditional institutions must be destroyed, and the land must be privatised' (Goldman 1998, p. 24 quoting Bentham cited in Polanyi 1944). 'Mineral deposits vary in the nature and extent; costs of working differ in each case, so it is quite impossible to use comparative sales or any method which is unrelated to the working of and returns derivable from the subject mine' (Syagga 1994, p. 206 citing Murray 1973, p. 389). 'In this respect,' accordingly, 'capitalisation is the most appropriate method' of valuation of mineral lands although it is possible to use the following six methods in mineral valuation: residual, comparative, cost, investment (or capitalisation), profit or going concern approach (Syagga 1994, pp. 207–208). Similarly sustainable development economist Mikesell (1993, p. 23) argues from the standpoints of governments and donors that: 'Whether a development project ... should ... be undertaken should normally depend on its Net Present Social Value (NPSV)'. Hence 'development' of a mining project 'should not be taken where the NPSV is not positive'. Consequently: 'To embody the principle of sustainable development into the process of project evaluation, any resource depletion arising directly or indirectly from the project should be treated as a social cost in calculating the NPSV. Thus a project that involves a large amount of natural resource depletion will tend to have a lower NPSV than one that has little impact on the environment' (Mikesell 1993, p. 24). 'Mining firms can be required to put up bonds to guarantee they will clean up the mining area including filling up an open pit at the termination of the operation' (Mikesell 1993, p. 61). And: 'Also, comprehensive surveys are necessary to determine the geologic stability of the mining area and its vulnerability to earthquakes and volcanic activities, and to slides generated by heavy rainfall.' Therefore 'environmental hazards need to be identified in the EIA and necessary measures for dealing with them should be embodied in the project design.' However, 'the EIA and environmental safeguards are often very expensive, running to many millions of dollars. These outlays could run to as much as 20 percent of the capital cost of the mine.' Therefore 'unless the ores are very rich and other mining and transportation costs relatively low' prospective companies are reluctant to prepare such EIA reports until and unless 'they have already signed a mining contract' with a host country 'to operate and develop the mine and they are also unwilling to commit themselves to make unlimited expenditures for environmental safeguards' (Mikesell 1993, p. 62). Assuming a mining project has a positive NPSV, then according to Mikesell (1993) a twofold problem nevertheless remains: First: 'How do you measure the value of the depletion?' Second, 'how much revenue do you need to save and reinvest to maintain the same income (after allowance for depletion) for future generations?'. He explains that: 'The revenue from a resource project that is attributable to the natural resource is the total receipts from the sale of the products less the associated capital and labour costs' (Mikesell 1993, p. 23). Hence: 'The sustainability criterion is satisfied by including in the social cost of the project any reduction in the value of the resource caused by depletion and degradation. This would include not just the natural resources used in the project but any adverse effects on the natural resources base.' The crux of the dispute between environmentalists and mining companies lies in the definition of environmental risk. Webster (1954) defines risk as 'hazard; danger; peril, exposure to loss, injury, disadvantage or destruction'. 'Risk is often voluntary while hazard is a product of chance. Mathematicians define risk as a fraction between zero and one that measures the chance something will happen. A probability of one is a sure thing while zero means it is impossible' (Lewis 1990). The problem with pollution risks, in his view, is that they comprise 'events whose probability is very low that they have never happened at all yet but whose prospective consequences are so awful that they deserve attention.' Lewis (1990) regrets that 'legitimate uncertainty provides an opening through which demagogues and technical charlatans can get into the act, and exert a disproportionate and ultimately destructive influence on public decision-making. The mischievous exploit uncertainty.'

To establish a guideline, to measure environmental depletion, Mikesell (1993) formulates an accounting equation. He postulates that natural resources comprise non-renewable capital inputs which upon mining are combined with labour to produce finished products. The sale of products of a mine generate revenue (*R*). The resource depletion (*R–X*) component is thus equivalent to the revenue, less the income component (*X*). He then explains how such environmental depletion is estimated. First, by estimating the lifespan of the mine (*n* years). Second, by estimating the annual interest (r) expected to accrue from a fund reinvested each year during the number of years of the mine's life. Third, resource sustainability therefore requires that these annual interest sums—upon deduction from revenue—are invested in an accumulated fund or are reinvested to replenish depleted capital resources. Then, fourth, compounded savings should accumulate into a sufficient fund so that upon the exhaustion of the mine, future annual withdrawals by successive generations, in perpetuity, remain constant—equivalent to the average annual incomes drawn by the mining company during the mine's lifespan. Effectively: 'It makes little difference to future generations when actual depletion takes place during the life of a project' (Mikesell 1993, p. 25). Yet, the problem with these estimates is that because of ambiguity in subjective interpretations of risk and given self-interest therefore mining companies tend to overestimate the length of the mine, its annual revenue stream and expected annual interest rate, thus underestimating environmental depletion. This results in higher estimated income. Conversely, inflated environmentalist estimates of depletion proceed vice versa to the detriment of mining income.

2.4 *From resource conflicts to community participation*

The conflict between economic feasibility studies and EIAs if unresolved, foments political risk. In the Papua New Guinea *case of Ok Tedi Mine*, the government simply sacrificed environmental interests by refusing to shut down an environmentally unsustainable mine in 1990. In a previous conflict involving inhabitants of Bougainville Island, the government preferred to reap considerable tax revenues while neglecting both environmental hazards and risks to the community (Mikesell 1993, pp. 120–121; Ghai & Regan 2000, p. 257). It precipitated a revolution by angry inhabitants who themselves closed the destructive mine and ejected a multi-million dollar copper mining company. Hence the PNG government learned how to simultaneously exercise high control over the copper mining developer while providing support in negotiations to compensate or relocate affected communities. 'The grievances in Bougainville which were multifaceted and entangled with the wider processes of identity,' in the 'Ok Tedi situation there was no social crisis ... no conflict within landowning groups, no challenge of local political authority and no intergenerational strife' (Linnett 2009, p. 28). Thus governments have a duty to 'prevent economic activity from overwhelming the pollution absorptive capacity of the environment, or depleting or degrading the natural resource base' (Mikesell 1993, p. 5).

'Autonomy is a device to allow ethnic or other groups claiming a distinct identity to exercise direct control over affairs of direct concern to them, while allowing the larger entity those powers which cover common interests' (Ghai 2000, p. 8). By 'involving' the community in administrative decisions to determine the best use of natural resources, one privileges the community participation interest. Hence Mikesell notes that: 'Resource depletion need not be confined to depletion of reserves in a mining project. It may take the form of environmental damage caused by a mine ... that reduces the productivity of natural resources' (Mikesell 1993, p. 25). Linnett quotes PNG islanders who say 'Land is our life. And is our physical life—food and sustenance. Land is our social life; it is marriage; status; it is security; it is politics; in fact it is our only world' (Dove *et al.* 1974 cited in Connell 1997, p. 137 cited in Linnett 2009, p. 14). Similarly, Okoth-Ogendo asks us to 'enrich our environmental ethos and management strategies with our own cultural traditions and norms' (Okoth-Ogendo 1999, p. 60). He urges that: 'The rediscovery of the virtues of common property (read *customary*) management regimes is a healthy indication ...' He warns that the notion that 'developing countries have no property in their environmental resources' is 'becoming a new programme for colonisation' (Okoth-Ogendo 1999, p. 61). The latter author continues, that although: 'There is a great deal in our own history to draw from,' nevertheless: 'A major limitation of territorial devolution is that it is restricted to circumstances where there is a regional concentration of an ethnic group' (Ghai 2000, p. 9). Instead the NSPV may be calculated to preserve—not only community health—but also their cultural heritage as part of environmental depletion, in addition to extraction of raw materials. Expressing the intergenerational problem in Bougainville, 'one former landowner has stated that: "I can't pass the land on now because most of it has been covered up by the mine ... The traditional system will never work again. The company has only paid the parents for this. ... (Yet) everybody, right down to the last born, should get compensation because our traditions have been broken, and we will not be able to pass anything down to them"' (Linnett 2009, p. 23 citing O'Callagan 2002). Thus for Linnett: 'natural resource extraction leading to violent conflict is dependent upon its surroundings and the ideological and identity related resources that exist to those ethnic entrepreneurs and other actors involved. As a result "the severity ... consequences, or other intrinsic qualities of the resource exploitation itself are less important than the concurrent narrative of identity and suffering."' (Linnett 2009, p. 28 citing Aspinall 2007). Conversely, the overriding free market principle is that commerce and contractual exchange seems to provide a valid basis for producing peace since underlying communitarian theory of group solidarity implies that tribal land was collectively, not individually, owned. Hence mineral income can compensate communities for their mineral-laden land as part of resource depletion of the mining project and nevertheless confer a positive NPSV (Mikesell 1993, p. 25).

2.5 *Development, compulsory acquisition and compensation*

Although Articles 55 and 56 of the 1945 UN Charter[4] provide the duty of states to co-operate for the pursuit of peace and development: 'The term right does not necessarily imply a distinct legal claim or a clear corresponding duty. Hence the right to development does not meet the criteria of positive law. It exists in the need of each human being to realise his essential humanity' (Olale 1995, p. 18). 'In the African conjuncture, it is submitted' by Shivji (1989, p. 83) 'that the *central* rights are "right to self-determination" and "right to organise". This rejects the capitalist notion that 'the ruling ideology in post-independence Africa ... developmentalism ... takes as its point of departure the real material conditions of underdevelopment, and argues that development takes priority over everything else ... Africa can ill afford the luxury of politics.' Instead 'as a matter of fact, in practice ... it is the politics of the people that is displaced while the politics of the ruling class is consolidated'. Hence, natural resources such as coral reefs, ocean shores or marine parks which may suffer salination by drilling of holes close to the seashores and leaving unsightly tailings thus depriving communities of recreation may render a negative NSPV such as to warrant denial of mining licence of a given location in favour of alternative uses (Mikesell 1993, p. 61). Moreover, community life may be disrupted and communities may require relocation and resettlement. Because of the wide disparities in capacities of parties to access information—leave alone undertake scientific investigations and further: 'Since all natural resources provide some services that go beyond those to the individual owner, some would argue that all natural resources should be subject to strict governmental control or even be government-owned and limited to uses imposed by the collective will' (Mikesell 1993, p. 76). However, 'experience has shown that government ownership and control can be highly destructive of natural resources, and that the benefits from their use have often gone to state bureaucracies rather than to society as a whole'. The previous Kenyan Constitution provided that: 'No property of any description shall be compulsorily taken possession of and no interest or right over property of any description shall be compulsorily acquired.'[5] Lawyer P.L. Onalo explained that first 'the taking over must be necessary in the interests of defence, public safety, public order, public morality, public health, town and country planning or the development or utilisation in such a manner as to promote the public benefit'[6] (Onalo 1986, p. 57, Syagga 1994, p. 96). Second, such 'necessity must be such as to justify any hardship to the owner or the person with interest.'[7] Third, 'provision must be made for the prompt payment of full compensation.'[8] Arising from injustices inflicted by the Tiomin titanium mining cases, the new Constitution significantly enshrines the right of persons with affected property interests to access to courts of law. The procedure required that specific written awards be made to individuals directly affected including landowners and the Registrar of Lands (58). In ensuring prompt and full compensation,[9] the Commissioner of Lands, was bound to 'determine, in accordance with the principles set out in the schedule, what compensation is payable to each of the persons who he had determined to be interested in the land.'[10] The schedule establishes principles on which full compensation is to be determined as comprising: 'market value, the damage sustained by reasons of severance of land acquired from the owner's land, damage sustained by reason of acquisition injuriously affecting the owner's other property or his income, and expenses a person may have had to incur in order to change his place of business or residence, and an additional fifteen percent (15%) *salotuim* to be considered for the compulsory nature' (Onalo 1986, p. 58; Syagga 1994, p. 98). The commissioner, under the old constitution, however was not bound to consider the 'degree of urgency in acquisition, the disinclination of interested persons, the damage which if caused by a private person would not render such person legally responsible, any increase in the value of the land of the interested person after the acquisition and any improvements effected during the acquisition proceedings'. Only if notices were issued and subsequently withdrawn are individuals entitled to compensation for disturbance. Bhalla (1996, pp. 75–76) criticised the old Constitution since: 'Whether or not compensation is adequate is an important matter in the social and economic well being of an individual, since acquisition, in whatever form, is always considered an intrusion upon private property. No matter how substantial the compensation paid, the satisfaction found in one's belongings can never be assessed in monetary terms. In fact compensation can never cover the self-esteem and reverence for his or her property nor the personal identity associated with property'. Similarly, for land economist Aritho (2000, p. 75): 'The main question is how is *full compensation* or *just indemnity* to be determined'. He argues that 'a declaration or recognition of principles of compensation to be paid is one thing, their practical application is something different altogether.' He quotes Denyer-Green (1985) that 'a measure of compensation is not fair,' just, adequate, efficient, or full because the statement of the rule defining it satisfies those criteria; it is only fair, just, adequate and efficient 'if it is broadly accepted by expropriatees through their experience of its implementation' (Aritho 2000, pp. 75–76 citing Denyer-Green 1985). Aritho (2000, p. 76) laments that 'where valuations have been carried out for the purposes of determining the constitutionality and statutorily prescribed full and fair compensation for lands that are compulsorily acquired in Kenya, genuine dilemmas in defining terms for valuing land so acquired has led to problems of ascribing value to land and property in a clear-cut and unequivocal way'. He also explains that 'Debates and arguments abound on how to account (separately) for each of the varied and concurrent cause-effects which influence changes in land values and particularly in the petty commodity production lands occupied by so called informal or *jua kali* sector'. Hence he diagnoses that 'the real problem is not compulsory acquisition *per se*. Rather it is the misconceived assumptions of the disarticulated or incoherent economies of rural society and the so called informal sector markets'. Because 'the abovesaid dilemmas and misconceptions have characterised the history of compulsory acquisition of lands in Kenya (consequently) rendering it socially vexing, incoherent, and lacking in social sensitivity towards and consideration of the people who are expropriated,' Aritho (2000) concludes that use of market value displays defects that it is 'too restrictive; it may not accommodate individual cases of equality,' unless market value is taken by claimants to be a minimum expectation. In short, because a compulsory acquisition can be 'equated with a free transaction only in respect of compensation, not compulsion, therefore adequate compensation balances acquisition and only the public good can balance the compulsion'. Kibugi (2008, p. 362) suggests that 'government should establish a legal framework to specify the rights and obligations of all persons involved in involuntary settlement. His view is that: 'Compensation should be land for land pegged to the productive value of the vacated land in addition to the costs of relocation'.

By comparison, Gordon (1995, p. 521) indicates that in the US 'Extensive government programmes affect mineral supply, including controls by land policy and controls by price regulation. Mineral land policy is part of the overall land use policy'. Before announcing that 'Kenya strikes Gold!' (Headline The East African Standard, 13 September 2010), the Kenya National Land Policy stated that: 'To sustainably manage land based natural resources, the Government shall: (a) Encourage preparation of participatory environmental action plans by communities and individuals living near environmentally sensitive areas to preserve cultural and social-economic aspects.'[11] A new Mining and Minerals Bill proposes a mining legal framework and policy with a clear direction on mineral exploitation ventures.

3 RESTORATIVE ENVIRONMENTAL GOVERNANCE

3.1 *Linking modern and traditional environmental governance through responsive law*

Hyden & Bratton (1992) argue that governance subsumes both 'government' and 'leadership'. In more practical terms, Hyden & Mugabe (1999, pp. 33–34) state that 'governance refers to measures to lay down the rules for exercising power and settling the conflicts emerging from such rules'. Sociologists define governance as 'the guidance or control of an activity in order to meet a specified objective' (Fox & Ward 2008, p. 524 citing Jewson & MacGregor 1997, p. 6 and Rosenau 1997, p. 146). Furthermore: 'For regional, national and international governments, governance describes the accountability processes used to shape economic and social activity and may involve legislation and regulatory processes to set standards, monitor or correct defined areas of activity' (Fox & Ward 2008 citing Salter & Jones 2002, p. 810). More particularly, to Mugabe & Tumushabe (1999, p. 15): 'Environmental governance can be defined as a body of values and norms that guide or regulate state-civil society relationships in the use of, control and management of the natural environment ... [It] provides a conceptual framework within which public and private behaviour is regulated in support of sound ecological stewardship ... the rules, rights and responsibilities may either flow from custom and practice or be codified'. Therefore firstly: 'It implies collective deliberation and decision-making in arranging the environmental. Secondly, management involves political issues and processes.' Moreover: 'When extending the principle of "responsive and responsible leadership" to environmental issues generally and to managing environmental resources in particular, we are concerned with the extent to which the government or the state generally is capable of formulating and willing to formulate and implement appropriate measures to cause environmental sustainability.' Modern environmental governance is distinguishable from the traditional model. The former 'is mainly composed of written and formal policies, environmental plans, legal instruments and informal laws, rules of practice and institutions that explicitly or implicitly impact on environmental management'. Conversely the latter 'is composed of unwritten, informal and systemic taboos, rituals and rules that regulate the interactions between individuals and the natural environment.' Traditional environmental governance systems are characteristically first, *evolutionary*, second *responsive* and *resilient* to the ecology on which they are based; and third are *participatory* (Mugabe & Tumushabe 1999, p. 17).

In this regard, ironically traditional governance can be seen as responsive law which to Nonet & Selznick (1978, p. 79) is the modern successor to autonomous law. Indeed according to the 'communitarian thesis' of restorative justice, consensus dispute resolution approaches with an objective of victim compensation rather than offender punishment, can learn from traditional customary mechanisms (Christie 1977; Zehr 1990, both cited in Dignan 2005). For Nonet & Selznick (1978, p. 95): 'Whereas autonomous law encouraged a restrictive view of official obligation and was concerned mainly with the restraint of authority, responsive law is purposive and concerned with substantive outcomes, paying due regard to implicit values in rules and policies which may be flexibly interpreted and applied to new institutional settings. The responsive form of law assumes a multiplicity of sources of law and law-making within the framework of modern government and implies participating in the function of making and interpretation of legal policy'. Responsive law proceeds by assigning roles to various state agencies allowing them 'initiative in selecting ways and means and emphasising their native organisational roles and responsibility to enlist participation by citizens.' Hence, it has been argued that 'governance cannot be reduced to legislation and regulation ... governance increasingly entails decentralisation, inclusivity, engagement with the public and bottom-up constructions of consensus and collaboration' (Fox & Ward 2008 citing Coglianese 1999; Dorf & Sabel 1998, Freeman 1997). This is what I understand Ojwang (1996) to mean when he says that "law that facilitates" could in the first place be law that established certain patterns of legality, such as might be done by tax relief laws, or monetary benefits made payable under some law, provided as a device for encouraging private initiatives in environmental protection. Such laws are not imperative in the normal sense; but they give rights only to those who come to satisfy the basic conditions prescribed. In this manner, they facilitate the goals sought. But laws that establish agencies to perform management and discretion-oriented duties for instance in the cause of environmental protection, may be said to be facilitative, albeit in a different sense. By establishing management machinery, such laws will facilitate the task of environmental protection' (Ojwang 1996, p. 47). He however concedes that 'it is nonetheless apparent from his (Okidi's 1984 discussion) that environmental law will remain a mixed bag for quite some time without the relatively sharp boundaries of property law, contract or tort' (Ojwang 1996, p. 48). It is therefore argued under the post-modern Kenyan law, that this responsive function linking environmental governance to development projects is achieved through contract law, underpinned by the equitable doctrine of estoppel, imported from England. Equity functions as a doctrine of general application. Hence, Lubbe (2006, p. 784), explaining the South African context, which is similar to Kenya's liberal democratic constitution, indicates estoppel's multiple functions. Although 'its *primary* application are in the fields of private and commercial law' ... 'it is thought to be regulated by a uniform set of requirements invariably applicable irrespective of whether the estoppel is relied upon in the areas of property, contract, agency, partnership, or company law.' The estoppel doctrine originated from the English case of *Freeman v Cook*.[12] It provides that: 'If whatever a man's real intention may be he so conducts himself that a reasonable man would believe that he was assenting to the terms proposed by the other party, and that other party upon that belief enters into the contract with him, the man thus conducting himself would be equally bound as if he had intended to agree to the other party's terms.' In addition to the objective test under 'reliance theory' (Lubbe 2006, pp. 754–755), The latter cites another example, the 'will theory' expressed as a rule of law incorporating the Latin maxim *consensus ad idem* as the source of reliance upon representations during contractual negotiations generating estoppel. Indeed: 'The South African Supreme Court has emphatically asserted that good faith and other so-called "abstract ideas" such as reasonableness, fairness and justice are not properly regarded as legal concepts or "rules" but amount to fundamental principles underlying the substantive law' (Lubbe 2006, p. 753).

3.2 *Resolving environmental crimes*

According to Albrecht (2001, p. 305), an understanding of a criminal offence ... is either regarded as a consequence of a basic problem in the interactions between victim and offender, or as a cause of conflict in terms of a dispute on the type and the amount of what should be compensated'. He notes that 'there is something in the field of control of economic and environmental crimes, in particular, which comes close to dispute resolution, argumentative discourses, bargaining and ultimately to the goal of replacing criminal punishment and the criminal process through mediation' (Albrecht 2001, p. 306). Restorative justice approaches to problem resolution resolve poly-dimensional conflicts. It 'is a process whereby parties with a stake in a specific offence collectively resolve how to deal wit the aftermath of the offence and its implications for the future' (Marshall 1999, p. 5). According to the 'civilisation thesis' (Dignan 2005) of restorative justice theory, on one hand, 'centralisation and the monopoly of (criminal prosecution) provided benefits for both the offender and the victim. This is because the immediate conflict resolution was characterised by imminent risk of escalation which was limited only by the resources available to both parties and their alliances' (Albrecht 2001, p. 306). On the other hand according to the 'communitarian thesis' (Christie 1977): 'Administrative control of the environment ... is based on the consideration that compliance-achieving mechanisms of enforcement of environmental laws should preferably be based on voluntary action and persuasion or positive incentives but not on coercion and criminal sanctions. The model relies heavily on co-operation and bargaining' (Albrecht 2001, p. 306). The characterisation of administrative decision-making by 'the pursuit of two differing and conflicting goals: protection and preservation of the environment on one hand and commercial use of the environment on the other' provides the 'rationale for empowering administrative agencies to use discretion ...' Albrecht decries the fact that 'the use of the criminal law evidently results in a zero-sum game likely to increase the problem of non-compliance as well as to aggravate the problem of other legal conflicts between companies and administrative authorities.' Therefore 'the criminal prosecution of environmental offences is counterproductive' (Albrecht 2001, p. 307). He concludes that: 'Handling all deviant acts by the criminal justice system is neither economically feasible nor functionally viable.' Instead he is critical that 'short term benefits in terms of successful criminal prosecution of environment offences would be exchanged for long term benefits in terms of achieving the goal of future compliance with the objectives of administrative law' (Albrecht 2001, p. 306). Significantly, however given that 'it is not known whether informal and immediate conflict resolution yields, on average, as satisfying results as formal processing of cases and offenders,' he cautions that 'we may assume that successful conflict resolution is dependent on the availability of formal procedures and potential of coercion' (Albrecht 2001, p. 308).

3.3 *The social disciple window*

First, they observe that: 'Every part of society faces choices in deciding how to maintain social discipline whether it be ... justice systems responding to criminal offences' (McCold & Wachtel 2000, p. 1). Second, applying Albrecht's (2001) reasoning, because: 'Mining and associated mineral processing is one of the most environmentally harmful activities' (Mikesell 1993, p. 61), under criminal justice policy, a mining company may be construed as an offender, of crimes against the environment. Third, upon further dissecting this bivalent construction into its constituent parts, McCold & Wachtel (2001) indicate that 'one can identify social discipline choices when one looks at the interplay of two continuums, control and support, which comprise the Social Discipline Window.' Quoting Black's definitions, on the horizontal axis: '"Support" is defined as the provision of services intended to nurture the individual' (Black 1990, p. 1070). In practical terms: 'Active provision of services and assistance and concern for well being characterise high support' (McCold & Wachtel 2000, p. 1). Simultaneously: 'On the vertical axis, '"control" is defined as the act of exercising restraint or directing influence over others' (Black 1990, p. 329). In effect: 'Clear limit-setting and diligent enforcement of behavioral standards characterise high social control' (McCold & Wachtel 2000, p. 1). In application of criminal justice policy, first, *neglectful* social control is characterised by complacency or an indifferent response to misbehavior or wrongdoing. *Not* doing anything in response to harmful behaviour is indicated not only by minimum of physical and emotional support, but is also compounded by an absence of limit-setting (McCold & Wachtel 2000, p. 2). Second, *permissive* social control is classified as rehabilitative or therapeutic given its tendency to mitigate the rigours of consequential losses logically and naturally flowing from wrongdoing. Instead, abundant nurturing combines with diminished constraints to mitigate damages. One asks little in return, while instead doing everything *for* the offender thus mollycoddling misbehaviour.

Third, a *punitive* criminal justice system hardly offers any support to rehabilitate ordinary criminals but exhibits high-handed degrees of control to impose order, sometimes prone to circumventing the rule of law and trampling upon fundamental rights of powerless classes. Administrators, police, courts and prison authorities respond by doing things *to* the offender while overzealously administering retributive punishment which inflicts stigmatisation. Fourth, conversely, responding through a *restorative* social control approach demands working *with* all the stakeholders in the process including the offender, victims, family, friends and community—anyone affected by the offence.

3.4 *The multi-layered governance template*

In its widest sense, social control may be conceived as comprising four levels of activities: 'Governance issues occur at a meta-level, above all other political activities. It deals with "meta policies," or what Kieser & Ostrom (1982) call constitutional choices' (cited by Hyden & Mugabe 1999, p. 35).

Their smallest sense of social control embraces: 'Project management … the … level—where the actual cost/benefit calculations are put into practice on a day-to-day basis.' In between these extremes: 'Policy-making is a second level … where specific choices are made on the development of particular sectors.' Of relevance here is mining. Such macro' social control is underpinned by 'meso': 'public administration … where co-ordination and implementation of programme activities take place.'

Thus, social control is a generic term which refers to governance, policy-making, administration or management. 'The nation-state,' according to the chief architects of new institutionalism, Keohane & Ostrom (1995) 'is hierarchical' (cited by Thompson 1998, p. 199). It is 'centralised and coercive—and the commons, which are voluntary and co-operative, are able to function at the two extremes' (The very large scale 'meta' and the very tiny scale 'nano') as table 1 illustrates.

However, because the nation-state's writ does not fully run into either the global or village levels, 'therefore international regimes are emerging to manage what they call "the global commons".' Yet their new institutional approach to the commons has been criticised by Thompson

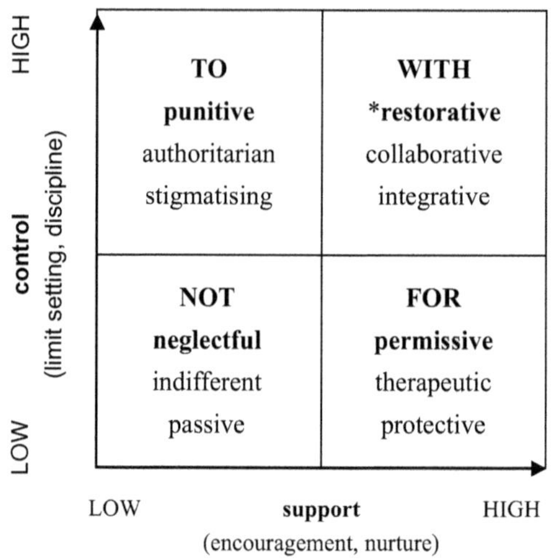

Figure 1. Social Discipline Window adapted from McCold & Wachtel (2000, p. 3).

Table 1. Governance and its relation to other relevant concepts adapted from Hyden & Mugabe (1999, p. 35).

Level	Activity	Concept
Meta	Politics	Governance
Macro	Policy	Policy-Making
Meso	Program	Administration
Micro	Project	Management

(1998, p. 198) for three reasons. His approach is based on Douglas's (1978) grid/group analysis. Because, first: 'Centralisation and coercion are *not* definitive of hierarchy; formal status and distinctions' *are*. Second: 'market relationships are often (indeed are usually) voluntary and many commons-managing institutions are hierarchical.' Hence: 'Co-operation and voluntariness are *not* definitive of the commons' (Thompson 1998, p. 199). Third: 'Hierarchical relationships are to be found at *all* levels (not just those that are beyond the reach of the nation-state). Therefore while: 'Scale, Cultural Theorists concede, does matter, but it plays no part in defining the commons.' Thompson's (1998) multi-layered template contains four typologies of styles at every level of the wider governance hierarchy. Where the new institutionalism has just a scale axis, with hierarchy in the middle and the commons at either extremity, cultural theory gives us a 'multi-layered template' in which all four styles are present at every level' (Thompson 1998, p. 200). He moreover cautions that: 'Order without predictability is what complex systems give in stead … when we are talking about institutions, we will be seriously wide of the mark if we assume (as pretty well all of current thinking does assume) that they are simple.' He explains that: 'What this means is that elegant institutions—solutions that begin with defining the problem in terms of just one of the myths of nature and then try to put all our transactions on to the pattern of social relationships that is supported by that myth of nature—are wholly inappropriate for complex systems' (Thompson 1998, p. 210, Fig. 2).

Thompson (1998) sorts out 'the various arrangements that are not commons' (he calls them 'privates') (p. 199) using a *typology of styles*. He asserts that: 'It is the distinctive combinations of these key predictions from Cultural Theory that define the various styles and their manifestations on all scale levels, in different proportions and patterns of interaction' (Thompson 1998, p. 200). Thus: 'Cultural Theory ... typology ... distinguishes those familiar arrangements, *markets* and *hierarchies* (the former promoting competition and instituting equality, the latter setting limits on competition and instituting inequality)—and then goes on to distinguish the less familiar arrangements—*egalitarianism* and *fatalism* (the former setting limits on competition and instituting equality, the latter enduring unfettered competition and inequality) that complete the permutational possibilities' (Thompson 1998, p. 199).

3.5 *Social discipline and communities of care*

One advantage of introducing cultural theory's multi-layered template is that while retaining the explanatory power of McCold and Wachtel's social discipline window, the template is more convenient in that it dispenses not only with McCold and Wachtel's additional use of three concentric circles to display what they call the stakeholders' needs structure, but also their reliance on three overlapping circles to depict their restorative practices typology (McCold & Wachtel 2000, pp. 5–6). It is perhaps preferable to introduce new institutionalism's perpendicular hierarchic axis so as to simultaneously examine variations of community participation in relation to both the mining company offender as well as the environment as a victim. Indeed, focus on 'macro' policy-making in the mining industry requires appreciation of different dimensions of mining policy activities which prescribe modalities for programme administration and in turn project management. Hence fatalism, individualism, hierarchy and egalitarianism ideologies give rise to neglectful, permissive, punitive and restorative justice social control responses respectively. Emphasising these distinctions Thompson asserts: 'Rather there is a double bifurcation: the commons come in two very different forms—hierarchy and egalitarianism—and the privates come in two very different forms—individualism (the form that is supported by market relations) and fatalism' (Thompson 1998, p. 228). So as to provide a more incisive working model by which to compare governance responses to polycentric mining conflicts, this contribution proposes to modify McCold and Wachtel's single-plane social discipline window—which is restricted to displaying micro projects in two-dimensions—by combining it with Thompson's multi-layered template to produce a three-dimensional governance diagram. Cultural theory's template, understood as a Social Discipline Cube, not only describes the relationship between an offending mining company and the environment or commons as victim, but also simultaneously defines their relationship to the third direct stakeholder, their 'communities of care' (Fig. 3).

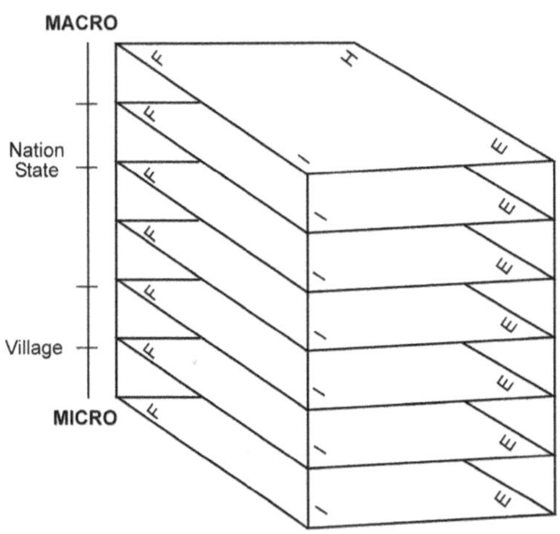

Figure 2. Cultural Theory's multi-layered template: four styles at every scale adapted from Thompson (1998, p. 200).

For Walgrave (2000, p. 1), the goal of reparation of harm is a necessary element of restorative justice. He asserts that restorative theorising should unpack the social values from their 'community' and find a way to combine these values with the principles of a democratic constitutional state. 'Under "community justice" community is advanced as the social environment of informal human understanding, as opposed to formal institutionalised society (sometimes called "the government" or "the state") with its rules and rigid communication channels'. Stakeholder needs *for* government support by way of rehabilitation relates specific social responsibilities to meet those needs to the injuries caused by offending behaviour (McCold & Wachtel 2000, p. 3). Nonetheless, they distinguish, on one hand, 'the injuries, needs and obligations of those indirectly affected, by a specific offence.'

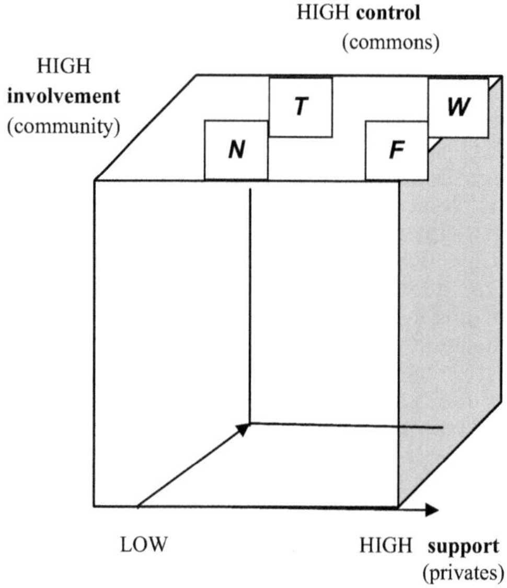

Figure 3. Social Discipline Cube showing intersection of three governance dimensions. Key: *N*: Not (Neglectful); *F*: For (Permissive); *T*: To (Punitive); *W*: With (Responsive).

They caution such groups of indirect stakeholders, 'those not emotionally connected to the principals' but 'who live nearby,' or 'are members or officials of the government, religious, social or business organisations whose area of responsibility includes the place or people affected by the incident,' against 'stealing the conflict' (Christie 1977 cited in McCold & Wachtel 2000, p. 4). Significantly, such theft of 'property' in offences is committed through usurping the responsibilities of those affected. Thus from a market perspective, it is plausible to argue that by insisting upon less efficient use of natural resources, environmentalists and communities reduce a mining company's expected income. Hence community participation in resource allocation decisions may offend against agreements between mining companies and environmentalists regarding the sufficiency of an EIA of a given mining project. Moreover another practical consequence of Thompson's (1998) stylistic depth is a fourfold typology of *ideas of fairness*. Every policy debate—local and global—is a debate about fairness, yet nearly all policy analysis is couched in the purported neutral terms of economic efficiency' (Thompson 1998, p. 212). If fairness is conceived from an egalitarian standpoint, the government as 'indirect stakeholders' have a responsibility of facilitating and supporting processes in which the direct stakeholders determine for themselves the outcome of the case (McCold & Wachtel 2000, p. 4). In this respect, interests of 'persons acquainted with the principals' include 'family and friends.' Such interests of members who comprise the 'community of care' are satisfied by acknowledging the wrongfulness, ensuring that something is done about it and taking concrete action to prevent recurrence. The most important stakeholders however, are the principals who are most directly affected. Offenders need to regain a sense of belonging to their community of care 'by accepting responsibility for their wrongdoing.' Citing Howard Zehr (1990), McCold & Wachtel note that: 'Victims need to regain a sense of personal power' (Zehr 1990 cited in McCold & Wachtel 2000, p. 3) to transform themselves into survivors. These processes will integrate both victims and offenders, build problem-solving communities and strengthen civil society.

3.6 *Clumsy institutions meet stakeholders' needs using restorative practices*

Regarding design of a restorative mediation institutional framework to implement multi-pronged mediation procedures to assess environmental impact, Thompson notes four advantages arising given that 'jurist Michael Shapiro (1988) has dubbed this sort of set-up (in which each conviction is given some recognition) as a *clumsy institution*.

This is in contrast to those more elegant, and more familiar, arrangements in which just one conviction holds sway. The terminology is deliberately counter intuitive: clumsy institutions have some remarkable properties that are not shared by their unclumsy alternatives' (Thompson 1998, p. 208 citing Shapiro).

In environmental governance, Okoth-Ogendo's first 'duty to refrain from injuring the environment, involving both abstention and positive action' (Okoth-Ogendo 1999, p. 47)—reflected in the social control and support axes of the social discipline window—which is testable using 'collaboration' thesis. (McCold & Wachtel 2000, p. 6). By accepting to make sufficient provision in EIAs which contain guarantees to refrain from ecological harm, take pre-cautions and implement clean-ups from mining projects, mining companies collaborate with environmentalists. Hence Thompson (1998, p. 210) indicates that one advantage of clumsy institutions is that they are '*highly adaptive* (because by covering all these bases, they allow us to continuously reassess the appropriateness of our initial distribution of transactions and to switch those that are not working well to more promising alternatives)'.

Because the very process of interacting is critical to meeting stakeholders' needs, McCold & Wachtel's (2000, p. 4) restorative practices typology assumes that the degree to which any form of social discipline may be termed as 'fully restorative' is equal to the 'degree to which the direct stakeholders are involved in meaningful decision-making and emotional exchange'. Only when their social discipline window deals with all three sets of direct stakeholders, such as in conferences or circles, is the process 'fully restorative.' In what Thompson's (1998) three-dimensional template for assessing appropriateness calls 'institutional style,' the proposed Social Discipline Cube, attains a 'fully restorative' outcome upon convergence of a triple restorative response (high social, support, high control and high involvement). 'Mostly restorative' describes double-highly restorative responses, together with one low response whether of neglectful, permissive or punitive variables. Thus the Social Discipline Cube, to the extent of non-involvement of communities in decision-making, responses are merely 'mostly restorative.' Finally, a 'partly restorative' social control response is characterised by but a single restorative variable with two non-restorative ones. In environmental governance, McCold & Wachtel's (2000) 'involvement' hypothesis which tests their restorative practices typology appears akin to Okoth-Ogendo's second environmental duty of 'systematic and sustained action, which implies preparation of environment impact assessment studies' (Okoth-Ogendo 1999, pp. 47–48). The difference is that while Ojwang (1990, 1996), recognises the potential advantages of community participation, the Kenyan environmental legislation does not expressly require NEMA to consider cultural knowledge as part of an EIA. Moreover the insistence—by human ecologists—upon so-called 'objective' scientific foresight of environmental harms appears to be privileged. Yet McCold & Wachtel's (2000) social discipline window blurs the distinction between environmental harm inflicted through extraction of non-renewable natural resources, on one hand, from wider harm to community health and safety by pollution, on the other. Nevertheless, their ordinal supposition about restorative practices assumes that practices which 'involve' more sets of stakeholders will produce outcomes which are more 'restorative,' 'empowering' and 'transformative.' Similarly, for Thompson: 'Policies designed in this way are *non-hubristic* (because they fully acknowledge the extent of our ignorance)' (Thompson 1998, p. 210).

The 'empowerment' hypothesis stakeholders' needs structure thus entails that to regain their sense of autonomy, victims directly affected by harmful behaviour must be part of the solution. It is further implied in their social discipline window that there is a diagonal dimension running from permissive to punitive where social discipline which is done *to* offenders, and responses done *for* offenders are both fundamentally disrespectful (McCold & Wachtel 2000, p. 6). Such minimalist duty of 'prevention of loss of environmental quality' (Okoth-Ogendo 1999, p. 48) necessarily unites both mining companies and environmentalists against communities in determination of what standard constitutes a floor of minimum environmental quality under Okoth-Ogendo's third duty. However, such community exclusion through paternalistic governance, although based on scientifically objective studies or economic feasibility studies to determine the NPSV, may be disrespectful to self-identity, autonomy and cultural integrity. Hence, 'collaboration' dimension links neglectful to restorative responses. In this regard, McCold & Wachtel's (2000) 'collaboration' hypothesis predicts that because healing in relationships and new learning are voluntary and co-operative processes, therefore, responses which are simultaneously high both on social support as well as social control require active collaboration of the offender. In this regard, Thompson asserts a third advantage of a 'clumsy' rather than 'elegant' approach to institutional responses is that it gives rise to '*quick learners* because they do not set off by discarding three quarters of the experience and wisdom from which they could learn' (Thompson 1998, p. 210).

Hardin (1968) contends that a society can be badly served if each member of that society makes decisions serving his own self-interest. Michael Goldman 'begins by boldly capturing the ... work done on the commons into a dialectic between those who accept the tragedy of the commons explanation and those who systematically reject it' (Goldman 1998, pp. 23–25). Section 2 above considered various policy choices, with one school of thought suggesting that we 'change human nature (solution built into Aldo Leopold's "earth ethic" and into the deep ecologists insistence that we must develop a new relationship with nature)' (Thompson 1998, p. 201 citing Goldman 1998, quoting Leopold). Bromley's (1992) ideology is that 'rights have no meaning without correlated duties, and the management problem with open access resources is that there are no duties on aspiring users to refrain from use' (Goldman 1998, p. 27, 49 fn 14 quoting Bromley 1992). Indeed, Ojwang (1996, p. 56) recognises that: 'There are opportunities to integrate environmental concerns into the constitutional debate'. Similarly, Okoth-Ogendo's (1999, pp. 52–56) fourth environmental duty discusses 'the framework of environmental legislation' and the 'global consensus building efforts', which have developed 'a coherent set of norms and principles constituting international environmental law'. We also considered arguments of another policy choice available to those who accept Hardin's (1968) explanation to 'privatise the commons (a solution that is favoured by many development economists and much touted libertarian think tanks)' (Thompson 1998, p. 201 citing Goldman 1998, pp. 27–30). Recommending a responsive approach, we further concurred that it is necessary 'to locate our project, reflexivity—the ability to take up a position from which it is possible to recognise the arguments of those for whom development is the solution *and* of those for whom it is the problem' (Thompson 1998, p. 201 citing Goldman 1998). It is in this new dialectic that 'McCold & Wachtel (2000, p. 5) define restorative justice as a process involving the direct stakeholders in determining how best to repair the harm done by offending behaviour'. This includes a supposition that restorative justice outcomes have a 'transformation' dimension transforming victims into survivors, conflict into co-operation, shame into pride and individuals into community, thus testing restorative outcomes. In the context of environmental governance, such secondary transformation, it is submitted, tallies with Okoth-Ogendo's aforesaid fourth environmental duty imposed upon officials to superintend over implementation of the three primary environmental conservation duties. The difference is that while human ecologists hold hierarchical ideologies prescribing punitive responses some criminologists espouse egalitarian procedures preferring restorative justice. Cultural theory thus calls for 'set-ups in which each myth of nature is granted some legitimacy, and in which transactions are tentatively distributed among the various institutional bases.'

Kibugi 2008, pp. 361–362) proposes that the: 'Department of mines and Geology obtain plan from an investor and forwarding them to the Ministry for presentation to Parliament ... the titanium mining project in Kwale District provides an example of the lack of a proper legal and policy framework to address involuntary displacement in Kenya'.

4 ENVIRONMENTAL PROTECTION AND PUNITIVE GOVERNANCE: TIOMIN I

4.1 *The facts*

In Nzioka and two others versus Tiomin Kenya Ltd.[13] (Tiomin I 2001) local inhabitants at Kwale district, along the Kenyan south coast, bordering Mombasa to the north and Tanzania to the south, sued a mining company on three grounds (Tiomin I, p. 425). First, alleging that the amount of consideration offered by the mining company, of Kshs 2000 (US$25) per acre per annum over a twenty-one year duration together with a Kshs 9000 (US$112) downpayment to facilitate relocation and resettlement, was unreasonably low (Tiomin I, p. 427). They expected their land-holdings to be purchased outright, to be resettled and schools and hospitals, constructed. Second, they accused the foreign company of not only lacking relevant presidential exemption,[14] but also neither obtaining change of user or obtaining requisite consent under the Land Control Act.[15] Third, the inhabitants alleged that because neither a project report nor EIA study and report had been submitted to the National Environmental Management Authority (NEMA) in the prescribed form or at all, therefore the company was not only in breach of its statutory duties *per se* but was also poised to commence mining, to the detriment of the environment, health and safety of the local farming communities. They relied on scientific research studies rendered by the Kenyatta University, to support their apprehension of exposure to titanium, given failure to 'redress radioactivity, sulphur dioxide pollution and dust pollution.' in absence of any comprehensive resettlement plan or neutralisation of toxicity. Initially, the local inhabitants prayed for the equitable remedy of a temporary restraining order pending the trial. At the full trial, they would ultimately claim an order declaring that intended mining was illegal, a permanent restraining injunction and general damages (Tiomin I, p. 428). In its defence, the mining company replied first, that the local inhabitants—being mere squatters without valid title deeds empowering them to assert proprietary rights—lacked *locus standi* (capacity to sue) and were therefore technically non-suited. Second, that having obtained relevant special mining licences from the Commissioner of Mines and Geology pursuant to the Mining Act,[16] it was duly authorised to prospect and was not proceeding unilaterally. In any event, third, it denied that any 'harmful effects have so far been experienced.' Fourth, instead it insisted that consultative discussions were ongoing with local government officials and communities to resolve both title deeds and compensation issues. Moreover, fifth, it emphatically claimed to have submitted an EIA, but was yet to commence mining activates because it was awaiting approval from the Environmental Authority.

4.2 *Geographical location and mineral deposits*

The inhabitants' EIA—commissioned by a coalition of non-governmental organisations interested in the project to mine titanium—specifically located Kwale as an administrative district along the Kenyan South Coast, bordering Mombasa to its North and North-Western Tanzania to its South, lying between longitude 38° and 39° East, and latitudes between 3° and 4° South' (Tiomin I, p. 425). It covers 8322 km^2 in area. About 0.73% of its area is covered with either fresh or salty water and from its waters fish and drinking water for humans and animals depend. On its coastline runs 3 to 5 km of living coral reef and a coastline with mangrove swamps (Tiomin I, p. 425).

A detailed mineral analysis of the inhabitants' EIA discovered that lying: 'In the Vumbu-Maumba area the titanium ore deposits constitute about 5.7% of the Magarini sediments the concentration reduces southwards to 3% in Nguluku area. The titanium deposits mainly occur in aliments and retile with specific gravity of 4.72 and 4.2 to 4.3 respectively. The zirconium containing mineral in this case is zircon, which has a specific gravity of 3.9 to 4.7. The specific gravity shows that these are heavy minerals and hence are deposited at similar sites through sedimentation in reverine, lacustrine and marine water' (Tiomin I, p. 426). Nearby: 'The Msambweni complex of mineral deposits has about 2.8 million tonnes of ilmenite. 1.0 million tonnes of rutile and 0.6 million tonnes of zircon. They occupy an area which is about 3 km long, 2 km wide and are generally 25 to 40 m deep. First the Iimenite contains up to 47.9% titanium oxide. Iron contents is also high being about 51.1% and there are low levels of calcium, magnesium and manganese. Secondly the native is a high-grade source of titanium containing about 96.2% of the metal, finally zircon in Msambweni contains about 66.0% of zirconium' (Tiomin I, p. 426).

4.3 *The preliminary considerations*

High Court Judge A. Hayanga first considered the procedural question of whether traditional common law rules and principles for the grant of temporary restraining injunction orders apply to environmental disputes or whether EMCA was an exclusive piece of legislation. Relying partly on American jurisprudence, the judge ruled that both the common law as well as the Mining Act—to the extent of any inconsistency with EMCA—have been superseded by EMCA (Tiomin I, p. 430). This follows because EMCA is enacted later in time. Therefore under rules of statutory interpretation, parliament must have intended an implied repeal of previous enactments or laws (Tiomin I, p. 433). Consequently, the court invoked EMCA[17] to confer *locus standi* upon the squatters. That provision expressly empowers any person to bring an action to safeguard the environment notwithstanding lack of proof of ownership of the affected land. Second, in addressing the question of which legal regime to apply, the judge distinguished the provisions of the Mining Act from EMCA. Essentially, the purpose of the Mining Act is to advance economic considerations which include vesting of minerals in the government and to confer authority upon developers to develop mines. Conversely, by introducing, *inter alia*, the health and safety of the public, through insisting that prospective developers must internalise the social costs or impact of their project upon the environment, EMCA recognised the principles of sustainable development. Quoting extensively from John Leeson's *Environmental Law* (Tiomin I, p. 435 citing Leeson 1995), the judge explained the principles of EMCA. One, 'every person in Kenya is entitled to a clean and healthy environment and has a duty to safeguard and enhance the environment[18] (Tiomin I, p. 429 citing Leeson 1995). Two that sustainable development 'requires that decisions throughout society are taken with proper regard to their environmental impact.' This entails 'development that meets the needs of the present without compromising the ability of the future generation to meet their own needs (hence intergenerational and intragenerational equity)' (Tiomin I, p. 431). Three, the court noted that reference must be made to international environmental law because 'conservation of natural resources extends beyond immediate environment to global issues so that (certain) principles (are) to be observed such as: (a) Decisions to be based on the best possible scientific information and analysis of risk. (b) Where there is uncertainty and potentially serious risks exist, precautionary measures may be necessary. (c) Ecological impacts must be considered, particularly where resources are non renewable or effects may be irreversible. (d) Cost implications should be brought home directly to the people responsible in the polluter-pays principle, are considered in the report because such assessment and assessment, of a ray (sic) of disparate factors require evidence from EIA to support sound judgment,' (Tiomin I, pp. 435–436, brackets mine). The judge concluded that 'most environmental disputes are resolved by application of the principles of common law like law of tort, property, injunctions and principles of administrative law, but the applicable law is statute law which in this case is EMCA.' Hence, the licence under 'the Mining Act does not legalise what is unlawful under EMCA.'

4.4 *The law*

In applying the above legal principles to resolve the substantive dispute, Judge Hayanga faced a dilemma. At one extreme, because environmental destruction is harmful to public health and safety, environmentalism *per se* would apparently prohibit pollution altogether. Yet at the other extreme, commercial progress and development although necessary, inevitably alter nature.

For Leeson, the question of whether or not a project attains sustainable development, should meet the test of 'best practicable means and best available techniques not entailing excessive cost' (Tiomin I, p. 431 citing Leeson 1995, p. 34). Consequently, 'the application of this principle to existing activities precludes cessation of the business or process because of its environmental impact ... the definition (sic) and interpretation of the phrase is therefore important in determining the extent of the obligation to remedy and the consequent degree of pollution permitted in a particular situation.' Judge Hayanga held that 'the law requires developers to take certain steps to consider the environmental impact of their activities.' Hence, the question was whether the mining company has taken all practical steps to protect against danger. Furthermore, EMCA criminalises failure to submit a project report in addition to an EIA 'at its own expense'[19] prior to commencement of mining. Such offence is 'punishable by twenty-four months imprisonment.'[20] By definition, 'EIA means a systematic examination conducted to determine whether or not a proposed activity or project will have any adverse effects on the environment.'[21] Further the court emphasised that: 'The purpose of such EIA study and report is to enable the Authority to determine the effect and impact of the project on the environment.' Then 'the Authority after studying the project report' may determine whether 'the impending project is likely to have or will have a significant impact on the environment.' Subsequently, 'the Director,' upon 'being satisfied as to the adequacy of the EIA study evaluation or review report, issues an EIA licence on such terms and conditions as may be appropriate to facilitate sustainable development and sound environmental management.'[22] Alternatively, but curiously, if the Director-General fails to reply within three months of submission of the requisite reports then 'the applicant may start his undertaking notwithstanding but may need circumspection.'[23] Thus, no reply from NEMA, is deemed to infer tacit, albeit conditional, approval. A developer is then at liberty to commence mining. Fatefully, Tiomin Kenya Ltd. confessed to the court that it was yet to commence mining notwithstanding lapse of the three month grace period from the date it purportedly submitted its EIA. Therefore the judge reasonably presumed that such failure to commence mining as permitted under the 'official failure to reply option' implied that no accompanying project report, by Tiomin, had indeed been submitted thereby confirming the squatters' allegations and apprehensions (Tiomin I, p. 435). Moreover, the judge reasoned that it was not possible for NEMA to assess an EIA in absence of the detailed project report. Therefore, he concluded that the company's omission to submit a report, coupled with its poised commencement to mine amounted to a breach of EMCA which violations were not only criminal offences, but also warranted restraining orders.

4.5 *Injunction procedure*

Judge Hayanga considered whether the mining company's offending behavior attracted temporary restraining orders under three common law principles introduced into East Africa through the famous case of *Giella v Cassman Brown.*[24] First, he held that environmental degradation *prima facie* constitutes irreparable harm. Hence, in event of any injury in the absence of sufficient precaution, then, pursuant to the polluter-pays principle, the Tiomin mining company was likely to lose the suit proper and in addition to being restrained permanently, pay general damages. The court noted that the polluter-pays principle imposes an 'obligation on any person to conduct their affairs in an environmentally sympathetic fashion .... anyone conducting activity ought to be aware of and accept responsibility for the environmental consequences of that activity, with regards to sustainable development' (p. 437 citing Leeson 1995, p. 34). However, in departure from the common law, under EMCA significantly, there is no requirement for a plaintiff to establish that the defendant's action was likely to cause him or her personal loss or injury to material legal rights. Instead the court is 'guided by cultural and social principles, and principles of international co-operation, principles of intergenerational and intragenerational equity, polluter-pays and precautionary principles' (Tiomin I, p. 438). Furthermore, the squatters' claim also satisfied the second test for grant of temporary injunctions, namely that the threatened harm could not be adequately compensated by money damages. The court held that whether or not a project was injurious could only be determined upon assessment of the EIA in relation to the project report. Yet the mining company omitted to avail the court with its project report (ibid.). The circumstances in the Kenyan Tiomin I case were thus identical with the American case of *Sierra Club National Audibon Society* (Friends of the Earth Inc Association of Game Fish and Conservation Commissioners versus William T Coleman Jr Norbert Tiemann)[25] where a US court found that the developers (Federal Highway Administration) 'had started building highway before complying with NEPA requirements [Similar to our EMC]. Decision had not been taken on statement and such work ought to have began after decision-makers had fully adverted to the environmental consequences of the action.' In *Sierra* the US Court concluded: 'That this court agrees that when ... federal statutes have been violated it has been a long standing rule that a court should not inquire into the traditional requirement for equitable relief.' Similarly, Hayanga asked and answered: 'So the question to be asked is what environmental factors has the proponent of the project taken into account? None.' The third test for grant of temporary injunction is whether the balance of convenience favours such grant. This approach is analogous to an economic cost-benefit efficacy criterion to determine whether either the squatters or the company would suffer greater inconvenience upon the grant or refusal of restraining orders, assuming that such injunction turns out to have been wrongly issued. However, the judge considered that such third test for grant of a temporary injunction was inappropriate since the intended project could potentially result in global environmental hazards which were against the public interest. Hence indirect public interests were an overriding factor in favour of the court's decision halting the mining project (Tiomin I, p. 440).

In summary, in Tiomin I, the environmental interests (which the community may be construed as representing) obtained a temporary injunction against a mining company, to restrain a threatened public nuisance citing its failure to submit both a project plan and EIA study. To prevent a *tort* of public nuisance, the court applied reasoning analogous to the *rights* model of Aritho (2000, p. 80) which furthers Okoth-Ogendo's (1991) utilitarian environmental degradation considerations which the court termed the public interest. Instead of relying on common law moreover, Judge Hayanga's reasoning is based on EMCA which statute, by ousting the Mining Act, effectively elevates environmental administrative law above both common law and liberal property rights. However the issue of compensation payable to the aggrieved community did not arise since the court's ruling was made at an interlocutory stage of the proceedings rather than at the conclusion of the case. Hence, arguably, had the compensation stage been reached, then the tortuous method of assessment of damages is likely to have been applicable in order to indemnify the farmers according to projected damages estimated according to their own EIA. Worse still, the Attorney-General may then have been faced with the embarrassing dilemma of having to decide whether or not to prosecute an investor which the government had itself invited to develop a titanium mine. It is submitted that EMCA places the burden of proof upon the mining company to prove that its prospective project remains economically feasible while satisfying the authority's duty to 'facilitate sustainable development and sound environmental management.' This burden may work both in favour and against mining companies depending on the discretion of NEMA. Curiously, section 59 of EMCA confers upon the Director near unfettered discretion which latitude is contrary to administrative law principles (Craig 1989: chapter 2). Such vague provisions express a standard which assumes sophisticated decision-making to administer them according to the circumstances of particular cases. To this end, it is not clear what is meant by a developer 'proceeding but needing circumspection' in lieu of NEMA's response. Hence Okoth-Ogendo (1999, p. 55) criticised the reluctance of central government bureaucrats to devolve power to specialised authorities so as to facilitate objective decision-making according to scientific standards. 'Such organs are generally viewed with suspicion and hostility since they tend to usurp the powers and functions of more established agencies and bureaucracies'. Moreover, 'Africa's tendency to subject any parastatal organisation to ministerial supervision creates additional problems. Unless such organs are located in the chief executive's or cabinet office, issues of hierarchy, supervisory authority and enforcement jurisdiction are not easy to resolve' (Okoth-Ogendo 1999, p. 56). It is submitted that because of the punitive decision in Tiomin I and specifically because the court's 'blanket' injunction failed to distinguish the components of environmental degradation from community health within the EIA, therefore there remained room for successive and/or simultaneous proceedings so as to clarify the community interest. At worst if Tiomin (K) Ltd. were to partly comply with EIA requirement by preparing an EIA Report which satisfies NEMA regarding the natural resources expected to be extracted from the environment as a stakeholder, but without sufficient community involvement in the process, a 'mostly restorative' response would result.

5 MINING MANAGEMENT AND RESTORATIVE JUSTICE: TIOMIN II

5.1 *The facts*

In Nzioka versus AG[26] (Tiomin II 2006) the facts were as follows: In 1995 Tiomin Resources Inc., a Canadian corporation, through a wholly-owned local subsidiary Tiomin Kenya Ltd. sought to invest in the Kenyan mining industry to mine titanium. Under an exclusive prospecting licence, it approached various occupiers to survey their land parcels along the coast, including 'the disputed land' at Kwale district south of Mombasa (Tiomin II, p. 5). Upon striking titanium, the company entered into leases with the squatters upon terms stated in section 4.1 above. Subsequently, the government issued a special mining licence

(Tiomin II, pp. 6–7). The government then felt insecure with the short-term leasehold interests. It instead insisted that it became necessary to either re-negotiate an outright purchase of the mining land from the occupiers or compulsorily acquire it. In the political background, prior to the 2002 general elections, title deeds were issued to the squatters. Should title to such private land become vested in the government, then this would facilitate granting a longer-term lease to the mining company. Hence, the government established a Cabinet sub-Committee and later a District Resettlement and Compensation Committee which engaged a landowners' Farmers Committee in re-negotiations. A contract of sale was entered into where the government offered Kshs 80,000 (approx. US$1000) per acre, lumpsum per farmer, together with an additional sum towards extras such as crops and other developments. Several farmers however, remained dissatisfied with the sale price. In 2006 the dissatisfied farmers petitioned the constitutional court challenging the government's move to negotiate a contract for purchase of the private land on behalf of a foreign mining company. The dissatisfied landowners alleged first, that the government and its agents, officials and servants were colluding with the Tiomin mining company to effectively compulsorily acquire their private landholdings without following the correct constitutional and legal procedure. They averred, second, that such clandestine maneuvers contravened their property rights.[27] In addition to invoking violation of property rights, they further alleged, third, that the government had forcibly entered upon their land thus trespassing and violating a litany of constitutional rights to liberty,[28] freedom from servitude[29] and from torture or other cruel inhuman and degrading treatment of punishment,[30] to privacy[31] and from protection against forcible entry and their right to movement.[32] They prayed for several reliefs including a permanent injunction restraining the purported sale, general and exemplary damages (Tiomin II, pp. 5–6). The government through the Attorney-General, Kwale district commissioner and his district officers, the Commissioner of Mines and Geology, the Commissioner of Lands as well as the mining company and its managing director in reply demonstrated that the majority—75% of the affected farmers—had duly executed the sale agreement. Furthermore, at no time prior to the constitutional reference did the petitioners lodge an official protest to disown their Committee's capacity to represent them.

5.2 *The law*

High Court Judge J. Nyamu rejected as an afterthought, the dissatisfied landowners' suggestion that the government was not entitled to buy the disputed land. Instead the eight petitioners, comprising a minority of the overall 60 farmers—45 of whom had already signed the sale agreements—were estopped from disputing their Farmers' Committee's legitimate mandate to represent their interests and negotiate and contract on behalf of all farmers on the land. In any case, the court held that even assuming that the government had forcibly expropriated their land, then the correct procedure to ventilate their grievances would lie in an application for judicial review of administrative action—not in a constitutional petition. Neither does the Land Control Act require that consent should be obtained from the Land Control Board to sanction land transactions by the government.[33] Not only were the minority landowners' property rights found to be fully respected and intact (Tiomin II, pp. 11–12), but also, judge Nyamu dismissed suggestions that the government violated any constitutional rights as alleged. Instead the judge recognised that 'the real objective of the Petitioner is to obtain for himself an award of a higher compensation than what is contained in the compensation package and if there is no agreement the petitioner would like the court to award more than what is being obtained hence the emphasis on the right of access to this court' (Tiomin II, p. 7). Indeed, the first petitioner, Rodgers Nzioka—owner of title deed Kwale/Maweche/232—said that: 'I instructed Mr. Aritho Gitonga, a valuer to undertake a valuation of my said property for the purposes of renegotiating the purchase price' (Tiomin II, p. 8). Even assuming that admission was true, then again the court held that appropriate legal remedies were available before the ordinary civil courts by way of a suit under statute law regulating the law of contract or common law tort—not in a constitutional reference. Moreover, if the petitioners did not consent to their representation by the Farmers Committee then they bore the burden of disassociating themselves from its actions at the earliest possible stage of negotiations. As it stood, the judge presumed that the minority farmers had constructively consented to be bound by the decisions articulated by their committee. For: 'Although the petitioner had two choices he elected to be bound by an out and out negotiation to purchase his land and he opted out of the compulsory acquisition choice (which apparently he is also now opposing in these proceedings)—he cannot approbate and reprobate' (Tiomin II, p. 8). Moreover, acting on the promise made by the Farmers' Committee, the Tiomin mining company had first, obtained development funding from a financier. Second, engaged a contractor. Third, conducted the relevant EIA. Fourth, obtained a special mining licence. Fifth, paid Kshs 3 billion (approx. US$37,500,000) to the Kenya government out of which, sixth, some of the majority farmers had already received payment. Seventh, the government located a host site to resettle the landowners (Tiomin II, p. 9). In sum the minority farmers were bound by the valid contractual commitments made by the majority and were estopped from injuncting, varying or rescinding the transfer process.

5.3 *The judicial rationale*

In Tiomin II, the government chose not to compulsorily acquire the farmers' land using administrative law. Instead, it entered into private contracts with the farmers by actualising Aritho's (2000) *purchase* model. Significantly, only after the court had punitively restrained the mining company, in *Tiomin I*, was the *purchase* model (Aritho 2000, p. 80) preferred. This confirms the abovestated critical 'reason why Kenya's ... Environmental Management and Co-ordination Bill (was) steeped in controversy' (Okoth-Ogendo 1999, p. 56). The government wrongly purported to divide NEMA's duties. Consequently, it separated the 'environmental management' component of NEMA's duty under section 59, from the community involvement component in attaining sustainable development. This aspect of the court's decision adopts a strong Kantian philosophy, similar to Ojwang's embracement of *self* as the person-centred basis of rights-creation, to the extent that Judge Nyamu upholds the maxim that '*pacta sunt servanda* (agreements must be honoured)' (Ojwang 1996, p. 44). On one hand, thus Judge Nyamu confers a narrow, restrictive reading to the constitutional protection against property violation. A feeble attempt at high social control is displayed through mild rebuke of the Commissioner of Lands for initiating compulsory acquisition notices during the pending petition proceedings. Yet, the court ultimately on one hand, firmly rejected the minority farmer's claim for any remedies regarding alleged violation of their constitutional rights. The court's jurisprudential interpretation appears to dismiss this constitutional claim because the dissatisfied landowners appear to base their assertion on Lockean *natural rights* (Locke 1823, section 27). Hudson (2003) commends Locke's idea of property for the advantage it offers because 'it does subject forms of property such as land, goods and wealth to a principle of equal compatible freedom, alongside life and liberty.' On the other hand, the court then departed from Kant's person-centred natural rights morality, and instead embraced economic utilitarianism, in stark contrast to Okoth-Ogendo's environmental determinism. This aspect of Nyamu J's decision resembles South African jurisprudence where he, categorically insists that 'fundamental rights and freedoms are guaranteed by the Government to individuals and are vertical not horizontal (following *Richard Nduati Kariuki v R*[34] citied in Tiomin II, p. 16).' Hence to defeat the minority landowners' constitutional claims to fundamental rights violations, and instead uphold a theory of economic developmental values conferred by the land purchase contracts, the court appears to endorse a utilitarian *rights-creation* position. The judge's cost-benefit analysis prioritised the public interest as the mining project would increase social welfare in a multiplicity of spheres. First: 'It is the largest investment in the minerals and mining sector.' Second through: 'Generation of Government income and increase of GDP of about 1%.' Third by: 'Creation of wage employment in the Coast Province of about 2–3%.' Fourth upon: 'Development of mining skills and competence.' Fifth through the: 'Improvement of infrastructure.' Sixth the: 'Development and creation of social services and a modern settlement scheme' (Tiomin II, p. 10). It is noted that the court's definition of sustainable development is broader than that of conventional economics which relies on increased GDP as the only criterion without reference to equitable parameters (Mikesell 1993, p. 5). To further oust the basis of the minority farmers' constitutional claims, the judge asserted various common law doctrines such as estoppel, limitation of actions, waiver, or compromise' (Tiomin II, pp. 12–3). Moreover, the minority landowners did not allege that their contracts were vitiated by factors which may render unfair contracts voidable, such as economic duress, misrepresentation or undue influence (Freidmann 1995). Hence the landowners were faulted for invoking the 'wrong' procedural jurisdiction as they may have correctly prosecuted such unfair contractual claim before ordinary civil courts. The aggrieved farmers appear to have experienced technical difficulty in identifying or articulating minority interests, where some farmers may have believed that they were forfeiting cultural heritage to ancestral land to which they had become emotionally attached. Perhaps their failure to expressly claim compensation for loss of heritage resulted from the fact that the current constitution did not expressly safeguard cultural rights as part of the public interest. Neither does Kenyan legislation enshrine the rights of indigenous peoples. Onalo and Bhalla each explain how section 75(1) of the previous Kenyan Constitution enshrined the eminent domain doctrine. Yet 'property as an absolute right has never existed' because 'the social control of property is exercised by putting restrictions on the use of property, and by compulsory acquisition' (Bhalla 1996, p. 67). The Schedule to the Land

Acquisition Act[35] does not permit Kenyan courts to consider subjective factors in computation of compensation awards. In addition to defining the meaning of full compensation[36] the Land Acquisition Act lays down procedures for giving notice and determining the public interest (Onalo 1986, pp. 57–59; Bhalla 1996, pp. 64–69). Hence Mikesell (1993, p. 76) argues that 'The issue of government regulation presents a problem for sustainable development economists, many of whom also believe in the efficiency of economic competition, including free trade and incentives, and share the distrust of big government and state enterprise' ... the dilemma arises in how to preserve a liberal economic order while at the same time preserving a natural resource base for future generations, and requiring private entities to take responsibility for external diseconomies that they generate.' estimates of harm to community health or cultural land values would reduce the NPSV of projects. Ultimately because the land transfer process in the Tiomin II case was deemed to be voluntary, rather than compulsory, therefore the government was not obliged to consider an award of 15% mark-up or *salotium* as payable to expropriatees in addition to the market price towards disturbance.

6 OVERCOMING STRATEGIC HOLD-OUTS THROUGH COMMON LAW AND EQUITY

6.1 *Strategic bargaining by environmental guardians*

Regarding s108 of EMCA, Kibugi (2008, p. 361) argues that politically: 'Interpreted liberally this provision empowers NEMA to require mandatory resettlement plan, where involuntary displacement is identified as an environmental and socio economic impact'[37]. Yet from an economic conservative perspective, one effect of the Kenyan court's decision in Tiomin I is that it delayed the mining company from initiating its investment enterprise and reduced its profitability since it entailed postponement of mining works pending preparation of an EIA, presumably at considerable expense. An appreciation of the post-Tiomin I bargaining position may be gained from environmental economist A. Alan Schmid's consideration of the US case of *Boomer versus Atlantic Cement Co*[38] where the court granted a temporary injunction which would be removed only after the plaintiffs' damages were paid. The court noted that third parties may have been damaged but since they did not join the suit, the court ignored them. Schmid (1995, p. 47) points out that the court further noted that a permanent injunction would have been too drastic since it would have resulted in the immediate closure of the plant. Yet, according to him, closure seems unlikely if the parties could negotiate since the plant could buy air rights from the neighbours if the value to the neighbours was less than the value to the plant. A dissenting judge argued for a permanent injunction on grounds that air pollution was a 'wrong.' That minority decision was based on a natural law. Significantly, only when the state indicated that it would enjoin the phosphate firm did the firm begin to buy citrus land at a price above its present net worth in citrus (Schmid 1995, p. 55 citing Crocker 1971). Schmid recognises that: 'Technological change often produces great benefits on the average but great costs (real or imagined) for some ... Calabresi calls these "gifts of the evil deity" because while producing benefits they may cause deaths and disability at random' (Schmid 1995, p. 51 citing Calabresi 1985). However, Mikesell (1993, p. 27) gives 'two reasons for not including technological progress in the calculation of the resource base. First, we do not know enough about future technology to assess its impact on the productivity of resources used in present projects. Second we should allow technological progress to have its full effect of improving the well being of future generations'. Consequently, 'the social costs of the depletion of natural resources attributable to a project is based on current technology.' Ultimately governments—through public policy must decide whose perception counts—the communal or commercial? In Tiomin I, when the court indicated that the Tiomin mining company was likely to lose at full hearing, from then on the company became indebted to the landowners who became its creditors. He would argue that it became necessary for the mining company to find the cheapest means of buying—not only the land—but also the injunction liability. This raised the price of the mining land above its pre-litigation value to the farmers (Schmid 1995, p. 55). Why would Tiomin pay a premium for the land? Schmid (1995) would answer that because after the Hayanga injunction, the landowners (in Tiomin II) became aware that the value of the land for mining is greater than for habitation and could bargain for part of this surplus. Moreover, they could take advantage of the economising effort by the mining company to avoid the high transaction costs associated with strategic bargaining (Schmid 1995, p. 56).

6.2 *Contractual consent*

'Cultural Theory's ideas of fairness ... predict a willingness, among individuals and hierarchists, to go along with the notion of *pareto optimality*, that is, an outcome in which winners can compensate losers and still be better off than they were, is seen as preferable to the *status quo*. But egalitarians, Cultural Theory suggests, prefer the opposite: a situation in which people are less well off but more equal' (Thompson 1998, p. 203). However, as Mikesell (1993, p. 77) notes 'there has been less acceptance of government control over the use of natural resources to protect utilities provided by private property that are commonly shared. 'Equally important is whether the state is exploiting natural resources in the public sector in a way that maximises social benefits, as contrasted with using resources in a way that serves the interests of private industry e.g. mining ...' (Mikesell 1993, pp. 77–78). If the miners of titanium accept to account for the social cost for their activities then the titanium output would be less. Yet decisions depend on the moral perspective of the maker. Hence in a private property system the owner of property rights determines user of objects by creating costs for others. The stronger parties, unless controlled, would tend to disrespect the weaker (Schmid 1995, p. 55). Therefore the latter proposes that one alternative available to the offending company if it chose not to reach a prior agreement with the farmers in bargaining over their strategic position would be to proceed with the mining operation on the pain of criminal punishment, including contempt of court. In *the Florida Phosphate case*, he argues that 'payment of court-assessed damages is an alternative to the permanent injunction.' In this case the phosphate firm does not have to reach a prior agreement with the farmers but makes use of the resource and pays afterwards as required. This can be referred to as "private condemnation" since it parallels public condemnation where the state need not meet a willing seller.'

Clearly, such punitive option in the *Tiomin cases* would prove counter-productive to all stakeholders' needs since, in monetary terms, upon the state 'stealing the conflict,' the damages awarded may be much less than what parties may agree or even the land's market value. The alternative of compulsory acquisition is also problematic. This is because, Schmid (1995) argues, that where regulation such as the compulsory acquisition power is a source of the mining company's opportunity to enjoy a right, 'it is fundamentally different from a tradable right as the non-right holder may not legally offer money in exchange for the right.' Moreover, the government is not a disinterested bystander and would be irresponsible if it failed to facilitate either the moral choice of parties or the maximisation of social welfare.

In McCold & Wachtel's (2000) restorative practices typology, the government is an indirect stakeholder in the dispute. The government may consider it to be against public policy for the landowners to take strategic advantage of their mineral-rich land and simply use high support for the mining company, by acquiring the land compulsorily in the public interest of advancing development and progress. According to Lesson's test, provided an EIA is forthcoming from the mining company following such high control, then the government may reciprocate with high support to allow the mining project to proceed since 'for the best practicable means one would like to consider whether one has or can do what is practicable in terms of prevention or reduction where the defendant has discharged the obligation bestowed on him the nuisance or pollution may be allowed to continue' (Tiomin I, p. 437 citing Leeson 1995, p. 34). Hence Lubbe (2006) would argue that in Tiomin II, an agreement appears to have been reached between the government and the mining company which involved the government taking over the liabilities owing from the company to the landowners and offering a price in consideration for the land which would obtain consent for release of the developers from their liabilities under the temporary injunction together with transfer of land to vest in the government for onward longer-leasing to the mining company (Lubbe 2006, p. 783). Such tripartite agreement involving transfer of contractual rights and obligations is called a *novation*, according to the decision of the court of Appeal for Eastern Africa in *Settlement Fund Trustees v Nurdin Ismail Nurani*.[39] Presumably, valuable consideration must have been paid to the government for its taking-over of Tiomin's liabilities. Yet a few disgruntled landowners objected. Thus while, on one hand, *the Tiomin cases* may have reflected a restorative process—according to Marshall's definition—on the other, the case may be seen as entailing a moral decision which may be evaluated according to how restorative the outcome was, under the restorative practices typology. Outstanding questions which arise are: Given that the project was worth Kshs 12 billion (*Tioimn II* 5). How were the non-monetary social costs estimated to arrive at Kshs 80,000 plus extras per farmer? Yet no information is available from the court judgment to indicate how the NPSV of the mine was arrived at, whether using estimates acceptable to the community valuer or scientifically objective values and if so then what proportions of revenue were allocated towards environmental damage and for community harm? Was this a case of brow-beating? Is Judge Nyamu's presumption that the petitioners had the election of either negotiating with the government or face compulsory acquisition warranted? It is true that the sale price in Tiomin II was double the Tiomin I offer. Nonetheless, by the very nature of compulsory acquisition, communities have no control over the process. In reality, by threatening to saddle communities with the statutory market value plus 15% salotuim, the government's hard bargain effectively pressured the community to settle for the 'negotiated' contracts rather than face the uncertainty of possible prolonged litigation to review the Land Commissioner's administrative action. To what extent did the Kenya government apply McCold & Wachtel's (2000) 'collaborative hypothesis' to resolve the successive *Tiomin cases*? Did the minority farmers voluntarily consent to sell their land as required by contract law? Or can common law estoppels extend from private law to upgrade mere negotiations into binding contracts thus defeating constitutional claims to property and freedom (Lubbe 2006, pp. 761–771)? Were the minority farmers—as victims of expropriation—entitled to strategically hold-out for a higher price? Or were they bound by majority collective consent? McCold & Wachtel's (2000) stakeholders needs structure suggests that victims need to be involved and empowered so as to become transformed into survivors?

In contract law, the principle of autonomy requires that contractual liability again depends on the consent of the parties. Going by the court's decision in Tiomin II it appears however, that the Kenyan courts are moving to recognition of the need to qualify the traditional notion that parties in general are free to withdraw from negotiations. In case of disagreement estoppel might result in liability even in the absence of agreement in order to protect induced reliance (Lubbe 2006, p. 755 citing Freidmann 1995, p. 399, 406). Yet under common law, where an actual intention to enter into an agreement is established, then liability accrues under contract law, there is no need for estoppels. Judge Hayanga presumed that non-commencement of mining by the company was tantamount to its failure to comply with EMCA conditions. In this respect, Leeson's test appears to balance the requirements of Mikesell's (1993) economic feasibility study and the environmental interest in an EIA study. Hence, mining interests hold sway in the Tiomin cases. To this extent the net effect resulting from the respective punitive and permissive decisions, is to partly accept but partly reject Ojwang's environmentally-friendly conclusion. He stated that 'bearing in mind the gravity of the present environmental crisis and the fact that it is likely to seriously undermine the material supports to human survival and comfort, ... the scales should be gradually titled against those rights and interests that rest mainly on property values and in favour of a more planned social life and use of natural resources' (Ojwang 1996, p. 58). Rather a valid sale contract between the government and landowners may be relied upon by third parties—including the mining company, financiers, contractors and others who can prevent the landowners from denying their intention to transfer—since such third parties are entitled to act on the basis of a valid contract. However, estoppels would operate only as between the government and landowners (Lubbe 2006, p. 784).

6.3 *Promissory estoppel and apparent authority*

The doctrine of promissory estoppel is found in English precedents, handed down by colonial court decisions and inherited into post-independent Kenya. The modern understanding was stated by Lord Denning in the famous case of *Central London Property Trust versus High Trees House Ltd.*[40] In Canada: 'The principle has been well defined in the case ... *Newborn versus City Mutual Life Assurance Society Ltd*[41] as follows: 'But, even so, yet another element is required to make out the estoppel. The reason for precluding a party from relying upon an actual state of affairs as the formulation of his rights lies in the injustice of permitting him to depart from some contrary assumption if another party has based his conduct upon it' (Tiomin II, p. 9). Estoppel is a rule of evidence and not substantive law. It functions as a shield, rather than a sword. Estoppel creates a legal fiction and generates uncertainty. Thus for natural law judges, such fictions operate as a rule, not a guideline (Khamala 2005, p. 109). Curiously, faced with two conflicting fictions, one that 'promises must be kept' the other that 'persons should not renege on a promise which another has relied upon even though the promise is not supported by consideration,' Nyamu upheld both. Recall that Judge Hayanga was earlier prepared to presume that a developer's failure to mine—in circumstances of official silence to its licensing application—is tantamount to 'failure to submit an accompanying project plan.' Because EMCA imposes a passive duty on NEMA, therefore the legislation effectively allows the authority to ignore applications for EIA licences in order to exercise hierarchic, high control over indecisive developers. Conversely Judge Nyamu's successive decision in Tiomin II reflects simultaneous high support for the individualistic mining company. The court's support for the mining company is evinced by facilitating low community 'involvement' thus preventing the third direct stakeholder from, 'collaborating' with the government—through the Hayanga injunction—using Christie's conceptualisation, to 'steal the conflict' from the two principal stakeholders, the environmentalists, as victims and the mining company, as offenders, the parties in Tiomin I.

Indeed, the principles of ostensible or apparent authority find important application in the law of agency and partnership. Agents and partners are generally regarded as having capacity to bind their fellows by means of contracts (Lubbe 2006, p. 785).

On the termination of an agency or partnership the fiduciary duty or mutual mandate, respectively is terminated. Thus, it was essential for Nzioka and the minority farmers to disassociate themselves from the Farmer's Committee. Instead Nzioka himself was a member of the Committee which brokered the agreement. It therefore appears that his remedy, if any, lay in suing his partners if they exceeded their mandate. Yet in order to enhance their bargaining capacity, the minority landholders apparently depended upon the wider community group for solidarity perhaps to enhance their negotiating capacity. Hence, the court held that the Mining Act provided an appropriate remedy for alleged violations to constitutionally enshrined private property rights. Judge Nyamu held that: 'Indeed, the right to receive compensation as per section 75 is still open and if aggrieved, the applicant can articulate his grievance under of the Mining Act to seek compensation therein. Any challenge must be within the scope of the Mining Act, Land Acquisition Act and s 75' (Tiomin II, p. 12). Moreover, 'the contested Agreements and Leases have arbitration clauses and the principle of party autonomy must be honoured by the court. The validity of the signed agreement is a matter for the arbitral tribunal and not for this court' (Tiomin II, p. 14). By inclusion of the Farmers' Committee it can be concluded that the government fully respected the autonomy, dignity and identity of all the farmers and indeed their voluntary choice was to effectively waive their injunction thereby releasing the mining company from injunction liability in acceptance of the purchase price offered by the government for their land. Finally the judge was ready to uphold the government's fall back onto compulsory acquisition procedure as being been validly invoked. In this regard Judge Nyamu recalled his earlier decision '*in Booth versus Mombasa Water Company*[42] this court observed (citing *Labshons versus Manula Haulers Ltd.*[43]).... in the articulation and enforcement of fundamental rights this court cannot disregard the fundamental principles of law such as *res judicata*, limitation of actions, laches, and if I may add to this list, estoppels, waive and compromise' (Tiomin II, p. 12). Lubbe (2006, p. 788) concludes that: 'In such cases, estoppel could be seen as a manifestation of delictual liability for misrepresentation entailing either a liability for damages or a making good of the representation as between the parties as an exceptional delictual remedy or a combination of both of these remedies'.

6.4 *Public policy and constitutional rights*

In Tiomin II the Nyamu court was of the view that: 'Our Constitution does assume that there are also fundamental principles of law in existence although the constitution does so to speak occupy the position of the superstructure. That superstructure has a foundation and one of the foundations is the existence of fundamental principles of law. It is reckless for applicants to ignore fundamental principles of law when filing or articulating constitutional applications. Many of the identified principles of law are based on public policy principles of fairness and justice' (Tiomin II, p. 13). On one reading, Judge Nyamu appears to embrace a position similar to Ojwang's ideology that the African condition is in a state of construction and therefore requires top-down development so as to enhance is material base. However, Nyamu J also insists that such development must be rooted in fundamental principles of law. Most significantly the judge indicated that he was not guided by what Okoth-Ogendo (1999) puts forward as customary rights or history alone. Instead the judge insisted that: 'To me we must give it much more.' Upon striking out the constitutional application he asserted that 'Our Constitution is not a cloud that hoover over this beautiful land of Kenya—it is linked to our history, customs, tradition, ideals, values and on political, cultural, social and *economic* situations. Its dynamics and relevance is deeply rooted in these values' (emphasis added). Hence it is evident that Judge Nyamu's ideology of utilitarianism expresses mainly liberal economics, albeit partly compatible with sustainable development. Relying upon the Privy Council's decision in the English authority of *Harriksoon versus Attorney General*,[44] he asserted that: 'The notion that whenever there is a failure by an organ of the government or a public authority or public officer to comply with the law this necessarily entails the contravention of some human right or fundamental freedom guaranteed to individuals by ... the Constitution is fallacious.... if it is apparent that the application is frivolous, vexatious or an abuse of the process of the court as being made solely for the purpose of avoiding the necessity of applying the normal way for the appropriate judicial remedy for unlawful administrative action which involves not contravention of any human right or fundamental freedom' (Tiomin II, pp. 13–14). He concluded that 'the court does frown upon any practice to trivialise this important jurisdiction. Sadly this appears to be the case here. The applicant's case is grounded on sinking sand. It is built upon an estoppel. No constitutional application can sustain itself in this way. The application cannot stand' (Tiomin II, p. 12).

6.5 *The 2010 new constitution*

The new Constitution of Kenya elevates the right to a healthy environment to an enforceable[45] fundamental freedom.[46] Moreover, it[47] further incorporates significant principles emerging from the Tiomin jurisprudence. Not only do minerals remain vested in the government,[48] but also, under, Parliamentary ratification of transactions for exploitation of natural resources is necessary. Principles[49] prescribe, *inter alia*, equitable sharing of accruing benefits. This requires that enabling legislation[50] to effectuate Chapter Five provisions, should introduce an obligation upon developers to prepare a Social Impact Assessment which evaluates cultural values where Community Land[51] is purchased by the government.

Institutions established pursuant to the new Constitution—appear likely to favour indigenous or group claims to native land rights over individual rights to private property. First, establishes a National Land Commission inter alia "(e) to initiate investigations, on its own initiative or on a complaint, into present or historical land injustices, and recommend appropriate redress."[52]

Second, a person who is not a citizen may hold leasehold land for a period not exceeding 99 years.[53] Ultimately, enshrines protection of right to property[54] subject to citizenship.[55]

7 CONCLUSION

This paper distinguishes three separate and complex conflicts emanating from mining projects. First, between mining companies and environmentalists, requiring limitations on private property rights by restoring the environment into its pre-mining state while allowing for 'fair wear and tear' by extraction activities. Prediction and assessment of damage and required repairs entail scientific EIA studies. The difficulty entails how to obtain objective estimation of environmental depletion variables given reluctance of mining companies to invest in expensive EIAs prior to contract. In answer to this dilemma we agree with Thompson (1998) that: '*Of course* we need the commons; *of course* we need the markets. It is never a matter of one or the other. Nor (and this is the crux of the complex—not simple realisation) is it ever a straightforward, two-fold allocation: this to the commons that to the markets. Indeed environmental crimes are polycentric. Hence the need to involve other stakeholders whose legitimate concerns, in all fairness, should be factored into the NPSV of mining projects.

Thus a second dispute involves environmentalists and communities regarding the nature of public interest to prioritise in the user of the land. Whereas the community may require their land for habitation, environmentalists, using Leeson's test may determine that once all practicable measures have been taken at a reasonable cost, by a mining company to manage environmental hazards, then mining interests hold sway. Communities may be re-located to prevent harm to their health but cannot prevent sustainable development from being undertaken in the wider public interest. Natural resource allocation may generate pressures for regional autonomy or recognition of self-identity, by communities seeking ecological stewardship. Yet the community interest is difficult to articulate given problems generally afflicting consumer organisations (Cranston 1978). They are disadvantaged not only in terms of lacking access to information but also devoid of organisational capacity. We join Ojwang's (1996) call for a freedom of information Act to facilitate the right-to-know thus enhancing community participation. It is necessary to interrogate what 'Dichiro calls 'colonial nature talk' or discursive strategies that separate peoples colonised histories from their naturalised environment. By contrast people-of-colour activists are bridging the ecological with social justice concerns, working through the organising of fragmented and alienated communities' (see Goldman 1998).

A third dispute pits mining companies against local communities whose land is required for purposes of access to minerals. In this regard, the: 'Intervention in land law, particularly that related to mineral rights, prevails throughout, the world. Private activity is restricted and in many cases prohibited' (Gordon 1995, p. 529). The Kenyan draft Mining Bill 2007 is no exception (Kibugi 2008, p. 369). Nevertheless a problem of valuation of the land for full compensation of expropriatees arises for determination either by agreement, under contract law or upon compulsory acquisition, under administrative law. In the latter scenario, compensation, at market value ignores the community's cultural attachments to land and may foment discontent by dissatisfied landholders. Hence for Calabresi (1985) the decision to conserve natural resources for future generations is a social decision not a private one. He observes that 'what is deemed unreasonable behavior, no less than who is the cheapest avoider of a cost, depends on the valuations we put on acts activities and beliefs by the whole of our law and not on some subjective scientific notion.' … 'What is efficient, or passes a cost-benefit test is not a "scientific" notion separated from beliefs and attitudes and always must respond to the question of who we wish to make richer or poorer'.

Tiomin I reflects high control of the offender while Tiomin II depicts high support. The net effect is restorative. The crucial question in both the Tiomin cases was whether some minority individuals could claim violation of their constitutional rights upon negotiation by the government with some community representatives to purchase communal land on the majority's behalf. At successive or parallel proceedings, Kenyan courts in Tiomin II rejected such notion, describing it as an opportunistic attempt to re-negotiate valid contracts through forum shopping. The court effectively implied sound that environmental governance demands that the 'project assessment processes should incorporate the *subsidiarity principle* that: anything that is being done (or could be done) at a lower level should not be done at a higher level' (Thompson 1998, p. 221). Hence, the 'micro' contracts were upheld and indeed, reinforced using the equitable doctrines of estoppel. In sum, because restorative justice addresses stakeholders needs in three dimensions, shown by the Social Discipline Cube, therefore, collaboration between mining companies and environmentalists to establish an EIA with respect to environmental management, though efficient, is not sufficient. Community involvement is integral to a fairer, more equitable process. Moreover, in the context of superior government bargaining power against relatively illiterate, vulnerable and poor communities which require equal protection to confer fair competition would require courts to recognise and correct the market distortion.

The design of NEMA conferring vague project application procedures without effective community participation is a hierarchic, bureaucratic framework. Such 'meso' institution should be democratised so as to require consultation with affected communities. Instead a *clumsy* institution has the merits of flexibility by not privileging any of the four myths of nature. However in the context of a developing country which may lack sophisticated administrators to delicately balance community interests against the competing environmental and economic interests. Rather than Parliament, being a legislative organ, involving itself in direct oversight, there is need to establish an oversight institution. By underpinning the community participation requirement through community consultation programmes (Thompson 1998, p. 223), it is possible to introduce an independent check over administrative bureaucracy. Therefore restorative justice procedures should provide for mediation not only in the EIA approval process but also for restorative outcomes which can transform stakeholder behaviour. To ensure decency in the character of negotiations, the right of appeal by dissatisfied individuals to court should be enshrined. Hence community organisations need not be restricted to negotiating lumpsum down payments to vacate their land from the anticipated mining revenues. Rather they may insist upon maximising benefit-sharing so as to share-benefits equivalent to the annual income of the mining company, in perpetuity.

8 RECOMMENDATIONS

It is recommended therefore that at the meta governance level, first, that during the constitutional review process,[56] Kenyan public law should be reformed to facilitate community involvement in mining negotiations to enable all affected individuals not merely to hold dissenting views, but to express them and act upon their opinions. This may entail express incorporation of contentious community land rights in the new constitution to reverse the colonial discrimination against African cultural values (Juma 1996, p. 373). In addition to an EIA, a distinct Social Impact Assessment (SIA) study (Barrow 2000) should be prepared using prescribed criteria such as making of provision for all individual community members, including non-landholders such as women and children who may have benefited from the communal lifestyle attached to the land. Such SIA may form part of the 'environmental depletion' component in evaluating the NPSV of a mining project. The National Land Commission under the proposed new Constitution appears to embrace this function.

Second, that if there is justification in the environmentalists' claim that liberal economic values necessarily relegate or threaten to extinguish environmental gains (Okoth-Ogendo 1999), then it is for restorativists to define clear boundaries between ecological depletion and community displacement. By involvement of the third direct stakeholder in natural resource disputes, the 'communities of care,' liberal, economic interests of mining companies may be preserved, albeit constrained, rather than mining projects being nullified or frustrated in entirety. Mining companies are able to predict depletion, take precautions and guarantee clean-ups as well as make provision for future generations while nevertheless remaining reasonably profitable ventures.

Third, at the 'macro' level, political will should be mobilised to establish community trust funds to ensure that community representatives on their part to equally guarantee that any accumulated funds from mining projects are applied for the benefit of future generations and not misappropriated. This complex institution-building can benefit from capacity-building through technology transfer from mining companies to facilitate future reversion of mines to indigenous communities upon expiry of mining leases, assuming the mine is not exhausted, or that new technology emerges to exploit other minerals at the mining location. The vesting of community land in county councils under the proposed new constitution seems predicated on increasing benefit-sharing.

Fourth, two assumptions must be debunked. One, even if leaving natural resources in their pristine form were not considered as 'stagnation' but 'hospitality,' (Thompson 1998, p. 203 citing Goldman 1998) nevertheless, such neglect involves a trade-off against any potential advantages derivable from potential present uses, such as sale of land for access to extraction. It is not possible to have both community developments while simultaneously remaining rooted to the land. Indeed throughout civilisation, mankind has always migrated and adapted new cultures in order to survive. Two, a policy of exclusivity over natural resources also has its risks. While Africans may benefit from retaining dominion over our own natural resources, we must consider whether by embracing fatalism we may simultaneously become disentitled to claim our fair share of the global commons. For example, suppose the global commons in the fullness of technological innovation, turn out to harbor more natural resources than historically known in the privates? Individualism inspires development of human capital, a prerequisite to sustainable development. Optimum resource use would imply facilitating access or use of natural resources to deserving applicants. Indeed, a liberal policy—by encouraging repatriation of profits—places premium on competition of ideas which, while increasing risks, provides the means of stimulating incentives for innovation and generates capacity to develop. If a flexible approach to resource exploitation is deemed as a valuable procedure then global stewardship would be required to take egalitarianism seriously. Environmental governance institutions should therefore not over emphasise hierarchic punitive approaches to environmental crimes but only invoke sanctions as a last resort. Focus should be directed towards facilitating wide community participation so as to accommodate minority, including customary, views in continuous selection of self-correcting options.

ENDNOTES

1 Act no 8 of 1999 (hereafter EMCA).

2 United Nations General Assembly, Universal Declaration of Human Rights (Paris, 10 December 1948) General Assembly resolution 217 A (III).

3 For example, the International Covenant on Economic Social and Cultural Rights Adopted and opened for signature, ratification and accession by General Assembly resolution 2200A(XXI) of 16 December 1966 entry into force 3 January 1976; International Covenant on Civil Political and Cultural Rights Adopted and opened for signature, ratification and accession by General Assembly resolution 2200A (XXI) of 16 December 1966 entry into force 23 March 1976; and African Charter on Human and Peoples' Rights Adopted 27 June 1981, OAU Doc. CAB/LEG/67/3 rev. 5, 21 I.L.M. 58 (1982), entry into force 21 October 1986).

4 Charter of the United Nations. 1945. 1995. In *The United Nations and Human Rights 1945–1995*. Department of Public Information. United Nations. New York.

5 Section 75(1) The Constitution of Kenya. Nairobi. The Government Printer. (1998). Modified by Article 40(1) in the new constitution 2010.

6 s 75(1) (a) ibid. new Art 40(3)(a).

7 s 75(1) (b) ibid. new Art 40(3)(b)(ii).

8 s 75(1) (c) ibid. new Art 40(3)(b)(i).

9 Land Acquisition Act (Chapter 295 Laws of Kenya).

10 regulation 9(3)(c) pursuant to s 8 ibid.

11 The Kenya National Land Policy, 2009: para 131.

12 18 LJ (Ex 119) cited in *Smith v Hughes* LR (1870–1) 6 QB 597.

13 [2001] KLR (E&L) 1, 423 (herein *Tiomin I*).

14 s 55, The Mining Act (Chapter 306 Laws of Kenya).

15 s 7(m), The Land Control Act (Chapter 302 Laws of Kenya).

16 Ibid.

17 s 11(3) supra note 1 cited in *Tiomin I*: 433.

18 s 3 (1) supra note 1.

19 s 58(2) cited in ibid 433–4.

20 s 138 cited in ibid.: 435.

21 s 2 cited in ibid. 434.

22 s 59 cited in ibid. 434.

23 Ibid.

24 [1973] EA 358 cited in ibid.: 438.

25 US District Court for the District of Columbia Civil Action. No. 75-1040 cited in ibid.: 438–9.

26 *Rodgers Nzioka Mwema v Attorney General and 8 Others* HCCC no 613 of 2006; [2007] eKLR consolidated with seven similar suits, (hereafter *Tiomin II*), pagination mine.

27 Supra note 5 modified under new Art 40.

28 s 72 ibid. new Art 28 human dignity, Art 29(a), (b).

29 s 73 ibid. new Art 30.

30 s 74 ibid. new Art 29 (c), (d), (e) and (f).

31 s 76 ibid. new Art 31.

32 s 81 ibid. new Art 39.

33 s 6(c) supra note 15.

34 H C Misc Civil Application no 7 of 2006 (unreported).

35 Supra note 8.

36 Ibid.

37 As read with Regulation 18 of the Environmental Impact Assessment and Audit Regulations 2003 (Legal Notice 101).

38 257 NE 2nd 870 [1970] (*the Florida Phosphate case*).

39 [1970] EACA 3.

40 [1947] KB 130.

41 (1935) 52 CLR 723.

42 Misc Civil Application 1052 of 2005, cited in Tiomin II at p. 8.

43 HCCC no 204 of 2005 (unreported).

44 [1979] 1 WLR 63.

45 Art 70 New Constitution of Kenya 27 August 2010.

46 Art 42.

47 Chapter five, entitled Land and Environment.

48 Art 71.

49 Art 69.

50 Supra note 15.

51 Art 63.

[52] Art 67.
[53] Art 65.
[54] Art 40(1).
[55] Art 65.
[56] Proposed Harmonised Draft Constitution of Kenya adopted unanimously by the 10th Parliament on 1st April 2010 due for publication by the Attorney General on 6th May 2010 and ratification at a national referendum due before 6th August 2010 (Proposed new Constitution).

REFERENCES

Albrecht, H.-J. 2001. Restorative Justice—Answers to Questions that Nobody has Put Forward. In E. Fattah & S. Parmentier (eds), *Victim Policies and Criminal Justice on the Road to Restorative Justice: Essays in Honour of Tony Peters.* Belgium. Leuven University Press: 295–314.

Aritho, G.M. Gitonga 2000. Constitutional and Statutory Basis of Land Acquisition and Compensation in Kenya. In Wanjala, Smokin, C. (ed), *Essays on Land Law: The Reform Debate*. 63–83. University of Nairobi. Faculty of Law.

Barrow, C.J. 2000. *Social Impact Assessment: An Introduction.* Arnold. Great Britain.

Bentham, J. [1781] 1988. *The Principles of Morals and Legislation.* Amherst. Prometheus Books.

Bhalla, R.S. 1996. Property Rights, Public Interest and Environment. In Juma Calestous & Ojwang J.B. (eds), *In Land we Trust: Environment, Private Property and Constitutional Change*: 61–81. Nairobi and London. Initiatives Publishers and Zed Books.

Black, H. 1990. *Black's Law Dictionary*. (6th edn). St Paul. MN West Publishing.

Bromley, D. (ed). 1992. *Making the Commons Work: Theory, Practice and Policy.* San Francisco. Institute for Contemporary Studies Press.

Bromley, D. 1989. *Economic Interests and Institutions*. Oxford. Basil Blackwell.

Calabresi, G. 1985. *Ideas, Beliefs, Attitudes and the Law*. Syracuse. NY/ Syracuse University Press.

Christie, N. 1977. Conflicts as Property. *British J of Criminology* 17: 1–26.

Coglianese, C. 1999. The Limits of Consensus. *Environment* 41(3): 28–33.

Cranston, R. 1978. *Consumers and the Law*. London. Weidenfeld and Nicholson.

Craig, P.P. 1989. *Administrative Law*. 2nd ed. Oxford.

Crocker, T.D. 1971. Externalities, Property Rights and Transaction Costs: An Empirical Study. *Journal of Law and Economics* 14: 49–70.

Denyer-Green, B. 1985. *Compulsory Purchase and Acquisition*. London. Estate Gazette Limited.

DiChiro, G. 1998. Nature as Community: The Convergence of Environment and Social Justice. In Goldman Michael (ed), *Privatizing Nature: Political Struggles for the Commons:* 120–143. Pluto Press in association with Transnational Institute.

Dignan, J. 2005. *Understanding Victims and Restorative Justice*. Maidenhead. England. Open University Press.

Dorf, M.C. & Sabel, C.F. 1998. A Constitution of Democratic Experimentalism. *Columbia Law Review* 98(2): 267–473.

Douglas, M. 1978. *Cultural Bias*. London. Royal Anthropological Institute.

Fox, N.J. & Ward, K.J. 2008. What Governs Governance and how does it Evolve: The Sociology of Governance in Action, *British Journal of Sociology* 59(3): 519–538.

Freeman, J. 1997. Collaborative Governance in the Administrative State. *UCLA Law Review* 45(1): 1–98.

Freidmann, D. (1995). Good Faith and Remedies for Breach of Contract. In Beatson J. & Freidmann, D. (eds) *Good Faith and Fault in Contract Law*: 399–426. Oxford. Oxford University Press.

Fuller, L. 1978. Forms and Limits of Adjudication. *Harv L.R.* 92: 353.

Gasper, D. 2004. The Ethics of Human Development: From Economism to Human Development. New Delhi. Vistal Books.

Ghai, Y. 2000. Ethnicity and Autonomy: A Framework for Analysis in *Autonomy and Ethnicity: Negotiating Competing Claims in Multi-Ethnic States:* 1–26. In Y. Ghai (ed.), Edinburgh, Cambridge University Press.

Ghai, Y. & Regan, A. 2000. Bougainville and the Dialectics of Ethnicity, Autonomy and Separation in *Autonomy and Ethnicity: Negotiating Competing Claims in Multi-Ethnic States:* 242–265. In Yash Ghai (ed.), Edinburgh. Cambridge University Press.

Goldman, M. 1998. Inventing the Commons: Theories and Practices of the Commons' Professional. In Goldman Michael (ed), *Privatizing Nature: Political Struggles for the Commons:* 20–53. Pluto Press in association with Transnational Institute.

Goodland, R., Daly, H. & El Serafy, S. (eds). 1991. *Environmentally Sustainable Economic Development: Building on Brundtland.* Paris. UNESCO.

Gordon, R.L. 1995. Minerals Policy. In Bromley, D.W. (ed) *The Handbook of Environmental Economics*: 521–539. USA. Blackwell Publishers Ltd.

Government of Kenya. 2009. *National Land Policy Sessional Paper*. Nairobi. The Government Printer, *adopted by Parliament on 8th December* 2009 http://www.lands.go.ke/index2.php?option=com_docman&task=doc_view&gid=9&Itemid=46 accessed on 19 February 2010.

Hardin, G. 1968. The Tragedy of the Commons. *Science*. Vol. 162, No. 3859, 1243–1248.

Headline. Kenya Strikes Gold. *East African Standard*. 13 September 2010.

Hudson, B. 2003. *Justice in the Risk Society: Challenging and Re-affirming Justice in Late Modernity*. London. Sage.

Hyden, G. & Bratton, M. 1992. *Governance and Politics in Africa*. Colorado. Boulder. Lynne Reinner.

Hyden, G. & Mugabe, J. 1999. Governance and Sustainable Development in Africa: The Search for Economic and Political Renewal. In Okoth-Ogendo H.W.O. & Tumushabe, Godber (eds) *Governing the Environment: Political Change and Management in Eastern and Southern Africa*: 29–38. Nairobi. Acts.

Jewson, N. & MacGregor, S. 1997. Transforming Cities: Social Exclusion and the Reinvention of Partnership.

In Jewson, N. MacGregor S. (eds) Transforming Cities. London; Routledge.
Juma, C. 1996. Private Property, Environment and Constitutional Change. In Juma Calestous & Ojwang, J.B. (eds). *In Land we Trust: Environment. Private Property and Constitutional Change*: 363–395. Nairobi and London. Initiatives Publishers and Zed Books.
Kant, I. 1996. A Definition of Justice, from *The Metaphysical Elements of Justice.* J. Westpal (edn). Indianapolis. Hacket.
Kenya Law Reports 2001. Nzioka & 2 Others v. Tiomin KenyaLtd.(2001)1KLR(E&L).1,p.423.www.kenyalaw.org/environment/content/case_download.php?go=876800085596677323188498&link=
Kenya Law Reports 2006. Rodgers Nzioka Mwema v. Attorney General and 8 Others. *eKLR. HCCC no 613 of 2006*. www.kenyalaw.org/CaseSearch/view_preview1.php?link=13463381344121577185091&words=%27%29
Keohane, R. & Ostrom, E. (eds) 1995. *Local Common sand Global Interdependence: Heterogeneity and Cooperation in Two Domains*. London. Sage.
Khamala, C.A. 2005. The Poverty of Lords: the Value of Forensic Evidence and the Role of Experts before Kenyan Courts. In Franceschi, Luis G. & Ritho Andrew M. (eds) *Legal Ethics and Jurisprudence in Nation building*: 103–148. Nairobi. LawAfrica and Strathmore University Press.
Kibugi, R. 2008. Mineral Resources and the Mining Industry in Kenya. In Okidi, C.O., Kameri-Mbote, P. & Aketch Migai (eds) Environmental Governance in Kenya: Implementing the Framework Law. 355–371. East African Educational Publishers.
Lewis, H.W. 1990. Technological Risk. New York. W.W. Norton & Co.
Linnett, J. 2009. Grievances in Bougainville: Analysing the Impact of Natural Resources in Conflict. *POLIS Journal*. 2. Winter.
Locke, J. 1823. Two Treatise of Government, London: Printed for Thomas Tegg; W. Sharpe and Son.
Lubbe, G. 2006. Estoppel: South Africa. *European Review of Private Law*, 5/6, Kluwer Law International. The Netherlands. 747–788.
Marshall, T.F. 1999. *Restorative Justice: An Overview*. London. Home Office.
McCold, P. & Wachtel, T. Restorative Justice Theory Validation: *Presented at the fourth International Conference on Restorative Justice for Juveniles, Tübingen Germany, October 1–4, 2000.*
Mikesell, R.F. 1993. *Economic Development and the Environment: A Comparison of Sustainable Development with Conventional Economics*. England. Mansell.
Mill, J.S. 1859. *On Liberty*. London. Republished in Warnock M. (ed.) Mill. 1962. *Utilitarianism*. 126–250. London. Collins.
Mugabe, J. & Tumushabe, G.W. 1999. Environmental Governance: Conceptual and Emerging Issues. In Okoth-Ogendo H.W.O. & Tumushabe, G.W. *Governing the Environment: Political Change and Management in Eastern and Southern Africa*: 11–28. Nairobi. Acts.
Nonet, P. & Selznick, P. 1978. *Law and Society in Transition: Towards Responsive Law*. London. Harper and Row.
O'Callaghan, M.L. 1992, Secession may Perish in a Rebel's Cradle, 22 October. *The Age*, Melbourne, Australia: 6.
Okidi, C.O. 1984. Reflections on Teaching and Research on Environmental Law in African Universities, *Journal of Eastern African Research* and Development 18, quoted in Ojwang, J.B. 1996. The Constitutional Basis for Environmental Management. In Calestous Juma & J.B. Ojwang (eds), *In Land we Trust: Environment, Private Property and Constitutional Change* 128–144. Nairobi and London. Initiatives Publishers and Zed Books.
Okoth-Ogendo, H.W.O. 1999. The Juridical Framework for Environmental Governance. In Okoth-Ogendo, H.W.O. & Tumushabe, Godber W. *Governing the Environment: Political Change and Management in Eastern and Southern Africa*: 41–62. Nairobi. Acts.
Okoth-Ogendo, H.W.O. 1991. *Tenants of the Crown: Evolution of Agrarian Law and Institutions in Kenya*. Nairobi, Acts Press.
Ojwang, J.B. 1996. The Constitutional Basis for Environmental Management in (eds) Juma Calestous & Ojwang, J.B. (eds), *In Land we Trust: Environment, Private Property and Constitutional Change*: 39–60. Nairobi and London, Initiatives Publishers and Zed Books.
Ojwang, J.B. 1990. *Constitutional Development in Kenya: Institutional Adaptation and Social Change*. Nairobi. Acts Press.
Olale, L.C.A. 1995. Development as Right in International law: An Overview *University of Nairobi Law Journal*: 18625.
Onalo, P.L. 1986. *Land Law and Conveyancing in Kenya.* Nairobi. Heinemann Books.
Polanyi, K. 1944. *The Great Transformation*. Boston. Beacon Press.
Rosenau, J.N. 1995. Collaborative Governance in the 21st Century. In Global Governance 1(1): 13–43.
Salter, B. & Jones, M. 2002. Human Genetic Technologies, European Governance and the Politics of Bioethics. *Nature* 3 October 2002: 808–14.
Schmid, A. 1995. The Environment and Property Rights Issues. In Bromley, Daniel W. (ed) *The Handbook of Environmental Economics:* 45–60. UK. Blackwell Publishers.
Shapiro, M. 1988. Judicial Selection and the Design of Clumsy Institutions. *Southern California Law Review* 61: 6.
Shivji, I.G. 1989. The Concept of Human Rights inAfrica. London. CODESRIA book series.
Stephens, T. 2009. *International Courts and Environmental Protection*. Cambridge University Press.
Syagga, P.M. 1994. *Real Estate Valuation Handbook: With Special Reference to Kenya*. Nairobi University Press. Nairobi.
Thomson, M. 1998. Style and Scale: Two Sources of Institutional Inappropriateness. In Goldman Michael (ed), *Privatizing Nature: Political Struggles for the Commons:* 20–53. Pluto Press in association with Transnational Institute.
Tiomin I (2001). Nzioka & 2 others v Tiomin Kenya Ltd. (2001) KLR 548; (2006) 1KLR (Environment & Land) 423. High Court Mombasa Civil Case no 97, ruling of Judge Hayanga, Andrew.21 September 2001.

Tiomin II (2006). Rodgers Mwema Nzioka v The Attorney General & 8 Others [2006] eKLR; High Court Nairobi Petition no 613 of 2006, judgment of Judge Nyamu, J.G. 19 December, 2006. http://www.kenyalaw.org/CaseSearch/view_preview1.php?link=1346338134412157718509l&words=%27%29 accessed on 8 March 2011 (different pagination).

Walgrave, L. 2000. In Search of Social Values for Restorative Justice: From Community to Dominion. *Outline for a Presentation at the Cambridge Symposium, 6–8 October 2000.*

Webster. 1952. *Webster's, New International Dictionary of the English Language*, (2nd edn.) Unabridged. Massachusetts. Springfield. G. and C. Merriam Company Publishers.

Zehr, H. 1990. Changing Lenses: A New Focus for Crime and Justice. Scottsdale, PA. Herald Press.

Geological Resources and Good Governance in Sub-Saharan Africa – Runge & Shikwati (eds)
© 2011 Taylor & Francis Group, London, ISBN 978-0-415-58267-4

Good governance, transparency and regulation in the extractive sector

S.V. Rungan, C. Musingwini & H. Mtegha
School of Mining Engineering, University of Witwatersrand, Johannesburg, South Africa

ABSTRACT: The mining industry is being increasingly targeted by society for its negative impacts on the land, environment and people affected by its operations. Mining companies attempt to address such concerns through Sustainable Development (SD) projects. The legislation and regulation SD arise due to slow, ineffective, irregular and lack of effective responses to SD by mining companies. This development is likely to lead to uniformity of SD practices and standards. There are many initiatives directed towards the implementation of SD internationally. These include the Extractive Industries Review; Mining, Minerals and Sustainable Development Project; and United Nations Commission on SD. There are also many methods of measuring compliance with SD, such as the Equator Principles, Global Reporting Initiative, and Extractive Industries Transparency Initiative. Using South Africa as a case study, this paper discusses the various legislative developments to regulate SD, such as the Mineral and Petroleum Resources Development Act, Mining Charter and environmental legislations. However, the proactive role of the government needs to be stressed; whereby government creates an enabling environment for mining companies to operate and engage in SD. Where SD is regulated, legal certainty is a necessity.

1 INTRODUCTION

Throughout history mining has had significant impacts on the land, the environment and the people living in and surrounding the area of operation. In some cases the impacts affect people living beyond the vicinity of the mining activities. Accordingly, all those affected by mining operations require, and often demand, significant action and compensation from mining companies for these impacts. Attempts by mining companies to cater for these and other demands have been couched with the concept of Sustainable Development (SD). Legislation and regulations reflect the compensatory desires of interested and affected parties. While such interventions may not be entirely welcomed by mining companies, legislation and regulation are often a response to slow, ineffective, irregular or even the lack of effective responses by mining companies.

While many would argue against the regulation of SD, regulation thereof, on the face of it, would lead to a uniformity of SD practices and standards.

SD has become a standard feature of most mining operations world-wide. It is perhaps lost on many individuals that the mining industry itself played a major role in the development of SD when "nine of the world's largest mining companies decided to initiate a project to examine the role of the minerals sector in contributing to sustainable development, and how that contribution could be increased". This project, which is now known as the Mining, Minerals, and Sustainable Development Project (MMSD 2002), represents one of the most, if not—the most, extensive consultative study of this nature to date. In stating the aim of the study, the MMSD (2002) states that:

"One of the greatest challenges facing the world today is integrating economic activity with environmental integrity, social concerns, and effective governance systems. The goal of that integration can be seen as 'sustainable development'. In the context of the minerals sector, the goal should be to maximise the contribution to the well-being of the current generation in a way that ensures an equitable distribution of its costs and benefits, without reducing the potential for future generations to meet their own needs".

The reason for the MMSD (2002) engaging in this study can be gleaned from the project report itself, which says that:

"Perhaps the greatest challenge of all is the fact that past practices and social and environmental legacies, combined with continuing examples of poor performance and inadequate accountability, have undermined trust among companies, governments, and some in civil society. The public's perception of what industry is doing is often very different from what company managers think they are doing. As far as some observers outside the industry are concerned, companies have been resisting or at best offering only token improvements: they are seen as failing to meet rising standards of accountability, transparency, and participation".

As with any initiative, the MMSD study proved to be only the beginning of a new way of conducting mining operations. The International Council on Mining and Metals (ICMM) which "serves as a change agent—not in areas affecting competitive positioning, but related to our members' social and environmental responsibilities where collaboration makes sense. Our vision is one of leading companies working together and with others to strengthen the contribution of mining, minerals and metals to sustainable development" (ICCM 2010).

While almost all companies now freely admit to engaging in SD practices, how has regulation influenced this interest in SD?

Mining companies, due to their massive corporate structures and impact, often find themselves subject to both international and local regulations. This paper therefore considers the international regulatory regimes and uses South Africa (SA) as a case study to consider local developments.

2 SUSTAINABLE DEVELOPMENT (SD) ON THE INTERNATIONAL ARENA

There have been many international developments that have resulted in SD gaining a greater prominence which has made it difficult for mining companies to ignore SD activities. The most well-known developments are discussed in the following sub-sections. The first world summit on sustainable development was held in Rio de Janeiro in Brazil in 1992 where the Rio Declaration on Environment and Development was made and Agenda 21 adopted for implementation.

2.1 *UN Commission on SD*

This Commission was set up by the UN General Assembly to review progress on the implementation of two important international SD instruments; namely, Agenda 21 and the Rio Declaration on Environment and Development (CSD 2010). The Commission consists of fifty-three (53) members, thirteen (13) of which hail from Africa (CSD 2010).

The Rio Declaration on Environment and Development consists of twenty-seven (27) principles providing for the protection of the environment and ensuring the effective implementation of SD (Rio Declaration 1992). Principle 8 (Rio Declaration 1992) states that:

"To achieve sustainable development and a higher quality of life for all people, States should reduce and eliminate unsustainable patterns of production and consumption and promote appropriate demographic policies".

This is a major commitment of the signing States to SD as, in addition to committing States to SD, it also requires States to eliminate all unsustainable practices.

Agenda 21 is a programme of action aimed at the environment and SD. The Preamble (Agenda 21 1992), *inter alia*, states that:

"Agenda 21 addresses the pressing problems of today and also aims at preparing the world for the challenges of the next century. It reflects a global consensus and political commitment at the highest level on development and environmental cooperation (sic). *Its successful implementation is first and foremost the responsibility of Governments.* National strategies, plans, policies and processes are crucial in achieving this. International cooperation (sic) should support and supplement such efforts. In this context, the United Nations system has a key role to play. Other international, regional and sub-regional organisations are also called upon to contribute to this effort. The broadest public participation and the active involvement of the non-governmental organisations and other groups should also be encouraged".

Agenda 21 is a document that provides the details in ensuring that the environment and developmental goals are met. The Preamble further states that (Agenda 21 1992):

"The programme areas that constitute Agenda 21 are described in terms of the basis for action, objectives, activities and means of implementation. Agenda 21 is a dynamic programme. It will be carried out by the various actors according to the different situations, capacities and priorities of countries and regions in full respect of all the principles contained in the Rio Declaration on Environment and Development. It could evolve over time in the light of changing needs and circumstances. This process marks the beginning of a new global partnership for sustainable development".

It is clear that Agenda 21 has made the Rio Declaration an integral part of it as well. As a result, the Rio Declaration and Agenda 21 together create the broadest and most in-depth UN initiative on SD to date.

2.2 *Mining, Minerals and Sustainable Development Project (MMSD)*

As mentioned earlier, the mining industry's involvement with SD culminated in the MMSD project, which was to "examine the role of the minerals sector in contributing to sustainable development, and how that contribution could be increased" (MMSD 2002).

This study confirms that the mining industry does regard the issue of SD as an important part of the modern mining business. Whether the industry has actually followed through on the recommendations of the MMSD project falls beyond the scope of this contribution save to say that it is unlikely that the industry would invest in a project of this nature without implementing the same. As was stated in the Report on the MMSD process for Southern Africa (2002):

"The momentum that has been created by the MMSD project must not be allowed to dissipate. If all the stakeholders in the sector bring their strengths to a multi-stakeholder forum to carry the process forward, the mining and minerals sector can make a real and lasting difference to ensure an equitable dispensation for all aspects of sustainable development—governance, society, economic growth and the environment".

Further, the SD portion of the report the MMSD project concluded that (MMSD 2002):

"There remains much to be done in improving the sector's contribution to all aspects of sustainable development. But the largest companies and their newest operations at least are now being held to higher standards. Indeed, the best mining operations are now in the sustainable development vanguard—not merely ahead of what local regulations demand, but achieving higher social and environmental standards than many other industrial enterprises".

One would be forgiven for thinking that the MMSD places the burden of SD on the mining industry alone. However, while it requires a great deal of commitment from the industry, it does not close the doors to all the other sectors. The MMSD (2002) states:

"Implementation of sustainable development principles in the minerals sector requires the development of integrated tools capable of bringing these diverse principles and objectives into focus in a manageable decision-making structure. A wide range of instruments is available, including regulatory, fiscal, educational, and institutional tools. Instruments need to be effective; administratively feasible; cost-efficient, with incentives for innovation and improvement; transparent; acceptable and credible to stakeholders; reliable and reproducible across different groups and regions; and equitable in the distribution of costs and benefits."

It is clear that the government is being referred to herein. The government plays an important role and if there is no commitment from government to the SD process then mining companies would find that being a sole player is a difficult task indeed. This government involvement is stressed in the Preamble of Agenda 21 (1992), discussed below, that its "successful implementation is first and foremost the responsibility of Governments".

It would seem that the work commenced by the MMSD has been assumed by the ICMM which was formed in 2001 "to represent the world's leading companies in the mining and metals industry and to advance their commitment to sustainable development" (ICMM, 2010). With the ICMM being a "CEO-led organisation" it would certainly have the necessary commitment from the mining industry to ensure that the MMSD study is not confined to history but would be expanded further.

It would be disastrous if an industry initiative, and the significant strides it has achieved, is forgotten or, even worse, ignored by the very industry it seeks to improve.

2.3 *The Extractive Industries Review (EIR)*

The World Bank Group (WBG) has considered the issue of SD with specific reference to mining and other extractive industries in the Extractive Industries Review (EIR 2003). The WBG has been financing projects involving the extractive industries in developing countries and undertook this study to address the issue: "whether WBG involvement in the [extractive] industries is consistent with its objective of achieving poverty alleviation through Sustainability Development".

The result of the study is encapsulated in the executive summary of the EIR:

"Based on more than two years of consultations and research, the answer is yes, the Extractive Industries Review believes that there is still a role for the World Bank Group in the oil, gas, and mining sectors—but only if its interventions allow extractive industries to contribute to poverty alleviation through sustainable development".

However, a study that was part of the EIR recommended that the "Extractive Industries sector should be a 'no-go zone' for the Bank" (Caruso, Colchester et al. 2003). It is fortunate that this study was not influential on the World Bank as it would not have worked to the benefit of the poorer countries. The World Bank's involvement, with the necessary appropriate management, could assist these poorer countries to effectively participate in mining operations within their borders and derive benefit from the operations.

This contention seems to accord with the World Bank as the World Bank Group Management Response (2003) to the EIR process has been summed up as follows:

"Our future investments in extractive industries will be selective, with greater focus on the needs of poor people, and a stronger emphasis on good governance and on promoting environmentally and socially sustainable development".

2.4 *World Summit on SD (WSSD)*

SD was brought to the fore in SA when the World Summit on SD (WSSD) was held in Johannesburg in 2002.

This Summit culminated in the adoption of the Johannesburg Declaration on SD which committed the signatories to SD practices (WSSD 2002). A Plan of Implementation of the WSSD was also adopted which consists of 170 Articles (WSSD 2002). Article 24 states that "Managing the natural resource base in a sustainable and integrated manner is essential for sustainable development" (WSSD 2002). Article 46 makes specific reference to the mining industry. While the Declaration acknowledges the importance of mining, minerals and metals to the economies and social development of many countries, and further that minerals are "essential for modern living" (WSSD 2002); Article 46 provides actions to enhance the contribution of mining, minerals and metals to SD which includes: supporting efforts to address the environmental, economic, health and social impacts and benefits of mining, minerals and metals throughout their life cycle, enhancing participation of stakeholders throughout the lifecycles of mining operations and the fostering of sustainable mining practices (WSSD 2002).

2.5 *The Natural Resource Charter*

The Natural Resource Charter, according to its Preamble (Natural Resource Charter 2010), has been drafted by "an independent group of economists, lawyers and political scientists" with the purpose to:

"(…) assist the governments and societies of countries rich in non-renewable resources to manage those resources in a way that generates economic growth, promotes the welfare of the population in general and is environmentally sustainable."

The Natural Resource Charter (2010) provides 12 precepts and, as per the Preamble:

"Ten of these offer guidance on core decisions that governments face—beginning with the decision to extract the resources, and ending with decisions about using the revenues they ultimately generate. The other two are addressed to other important actors and their responsibilities."

The twelve precepts are: Precept 1: The development of a country's natural resources should be designed to secure the *greatest social and economic benefits* for its people. This requires *a comprehensive approach* in which every stage of the decision chain is understood and addressed.

Precept 2: Successful natural resource management requires *government accountability* to an *informed public*.

Precept 3: Fiscal policies and contractual terms should ensure that the country gets *full benefit* from the resource subject to attracting the investment necessary to realise that benefit. The long term nature of resource extraction requires policies and contracts that are *robust to changing and uncertain circumstances*.

Precept 4: Competition in the award of contracts and development rights can be an effective mechanism to secure value and integrity.

Precept 5: Resource projects can have significant positive negative local *economic, environmental and social effects* which should be identified, explored, accounted for, mitigated or compensated for at all stages of the project cycle. The decision to extract should be considered carefully.

Precept 6: Nationally owned resource companies should operate transparently with the objective of being commercially viable in a competitive environment.

Precept 7: Resource revenues should be used primarily to promote *sustained, inclusive economic development* through enabling and maintaining *high levels of investment* in the country.

Precept 8: Effective utilisation of resource revenues requires that *domestic expenditure and investment be built up gradually* and be *smoothed* to take account of revenue volatility.

Precept 9: Government should use resource wealth as an opportunity to increase the *efficiency and equity of public spending* and enable the private sector to respond to structural changes in the economy.

Precept 10: Government should *facilitate private sector investments* at the national and local level for the purposes of diversification, as well as for exploiting the opportunities for domestic value added.

Precept 11: The *home governments* of extractive companies and *international capital centres* should require and enforce best practice.

Precept 12: *All extraction companies* should follow *best practice* in contracting, operations and payments (Natural Resource Charter 2010).

From these precepts, it is clear that, while there is a lot of expectation placed on mining companies, it is also expected that governments must equally play an essential part in the development of their natural resources. These precepts would form the basis of a country specific charter and it would seem appropriate that each country develop its own charter that provides for the unique nature of its mining industry. Seeing that this is a new development, only time would tell how effective an initiative this is.

2.6 *Measuring compliance with SD*

The development of SD has been matched by developments in the measuring of compliance with SD. The instruments discussed hereunder indicate the wide range of such developments.

2.6.1 *The Equator Principles*

These principles apply to Financiers of mining transactions. The reason that financiers would involve themselves with SD can be gleaned from the Preamble to the Equator Principles (2006) which, *inter alia*, provides that:

"In providing financing, particularly in emerging markets, project financiers often encounter environmental and social policy issues. We recognise that our role as financiers affords us significant opportunities to promote responsible environmental stewardship and socially responsible development. In adopting these principles, we seek to ensure that the projects we finance are developed in a manner that is socially responsible and reflect sound environmental management practices".

The Equator Principles (2006) consist of 9 statements to which the participating financing institutions must comply before providing finance. In this regard, a mining company that requires funding from such a financial institution would have to comply with SD principles. If not, then the financial institution would not be providing the required funding to enable the company to commence or continue operations.

2.6.2 *The Global Reporting Initiative (GRI)*

There also seems to be an initiative to measure a company's compliance with SD. The Global Reporting Initiative (GRI) is a body that provides a framework for the reporting of sustainability (GRI 2010) with a mission to provide "a trusted and credible framework for sustainability reporting that can be used by organisations of any size, sector, or location".

In stating the purpose of sustainability reporting, the Sustainability Reporting Guidelines state that:

"Sustainability reporting is the practice of measuring, disclosing, and being accountable to internal and external stakeholders for organisational performance towards the goal of sustainable development. 'Sustainability reporting' is a broad term considered synonymous with others used to describe reporting on economic, environmental, and social impacts (e.g. triple bottom line, corporate responsibility reporting, etc.)" (GRI 2010).

While the Guidelines provide for standards of sustainability reporting for all participants, they also provide reporting guidelines for various sectors. Provision is made for the mining sector in the Mining and Metals Sector Supplement (2010), which gives those that report on SD "the opportunity to describe their own scope of operation, particularly in the boundary-setting and materiality disclosures" (Sector Supplement 2010).

In mentioning the need to report on SD, the Supplement (2010) states that:

"Sustainability reporting is a living process and tool, and does not begin or end with a printed or online publication. Reporting should fit into a broader process for setting organisational strategy, implementing action plans, and assessing outcomes. Reporting enables a robust assessment of the organisation's performance, and can support continuous improvement over time. It also serves as a tool for engaging with stakeholders and securing useful input to organisational processes."

The Supplement leaves no doubt that, in addition to practicing SD, mining companies would do well to also report thereon. While the Supplement provides a detailed reporting mechanism tool, a detailed assessment thereof falls beyond the scope of this paper. Practitioners would be well advised to familiarise with the principles enunciated in the Supplement (Sector Supplement 2010); such as, principles for defining content (materiality, stakeholder inclusiveness, sustainability context and completeness) and principles defining quality (balance, comparability, accuracy, timeliness, clarity and reliability).

2.6.3 *The Extractive Industries Transparency Initiative (EITI)*

The Extractive Industries Transparency Initiative (EITI) was proposed by the then British Prime Minister, Tony Blair, in 2002 at the WSSD in Johannesburg which culminated in the establishment of the Board in 2006 (EITI 2010). It is stated that the EITI "supports improved governance in resource-rich countries through the verification and full publication of company payments and government revenues from oil, gas and mining" (EITI 2010).

The clearest notion of what the EITI, according to its Fact Sheet, is stated as follows:

"The ... (EITI) aims to strengthen governance by improving transparency and accountability in the extractive sector. The EITI is a global standard that promotes revenue transparency. It has a robust yet flexible methodology for monitoring and reconciling company payments and government revenues at the country level. The process is overseen by participants from the government, companies and national civic society. The EITI Board and the International Secretariat are the guardians of the EITI methodology internationally" (EITI 2010).

It seems that, as at 17th December 2010, according to the EITI Compliant Country list, only five countries (Azerbaijan, Ghana, Liberia, Mongolia and Timor-Leste) were EITI compliant. According to the EITI Candidate Country list there are twenty-seven candidate countries with four Other Countries signalling their intent to implement EITI (EITI 2010). It seems that there are also sixteen other countries, according to the Supporting Countries list, that support the EITI by providing political, technical and financial support (EITI 2010).

It would seem that the EITI shares commonalities with the GRI. It is reported by Moberg et al. (2010) that:

"Both the ... (GRI) and the ... (EITI) are tools for voluntary disclosure ... Both initiatives are global in nature, but while the EITI advances transparency in public payments of the extractive industries per country of operation, the GRI advances corporate-level transparency on sustainability topics including economic, environmental and social performance in all types of companies and organisations in all sectors."

It is therefore clear that there is a need for the GRI and EITI to jointly ensure that the mining industry engages in SD without accusations of engaging in such practices in name only.

In expressing its support for the EITI, the ICMM (Mitchell 2010) states that:

"It is generally accepted that development outcomes are enhanced by stronger economic and legal institutions, and the EITI is often seen as a component of governance-strengthening, offering particular value as a means of initiating broader reform."

The ICMM goes further to state that:

"The case for mining companies to support the EITI is clear: better governance standards support development efforts and improve the business environment for mining investment. Simply put, business support for the EITI could be viewed as enlightened self-interest" (Mitchell 2010).

Engaging in SD practices is not always easy for mining companies and it seems that the EITI would assist in ensuring that SD is a success. Darby and Lampa (2010) state that:

"While the EITI does not have an explicitly forensic anti-corruption focus, there is emerging evidence that it serves as a useful component in corruption prevention by increasing scrutiny ... of payments and revenues. The EITI is also increasingly being used to identify poor administration by providing a diagnostic of the efficacy of revenue assessment, collection, and ... redistribution systems. Most intangibly, but possibly of greatest value of all, is the EITI's ability to reduce political tensions and risks to extractive industry investments by creating a forum in which all parties (government, companies and civil society) regularly meet and come to better understand each other's position and concerns. This confidence-building aspect of the EITI is the most difficult to establish, but also has the potential to deliver long-term benefits by reducing the risk of conflict."

2.6.4 *The IASB Extractive Activities Project*

In 2004, the International Accounting Standards Board (IASB) embarked on a project to explore the possibility of introducing an International Financial Reporting Standard (IFRS) to adequately cover financial reporting by firms engaging in extractive activities including, minerals, oil and gas extraction (IASB 2010). This project culminated in the production of a Discussion Paper, DP/2010/1, that was released in April 2010 for public deliberations. One of the key issues explored in this research project is the issue of Publish What You Pay (PWYP) campaign. The IASB Extractive Activities Project complements the Extractive Industries Transparency Initiative (EITI) discussed in the preceding section. The aim of PWYP is to enable citizens of resource-rich developing countries hold their governments accountable for the management of revenues from the minerals, oil and gas industries. Additionally, investors and debt financiers would benefit from PWYP inclusion in financial reporting because information is useful for assessing an entity's exposure to country risk and reputational risk (IASB 2010). This initiative would require companies to disclose in their financial statements, payments made in cash or kind to host governments on a country-by-country basis.

From a SD point of view PWYP is useful in that it keeps a company's reputation in check. No company would want its reputation to be harmed by perceptions that it is associated with, or complicit in, corrupt government practices that have adverse social or environmental consequences (IASB 2010). The company also wants to be seen to be adequately compensating the host country for its extraction of the country's resources. The IASB Extractive Activities Project can therefore be seen as promoting the goal of good governance and transparency in the extractive sector.

3 CASE STUDY: SD IN SOUTH AFRICA—CHANGE THROUGH REGULATION

From the above discussions, one can see that there are various governance mechanisms to ensure that mining companies engage in SD practices. Some initiatives are voluntary while others appear to be voluntary but are actually peremptory. These developments do impact on countries in Africa as well.

South Africa's history of racial discrimination is well-known. Some countries look to South Africa to set an example as regards a smooth transition to democracy and, especially in Africa, the transformation of the mining industry by virtue of Black Economic Empowerment (BEE) and appropriate legislative instruments. While there have been many legislative interventions to remedy the effects of the past without "upsetting the applecart", the MPRDA and the Broad Based Socio-Economic Empowerment Charter for the Mining Industry (the Mining Charter), which have had a direct impact on the mining industry of South Africa, serve as another lesson that can be learned from South Africa as far as SD is concerned.

3.1 *MPRDA*

The Long Title of the MPRDA states that the Act is to "make provision for equitable access to and sustainable development of the nation's mineral and petroleum resources" (MPRDA 2002). This is affirmed in the Preamble as follows:

"The State's obligation to protect the environment for the benefit of present and future generations, to ensure ecologically sustainable development of mineral and petroleum resources and to promote economic and social development" (MPRDA 2002).

The objects of the MPRDA that deal with SD are:

"(...) (e) promote economic growth and mineral and petroleum resources development in the Republic; (f) promote employment and advance the social and economic welfare of all South Africans; (...) (h) give effect to section 24 of the Constitution by ensuring that the nation's mineral and petroleum resources are developed in an orderly and ecologically sustainable manner while promoting justifiable social and economic development ..." (MPRDA 2002).

Section 3 places a positive duty on the Minister of Mineral Resources to ensure SD of the nation's mineral resources:

"The Minister must ensure the sustainable development of South Africa's mineral and petroleum resources within a framework of national environmental policy, norms and standards while promoting economic and social development" (MPRDA 2002).

Though the MPRDA makes an effort to cater for SD, it is disappointing that SD does not form part of the considerations for granting assistance to Historically Disadvantaged People (HDPs) in terms of Section 12 (MPRDA 2002). The provisions of the National Environmental Management Act are incorporated into the MPRDA by Sections 37 and 38 with the environmental provisions extended by Section 39 (dealing with the environmental management programme and environment management plan), Sections 41, 45 and 46 (dealing with environmental damage) and Section 43 (dealing with mine closure) (MPRDA 2002).

Section 57 established the Minerals and Mining Development Board which in terms of Section 58, *inter alia*, acts as an advisor to the Minister on various aspects affecting mining and minerals and ensures human resources development in the sector (MPRDA 2002). Specifically, Section 58 further provides that the Minerals and Mining Development Board is also required to advise the Minister on "the sustainable development of the nation's mineral resources" (MPRDA 2002).

However, while this was the first legislation to regulate SD and mining in SA, international developments, as discussed above, suggest that it does not effectively deal with SD issues. An amendment to the MPRDA was passed in 2008 (Mineral and Petroleum Resources Development Amendment Act 2008); however, the Amendment seems to have been abandoned by the Department of Mineral Resources (DMR) as a new amendment is in the process of being drafted. This new amendment would hopefully take proper account of developments in SD.

SD in the MPRDA is further enhanced by the MPRDA Regulations which, in addition to providing the administrative mechanism to the MPRDA, requires mining companies to provide particular plans with their applications and to adhere thereto. These Regulations provide for SD by requiring a Social and Labour Plan in Part II, Environmental Regulations For Mineral Development, Petroleum Exploration and Production in Part III and Pollution Control and Waste Management Regulation in Part IV (MPRDA Regulations 2002).

While one could have many criticisms of the MPRDA, at a recent mining law workshop[1], delegates were admonished to look at the intent of the MPRDA and the improvements it seeks to introduce and not to criticise it. Unfortunately, such comments are disingenuous—the law must be certain and it is for the judiciary to consider the intent and other issues when disputes are referred to it. Until then, it is the wording of the legislation that must pass muster—if the wording is unclear, confusing or appears to be contradictory to other provisions or other laws; then the law is not a good one. Admonishments not to criticise the MPRDA only leads to a static law whose defects are not remedied—this makes the law uncertain and severely impacts on the Rule of Law.[2]

3.2 *Stakeholders' declaration on strategy for the sustainable growth and meaningful transformation of South Africa's Mining Industry*

Prior to the completion of the review into the Broad-Based Socio-Economic Empowerment Charter for the Mining Industry (Mining charter), all stakeholders in the South African mining industry concluded a Stakeholders' Declaration On Strategy For the Sustainable Growth And Meaningful Transformation of South Africa's Mining Industry (Stakeholders' Declaration 2010). This Stakeholders' Declaration contains various provisions that impact on the reviewed Mining Charter, yet the most significant development was a firm commitment to SD in Commitment 4 (Stakeholders' Declaration), where the parties have committed to:

- Develop and implement a National Action Plan for the management of acid mine drainage;
- Adopt a regional approach in dealing with integrated and cumulative environmental impacts resultant from mining;
- Embark on research and development initiatives directed towards the sustainability of mine closure and mining environmental legacies;
- Establish a multi-stakeholder forum on derelict and ownerless mines;
- Implement Mine Health and Safety Tripartite Action Plans;
- Establish a task team to develop mechanisms of accelerating exploration investment;
- Strengthen linkages of mining with other industries, such as infrastructure, upstream and downstream value addition, technology, services and manufacturing, to ensure sustainable mining 'beyond a hole in the ground';
- Work towards the development and effective implementation of a South African 'Mining Vision 2030' informed by sustainable development principles; and
- Adopt an integrated development approach through pooling of resources.

This provision seems to commit the stakeholders to international best practice as far as SD is concerned. However, at the above-mentioned mining law workshop,[3] reservation was expressed as regards the Stakeholder Declaration. In fact, one delegate stated that it was "not worth the paper it was written on". Such a negative perception, to what is essentially a radical declaration of particular behaviour cannot simply be pushed aside. This Stakeholders' Declaration indicates a firm commitment to behave in a particular manner and it is hoped, as far as SD is concerned, that the stakeholders do not adopt such a cavalier attitude as what some people might expect.

3.3 *The Broad-Based Socio-Economic Empowerment Charter for the Mining Industry*

The first time news of the Broad-Based Socio-Economic Empowerment Charter for the Mining Industry (Mining Charter) surfaced, it triggered disinvestment in the South African mining industry. The government then initiated meetings with the various role players in the industry that resulted in the refining of the Mining Charter.

The Mining Charter was established under the auspices of Section 100 of the MPRDA (2002) which provides:

"(1) The Minister must, within five years from the date on which this Act took effect-

(a) and after consultation with the Minister for Housing, develop a housing and living conditions standard for the minerals industry; and (b) develop a code of good practice for the minerals industry in the Republic.

(2) (a) To ensure the attainment of Government's objectives of redressing historical, social and economic inequalities as stated in the Constitution, the Minister must within six months from the date on which this Act takes effect develop a broad-based socio-economic empowerment Charter that will set the framework, targets and time-table for effecting the entry of historically disadvantaged South Africans into the mining industry, and allow such South Africans to benefit from the exploitation of mining and mineral resources. (b) The Charter must set out, amongst others how the objects referred to in section 2(*c*), (*d*), (*e*), (*h*) and (*i*) can be achieved."

From this Section, there are three things that have been required from the Minister: develop a housing policy, develop a code of good practice and develop the Mining Charter. The Mining Charter was recently subjected to a review and a new Mining Charter was developed and finalised in 2010 as the Amendment of The Broad-Based Socio-Economic Empowerment Charter for The South African Mining and Minerals Industry (New Mining Charter 2010).[4]

SD is brought to the fore in the Vision of the New Mining Charter (2010) which states:

"To facilitate sustainable transformation, growth and development of the mining industry".

It is further emphasised in the objects (New Mining Charter 2010) which provide:

- To promote equitable access to the nation's mineral resources to all the people of South Africa;
- To substantially and meaningfully expand opportunities for HDSA to enter the mining and minerals industry and to benefit from the exploitation of the nation's mineral resources;
- To utilise and expand the existing skills base for the empowerment of HDSA and to serve the community;
- To promote employment and advance the social and economic welfare of mine communities and major labour sending areas;
- To promote beneficiation of South Africa's mineral commodities; and
- Promote sustainable development and growth of the mining industry.

While the latter four objectives can be attributed to SD, the former two do not—they are important to redress the impacts of past discrimination but cannot rightly be included as being part of SD. The New Mining Charter was occasioned to increase the compliance of the mining industry to transformation of the South African mining milieu. It commits the mining industry to specific targets by 2014 with SD, as indicated, playing a much more significant role than before.

To measure the compliance of mining companies with the Mining Charter, a Scorecard for the Broad Based Socio-Economic Empowerment Charter for the South African Mining Industry (the Scorecard) was created. It contains a series of questions relating to the particular issues covered by the Mining Charter with a provision for ticks on either "yes" or "no" answers. One can be obligated to think that it constituted no more than a checklist. In fact, most companies have used the Scorecard as a format of responses but provided more details than the Scorecard required. In that regard, one could say that the Scorecard was superfluous, yet it was an integral part of testing the compliance of the mining company with the Mining Charter. The New Mining Charter does not seem to have replaced the reporting mechanism encapsulated in the Scorecard; however, the specific nature of the provisions of the New Mining Charter seems to make it difficult for the mining company to simply provide "yes" and "no" answers. Substantial reporting seems to be warranted.

On reading the New Mining Charter, as compared with the Mining Charter, all of the government responsibilities have been removed, making the mining industry solely subject to the requirements of the New Mining Charter.

3.4 *Codes of Good Practice for the South African Minerals Industry*

Section 100 of the MPRDA, as quoted above, requires the Minister of Mineral Resources to implement Codes of Good Practice for the South African Minerals Industry (the Codes 2009) which were published in the Government Gazette on the 29th April 2009, shortly after the 22nd April presidential elections. One could assume that the Codes would have been implemented in much the same manner as the Mining Charter, yet this was not so. It was published without consultation with the mining industry, much to the chagrin of many such as Cohen (2009) who stated that:

"You wouldn't think it possible, but the Department of Minerals and Energy has done it again; a big legislative overreach has again threatened the viability of SA's core economic sector."

Cohen (2009) further states that:

"Instead, days after the election, [the DMR] published something called the Code of Good Practice for the SA Minerals Industry, amazingly without informing the Chamber or anyone else it was doing so, even though it was in discussion with the Chamber at the time on the very topic."

Cohen (2009) believed that the Codes were an attempt to harmonise the BEE codes of the Department of Trade and Industry with the Mining Charter. However, the Codes (2009) themselves provide a purpose which states that:

"The purpose of this document is to set out administrative principles in order to facilitate the effective implementation of the minerals and mining legislation and enhance the implementation of the Broad-Based Socio-Economic Charter applicable to the mining industry and to give effect to section 100 (1) (b) of the Mineral and Petroleum Resources Development Act (2002) by developing a Code of Good Practice for the minerals industry in the Republic."

While the Purpose of the Codes does not seem to indicate anything untoward, the Introduction and Scope (Codes 2009) provide:

"The Code does not replace the key legislation and laws relating to the minerals and the petroleum industry but serves as a statement of present policy providing an overview and confirmation of the existing mineral and mining policy that is in place".

A detailed discussion on these Codes is not warranted as the Codes appear to have been abandoned by the DMR. The South African mining industry also has access to professional codes on good governance and reporting that complement SD requirements. These codes are voluntary, regulatory and provide guidance to mining companies, particularly those reporting to the public domain. These codes are the King II Code on Good Governance; The South African Code for Reporting of Exploration Results, Mineral Resources and Mineral Reserves (SAMREC); The South African Code for the Reporting of Mineral Asset Valuation (SAMVAL). Since these codes are professional codes of conduct and not directly related to SD, their detailed discussion cannot constitute a part of this contribution.

4 ENVIRONMENTAL LEGISTLATION

While it is clear that South African legislation does provide for SD, for information a few environmental legislation impacting on mining is mentioned below.

4.1 *National Environmental Management Act 107 of 1998 (NEMA)*

The Long Title of the National Environmental Management Act (NEMA 1998) states that it is to:

"Provide for co-operative environmental governance by establishing principles for decision-making on matters affecting the environment, institutions that will promote co-operative governance and procedures for co-ordinating environmental functions exercised by organs of state; to provide for certain aspects of the administration and enforcement of other environmental laws ..."

While the Preamble makes reference to SD, NEMA (1998) contains a list of principles in Section 2 that make a clear reference to SD:

"Development must be socially, environmentally and economically sustainable; Sustainable development requires the consideration of all relevant factors including the following:

i. That the disturbance of ecosystems and loss of biological diversity are avoided, or, where they cannot be altogether avoided, or minimised and remedied;
ii. That pollution and degradation of the environment are avoided, or, where they cannot be altogether avoided, are minimised and remedied;
iii. That the disturbance of landscapes and sites that constitute the nation's cultural heritage is avoided, or where it cannot be altogether avoided, is minimised and remedied;
iv. That waste is avoided, or where it cannot be altogether avoided, minimised and re-used or recycled where possible and otherwise disposed of in a responsible manner;
v. That the use and exploitation of non-renewable natural resources is responsible and equitable, and takes into account the consequences of the depletion of the resource;
vi. That the development, use and exploitation of renewable resources and the ecosystems of which they are part do not exceed the level beyond which their integrity is jeopardised;
vii. That a risk-averse and cautious approach is applied, which takes into account the limits of current knowledge about the consequences of decisions and actions; and
viii. The negative impacts on the environment and on people's environmental rights be anticipated and prevented, and where they cannot be altogether prevented, are minimised and remedied."

NEMA can be regarded as the basis for all environmental legislation in SA. Thc fact that NEMA makes such concise provision for SD, which is incorporated into the MPRDA (2002), indicates that mining companies would have to take note of these principles in their operations as well.

4.2 *National Environment Management: Air Quality Act 39 of 2004*

As with NEMA, SD is specifically mentioned in this Act. The Long Title of the Air Quality Act (2004) states that the Act is to "reform the law regulating air quality in order to protect the environment by providing reasonable measures for the prevention of pollution and ecological degradation and for securing ecologically sustainable development while promoting justifiable economic and social development ..."

This is mirrored in the objects clause of the Act (2004) which provides, *inter alia*, that the objects of the Act are:

"To protect the environment by providing reasonable measures for (...) securing ecologically sustainable development while promoting justifiable economic and social development".

While the language of the Act is general and applicable to SA as a whole, it would appear that air pollution has been specifically included as a responsibility for mining companies to take on board as part of their SD programmes.

4.3 *National Water Act 36 of 1998*

This Act also makes reference to SD. Section 2 states the purpose of the Act (National Water Act 1998) which is, *inter alia*: (a) "meeting the basic human needs of present and future generations (...) (d) promoting the efficient, sustainable and beneficial use of water in the public interest; (e) facilitating social and economic development".

The Act regards the Minister of Water Affairs as the trustee of the Nation's water resources and in Section 3 (National Water Act 1998) provides that he or she must "ensure that water is protected, used, developed, conserved, managed and controlled in a sustainable and equitable manner, for the benefit of all persons".

It seems that licences, in terms of Sections 27 to 55, would be needed for water usage in certain circumstances and it would also seem that, the mining industry is subject to this licence requirement (National Water Act 1998).

5 CONCLUSION

SD is an important facet of the mining business. While the mining industry may have initiated the development of SD, it is a partnership between all role players and is not limited to just the mining industry.

As was stated in the Report on the MMSD (2002) process for Southern Africa:

"The momentum that has been created by the MMSD project must not be allowed to dissipate. If all the stakeholders in the sector bring their strengths to a multi-stakeholder forum to carry the process forward, the mining and minerals sector can make a real and lasting difference to ensure an equitable dispensation for all aspects of sustainable development—governance, society, economic growth and the environment".

The benefits of mining cannot be looked at simply as a means of revenue for a country. It must deliver that which is necessary to ensure that the industry is viable and also that those affected by a mining operation also benefit from the operation.

However, the government should play a proactive role—while they cannot proclaim a law and expect the mining industry to follow; they can negotiate a coherent policy with all the stakeholders that would benefit from a SD mining operation. An important consideration would be that all participants comply with their respective undertakings. It would not bode well if the industry performs their obligations while the government shirks its responsibility.

While it may seem that the implementation of a Mining Charter is advocated due to its use in SA, it should be noted that an international Mining Charter has been developed to assist developing countries with mineral resources to effectively exploit such resources. This international charter does indicate that a form of agreement between the mining industry and a government would be necessary to ensure effective implementation of SD.

SA has opted for a regulatory approach to the introduction of SD. This approach places everyone on a footing of equality, in that all mining companies would have to comply. This alleviates the situation of having one company engaging in SD practices while others are not. However, due to the strategic importance of the mining industry to the economy, the government has also engaged in consultations with the various role players to ensure an economically sustainable mining industry and SD.

The Rule of Law is an essential aspect of any investment seeking country and the certainty of laws is an essential component thereof. By SA having two legal instruments being made ready for implementation and then abandoned does not auger well for legal certainty. Regulation of SD or an improvement thereto, is hindered and mining companies are left to their own devices.

SA's lack of effective regulation does not and should not affect the further development of SD practices in the country. As the MMSD (2002) project concluded in the SD portion of the report:

"There remains much to be done in improving the sector's contribution to all aspects of sustainable development. But the largest companies and their newest operations at least are now being held to higher standards. Indeed, the best mining operations are now in the sustainable development vanguard—not merely ahead of what local regulations demand, but achieving higher social and environmental standards than many other industrial enterprises".

While engaging in SD practices places mining companies in a positive light and is capable of improving the images of the mining industry, this should be considered in light of a comment by an ex-Chief Economist of a major mining company:

"Mining companies cannot be expected to step into the shoes of governments and do that which is rightly the responsibility of governments. Mining companies must accept social responsibility for those whose lives it impacts, but not for the whole country—that is the government's duty."[5]

Governments do require greater social investment from mining companies, but at the same time government cannot shirk its own social responsibilities. While the Mining Charter may have been the subject of much debate, it has had a positive impact on mining companies' SD programmes.

With all the requirements placed on governments and mining companies in terms of SD, one would be tempted not to conduct mining operations. However, it should be borne in mind that minerals that remain in the ground are not useful at all. Using SD requirements as an excuse to avoid commencing an operation would do more harm than good. After all, SD practices now, whether regulated or not, would lead to a sustainable future mining industry.

ENDNOTES

[1] Mineral Law for Industry Practitioners, held on 5th November 2010, jointly hosted by the School of Mining Engineering and The Mandela Institute.

[2] For a reaction to criticisms to the MPRDA see Makupula and Bonga (2010).

[3] Mineral Law for Industry Practitioners, held on 5th November 2010, jointly hosted by the School of Mining Engineering and The Mandela Institute.

[4] A comparison between the two Mining Charters falls beyond the scope of this analysis. As such, only the New Mining Charter would be considered herein.

[5] Personal communication with first author

REFERENCES

Agenda21, www.un.org/esa/sustdev/documents/agenda21/english/Agenda21.pdf Accessed 20.12.2010.

Amendment Of The Broad-Based Socio-Economic Empowerment Charter For The South African Mining And Minerals Industry http://us-cdn.creamermedia.co.za/assets/articles/attachments/29578_100908policy.pdf Accessed 20.12.2010.

Caruso, E. et al. 2003. *Extracting Promises: Indigenous Peoples, Extractive Industries And The World Bank.* http://bankwatch.ecn.cz/eir/reports/vol6_3.pdf Accessed 25.09.2008.

Codes 2010. *Codes of Good Practice For The South African Minerals Industry.* www.pmg.org.za/files/docs/090329goodpractice.pdf Accessed 20.12.2010.

Cohen, T. 2009. *Nothing innocuous about new mining BEE codes.* Business Day, 25 May 2009: 12.

Darby, S. & Lampa, K. 2009. Advancing The EITI In The Mining Sector: Implementation Issues. In Eads, C., Mitchell, P. & Paris, F. (eds.). *Advancing The EITI In The Mining Sector* 117. http://eitransparency.org/files/MINING%20Compressed.pdf Accessed 20.12.2010.

Extractive Industries Transparency Initiative. *Candidate Country.* http://eitransparency.org/countries/candidate Accessed 17.12.2010.

Extractive Industries Transparency Initiative. *Compliant Country.* http://eitransparency.org/countries/compliant Accessed 17.12.2010.

Extractive Industries Transparency Initiative. *EITI Fact Sheet.* http://eitransparency.org/files/2009-05-27%20EITI%20Fact%20Sheet.pdf Accessed 20.12.2010.

Extractive Industries Transparency Initiative. *History of EITI.* http://eitransparency.org/eiti/history Accessed 20.12.2010.

Extractive Industries Transparency Initiative. *Other Countries.* http://eiti.org/othercountries Accessed 17.12.2010.

Extractive Industries Transparency Initiative. *Supporting Countries.* http://eitransparency.org/supporters/countries Accessed 17.12.2010.

Extractive Industries Transparency Initiative. *What is EITI?* http://eitransparency.org/eiti Accessed 20.12.2010.

Global Reporting Initiative. *Preface—Sustainability Reporting Guidelines.* http://www.globalreporting.org/NR/rdonlyres/B52921DA-D802-406B-B067-4EA11CFED835/3882/G3_GuidelinesENU.pdf Accessed 20.12.2010.

Global Reporting Initiative. *Sustainability Reporting Guidelines & Mining and Metals Sector Supplement.* http://www.globalreporting.org/NR/rdonlyres/E75BAED5-F176-477E-A78E-DC2E434E1FB2/4162/MMSSFINAL115 NEW.pdf Accessed 20.12.2010.

Global Reporting Initiative. *What is GRI?* http://www.globalreporting.org/AboutGRI/WhatIsGRI/ Accessed 20.12.2010.

Internal Council on Mining and Metals. *About us.* http://www.icmm.com/about-us. Accessed 20.12.2010.

International Accounting Standards Board (IASB) 2010. Discussion Paper DP/2010/1, Extractive Activities. http://www.ifrs.org/NR/rdonlyres/735F0CFC-2F50-43D3-B5A1-0D62EB5DDB99/0/DPExtractiveActivitiesApr10.pdf Accessed 09.04. 2010.

International Council on Mining and Metals. *Our History* www.icmm.com/about-us/icmm-history Accessed 20.12.2010.

International Council on Mining and Metals 2009. Practical Implementation Challenges For Mining Companies. In Eads, C., Mitchell, P. and Paris, F. (eds.), *Advancing The EITI In The Mining Sector*: 92. http://eitransparency.org/files/MINING%20Compressed.pdf Accessed 20.12.2010.

Makupula, T. & Bonga, M. 2010. *SA's mining regulatory framework is unfairly criticised.* The Star Business Report, 20th September 2010: 16.

Mineral and Petroleum Resources Development Act 2002. *Act No. 28 of 2002. Mineral and Petroleum Resources Development Regulations.*

Mineral and Petroleum Resources Development Act 2008. *Act No. 49 of 2008.*

Mining and Minerals for Sustainable Development. *Breaking New Ground—Executive Summary.* International Institute for Environment and Development and World Business Council for Sustainable Development, London: Earthscan Publications.

Mining and Minerals for Sustainable Development, *Breaking New Ground.* http://www.iied.org/mmsd/mmsd_pdfs/finalreport_01.pdf Accessed 18.09.2008.

MMSD Southern Africa 2002. *The Report of the Regional MMSD Process.* School of Mining Engineering, University of Witwatersrand.

Moberg, J., Quiroz, J.C. & Fleur, M. 2009. EITI—One Of Many Efforts: Other Initiatives For The Extractive Sector. In Eads, C., Mitchell, P. & Paris, F. (eds.). *Advancing The EITI In The Mining Sector* 92. http://eitransparency.org/files/MINING%20Compressed.pdf Accessed 20.12.2010.

National Environment Management Air Quality Act 39 of 2004.

National Environmental Management Act 107 of 1998.

National Water Act 36 of 1998.

Natural Resource Charter. www.naturalresourcecharter.org/sites/default/files/NaturlaResourceCharter_Levels1and2_Oct2010.pdf Accessed 17.12.2010.

Paris, F. & Bartlett, S. 2009. EITI And The Mining Sector Today. In Eads, C., Mitchell, P. and Paris, F. (eds.), *Advancing The EITI In The Mining Sector*: 92. http://eitransparency.org/files/MINING%20Compressed.pdf Accessed 20.12.2010.

Stakeholders' Declaration On Strategy For the Sustainable Growth And Meaningful Transformation of South Africa's Mining Industry 2010. http://www.info.gov.za/view/DownloadFileAction?id=126562 Accessed 20.12.2010.

The Equator Principles 2006. www.equator-principles.com/documents/Equator_Principles.pdf Accessed 20.12.2010.

United Nations, *Commission on Sustainable Development*, http://www.un.org/esa/dsd/csd/csd_csd19_membstat.shtmll Accessed 20.12.2010.

United Nations 2002. *Report of the World Summit on Sustainable Development.* New York: United Nations Press.

United Nations, *The Rio Declaration.* www.sovereignty.net/p/sd/riodecl.htm Accessed 20.12.2010.

United Nations, *UN Department of Economic Affairs—Division of Sustainable Development.* www.un.org/esa/sustdev/csd/aboutCsd.htm Accessed 20.12.2010.

World Bank Group Management Response 2004. *Striking A Better Balance—The World Bank Group And Extractive Industries: The Final Report Of The Extractive Industries Review.* www.ifc.org/ifcext/eir.nsf/AttachmentsByTitle/FinalManagementResponse/$FILE/finaleirmanagementresponse.pdf Accessed 25.09.2008.

World Bank Group 2003. *Striking a Better Balance Vol. 1: The World Bank Group and Extractive Industries.* http://bankwatch.ecn.cz/eir/reports/vol1_eng.pdf Accessed 25.09.2008.

Geological Resources and Good Governance in Sub-Saharan Africa – Runge & Shikwati (eds)
© 2011 Taylor & Francis Group, London, ISBN 978-0-415-58267-4

The impact of EITI and the role of civil society in Africa in promoting and advancing transparency in the extractives sector

M.-A. Kalenga & members of the EITI International Secretariat
EITI International Secretariat, Oslo, Norway

ABSTRACT: Between 2003 and 2008, over 150 companies and their affiliates have reported under the EITI process in Africa, covering over US$130 billion in revenues paid to governments. Accurate, reliable data is an essential tool for sound economic management and policy making. It helps governments make policy decisions and helps reduce mismanagement, red tape and the opportunities for corruption. Increased transparency also strengthens accountability and can promote greater economic and political stability. EITI training activities carried out by the World Bank and bilateral development partners help to build management capacity in government institutions involved in the extractive sectors, as well as inform civil society groups about key aspects of the sector. These factors can enhance the investment returns from companies operating in EITI countries for investors, boosting a country's attractiveness as an investment destination.

1 INTRODUCTION

The beginning of the 21st century coincided with a growing awareness of the need for greater transparency and accountability in the management of revenue from extractive industries. Several initiatives such as the Kimberley Process, the international campaign of Publish What You Pay (PWYP) and the Extractive Industries Transparency Initiative (EITI).

The extractives sector has often become associated with mistrust, suspicion and hostility. Affected communities and ordinary citizens often assume that the government and companies are trying to keep the resource wealth for themselves and are undermining the economic development of the country through corruption and mismanagement. Conversely, extractive companies sometimes believe that governments and citizens are unaware of just how much they contribute to the greater economy through paying taxes and royalties, job creation and infrastructure development. Despite the stakes being high, dialogue about these problems is often lacking between stakeholders, contributing to a climate of suspicion, distrust and political tension. The EITI helps to address these problems through creating a platform for dialogue and information-sharing for governments, companies and civil society. We can see from the story below how in Liberia the EITI is providing a tool for building confidence and trust inside "this curious coalition" and in communities affected by oil, gas and mining activities. Community meetings between local leaders, civil society representatives, civil servants and government representatives in Liberia are providing a means to include citizens and disadvantaged communities in debates about the extractives sector which they never had before. The EITI is also contributing to wider peace and reconciliation processes in post-conflict countries through creating transparency in the management of the natural resources which are often at the centre of conflicts.

2 ORIGIN OF THE PWYP MOVEMENT

Publish What You Pay was launched in 2002, initiated by a group of international NGOs to require the publication of all payments made by extractive companies to host governments. The creation of this campaign is largely due to a growing global concern that many countries with significant natural resources are poor and have the lowest human development indices in the world. That is what is commonly called the 'resource curse' or the 'paradox of plenty'.

Angola is a Central African oil-rich country that sparked the launch of the PWYP campaign. Despite significant investments in the oil industry and the considerable income generated, Angola is one of the poorest countries in the world. In December 1999, the British NGO Global Witness, published a report entitled 'A crude awakening', which revealed the complicity of banks and oil companies in the plundering of Angola's oil resources with the blessing of government officials during the forty years of war experienced by the country and the confusion and lack of accountability in the exploitation of these resources. Also, Global Witness decided to launch an appeal to oil companies operating in Angola to 'Publish What They Pay'.

Over the past seven years, the PWYP coalition has grown considerably and is today a global network of over 350 civil society organisations which believe that transparency is an essential step for the management of revenue from petroleum, mining and gas industries. The campaign is firmly rooted in Africa, where it is an important platform for dialogue and discussions on issues of transparency.

However, PWYP is aware that it is impossible to ensure proper and accountable management of natural resources by looking exclusively at the payments made by companies (Publish What You Pay), requiring the disclosure of payments that companies make to governments in each country (taxes, fees, royalties, etc.) and revenue received by governments (Publish What You Earn) by asking governments of resource-rich countries to publish revenues.

PWYP's campaign scope has expanded to include new advocacy demands upstream and downstream. "Publish what you spend" for a transparent and accountable management of revenue from natural resources and "Publish what you didn't pay/should have paid" is a call for the maximisation of a country's revenue by establishing transparent, fair and equitable contracts.

3 HOW TO PROMOTE AND FOSTER TRANSPARENCY EFFORTS IN EXTRACTIVE INDUSTRIES?

For the civil society, the need for transparency in the extractives sector has two main objectives:

- Allow citizens to hold their government to account for the management of revenue from natural resources;
- Ensure that governments receive a fair share of their natural resource wealth.

The expertise and advocacy efforts of PWYP members spanned the entire value chain and in many cases they preceded the implementation of the EITI in their countries. In Chad and Cameroon, the construction of the Chad/Cameroon pipeline between 1999 and 2003 brought together civil society organisations in both countries to draw attention on the use of oil revenues (see contribution by G. Ngarsandje in this volume). They successfully led a campaign for the implementation of an environmental management plan for the project. The monitoring of the pipeline construction developed into a dynamic control of the management of oil revenue and budget tracking in Chad in particular, where human rights associations, religious organisations and trade unions continue to conduct advocacy on these issues.

Another example of "upstream" advocacy is the succesful campaign of the Gabonese PWYP coalition, in denouncing the detrimental effects of a contract between the Gabonese government and a Chinese mining company for the extraction of Belinga iron ore in the north of the country.

With regard to the EITI, the involvement of civil society and national coalitions of PWYP in particular is threefold:

- They participate actively in diverse national committees involved in implementing their initiatives and contributing their expertise and experience to the process at the national level.
- Members of PWYP play an important role in decision-making and governance of the EITI at the international level. CSO representatives from Africa and Europe sit on the International EITI Board, which gives them credibility and a platform to advance their advocacy goals.
- Finally, PWYP has the duty to monitor and closely follow-up the implementation of the EITI which must be done in accordance with a number of principles and criteria, including highlighting challenges and obstacles. Currently, Central African countries are under pressure to complete validation, an independent evaluation of the national EITI process, by the agreed deadlines. They may be tempted to bend some fundamental rules that the civil society will be obliged to remind them of.

The EITI is an important step towards creating a platform for tripartite communication between the government, the private sector and the civil society. It is sometimes the only avenue available to CSOs in many countries to talk about transparency and accountability in extractive industries, which was long considered taboo and even dangerous. The critical questions posed by civil society sometimes go beyond the narrow scope of the EITI and include upstream concerns such as whether contracts guarantee a fair share of revenue to the state, and downstream issues by questioning if revenue from extractive resources is used transparently and responsibly.

Moreover, it should be noted that the multiple discussions prompted by the initiative have enabled members of the civil society to strengthen their capacity and expertise on issues that are often very complex and technical.

PWYP is convinced that EITI is an important and appropriate mechanism and that additional steps are necessary. Indeed, many natural resource rich countries like Angola and Algeria have not yet joined the initiative. Its voluntary nature often depends on the goodwill of governments. Mandatory systems are therefore necessary to allow greater transparency of revenue from the exploitation of natural resources. Some countries have already embarked on this path, including Liberia and Nigeria, which have adopted specific national legislation on the EITI, thereby compelling companies in the extractive industry to comply with the EITI.

At the sub-regional level, the Economic Community of West African States (ECOWAS), in April 2009 adopted a directive to foster greater harmonisation of policies in the mining sector of member states. This measure will also help ensure greater transparency in the management of revenue generated by mining activities and better support the interests of the local people and the protection of the environment by mining companies. In turn, civil society organisations have called on ECOWAS member states to immediately establish mechanisms for implementing the clauses of the mining directive. The ultimate goal of the civil society is to strengthen its campaign, raise awareness and bring ECOWAS states to adopt a community mining code by 2011.

PWYP members in the Northern Hemisphere are also striving to adopt measures to strengthen revenue transparency in the extractives sector. In the United States, the civil society is working with some members of Congress to introduce a legislation that would require companies registered with the Securities and Exchange Commission (SEC) with to publish all payments made to governments. PWYP is also working with the Council of International Accounting Standards Board (IASB) for the adoption of new financial reporting standards for activities in the extractive sector.

All these developments should not make us lose sight of harassment, intimidation and restrictions faced by civil society because of its commitment to transparency and good governance. Some members of PWYP were imprisoned, others have witnessed the suspension of their activities and others have received death threats.

According to Criterion 5 of the EITI, civil society participation should be effective at all stages of the implementation process. The outcome is unfortunately mitigated in many countries where this participation has either been minimised or is marginal.

In a nutshell, it is important to recall that the imperative resolution of the paradox of plenty and of its major related problems (macro economic instability, the resurgence of conflicts within states and interstate; the Dutch disease whereby the extractives sector inhibits growth in other economic sectors; multiform practices of bad governance, and endemic corruption) require increased transparency and accountability.

To this end, civil society groups in Africa are formulating the following recommendations to the governments of Central African States and the CEMAC zone for:

- The harmonisation of mining policies to better meet the challenges of transparency and accountable management of revenue from the extractives sector, in light of ongoing efforts by ECOWAS and SADC;
- Joining and effectively implementing the EITI within given deadlines, including fostering genuine consensus and committing to abide by the EITI Principles and Criteria.
- Also, constructive participation in the EITI and expanding its scope to other resources such as timber and fisheries, and strengthening of the legal framework to entrench transparency requirements;
- Full participation of civil society in efforts to improve transparency and governance in extractive industries by strengthening the capacity of civil society and understanding its critical mission;
- Ensuring better security and protection civil society activists working on transparency and good governance in the extractive sector.

4 SHOW CASES: IMPACT OF THE EITI IN AFRICA

4.1 *Nigeria*

The Nigeria Extractive industries Transparency Initiative (NEITI) has become a catalyst for the application of the principles of transparency and accountability in the management of extractive industries in Nigeria. It is today an ethical reference point for the ongoing reform in Nigeria's Petroleum Industry: the Petroleum Industry Bill provides that all institutions in the industry as well as the National Oil Company, in achieving their functions and objectives, shall be bound by the principles of the NEITI Act 2007. Already, and specifically, mostly as a result of the valuable work of NEITI in identifying institutional and process deficiencies, and suggesting remedies, such key agencies as the Department for Petroleum Resources (DPR) and the Federal Inland Revenue Service (FIRS) have undertaken far reaching reforms aimed at building their capacity to perform their tasks more effectively. NEITI has had a significantly beneficial impact on the Nigerian economy, state and society (including positive contributions to policy debates and reforms relating to governance of the oil sector). Its periodic audits have opened up, widely, a hitherto opaque industry to public scrutiny. The NEITI audit reports have placed immensely rich data and information in the public domain thereby strongly empowering civil society to hold government to account. The reports have facilitated the process of recovery of significant amounts of underpaid revenue for the state and helped to improve the domestic climate for foreign direct investment. By placing embarrassing facts and figures about the bulk of Nigeria's public revenue in the public domain, NEITI has become both an instigator of civic interrogation of public officers and a social safety valve, redirecting youthful energies from resorting to violent conflicts to engagement in civil debate on sensitive issues. The audit reports of NEITI will have even greater impact on the nation in the future, given the provision of the NEITI Act that the Auditor-General for the Federation shall, within three months after the submission of the report to the National Assembly, "publish any comment made or action taken by the Government on the audit reports" (contribution by Humphrey Asobie, NEITI Chairman and International EITI Board Member).

4.2 *Liberia*

In Liberia people are excited and intrigued by the simple but very insightful EITI process of publishing mineral and forest taxes and revenues. Finding in one document a detailed comparative data of what mining and logging companies paid to the government and what the government acknowledged receiving is like a dream to many Liberians who for several decades had no idea of the hundreds of millions of dollars that came to the government from exploitation of the country's vast natural resources. The LEITI 1st Report, especially the summary version, has therefore emerged as a prized document that is being read and discussed all over the country. Discrepancies in the report are cited by some as evidence of corruption and fraud while many Liberians believe that the establishment of LEITI is in itself strong indication that the current government is committed to accountability and transparency. This created a significant amount of interest and discussion within the communities. Who is operating the mine down the road? What and who are they paying? Why was this payment made? How is this having an impact on our community? How can we raise our concerns with government and with the companies themselves? These questions are being discussed openly with the companies and government through local community meetings organised by LEITI all across the country. Communities are also using this opportunity to raise questions about how the money is being allocated and used, and whether the communities are receiving a fair return for their resources. Prior to the existence of LEITI, there was no real forum where these types of discussions could take place. However, now, through this process, suspicion and distrust are being reduced, helping to diffuse the tensions that led to conflict in the past. Sustainable implementation of LEITI therefore promises to bring radical positive changes to the way forest and mineral revenues are collected, used and accounted for in Liberia (contribution by Negbalee Warner, LEITI National Coordinator and International EITI Board Member).

4.3 *Cameroon*

The EITI process in Cameroon has been helpful in improving the monitoring and management capacity of the relevant government agencies. Through the EITI programme, the capacity of government officials and NGOs, and to some extent the general population to understand oil operations, taxation, accounting, and audit and control has increased significantly. Furthermore, there has been an increased appetite for getting detailed information and explanations on the mobilisation and the use of oil revenue. Through these activities, the EITI complements other capacity development efforts in Cameroon, especially public financial management. Before the EITI process, petroleum activity in Cameroon was completely opaque. Now in 2010 Cameroon has seen the publication of EITI Reports for the years 2001–2004 and for 2005. Reports for 2006 and 2007 are soon to be produced (contribution by Faustin-Ange Koyassé, Senior Program Officer, World Bank–Cameroon).

5 CONCLUSION AND OUTLOOK

The natural resource wealth of a nation belongs to its citizens. Exclusion of civil society in making decisions regarding natural resources has contributed to mismanagement, a lack of accountability and increased opportunities for corruption in public institutions. By requiring that civil society organisations play and integral role in the design, monitoring and evaluation of EITI reporting processes, the EITI provides a platform where civil society can engage with the companies and government institutions involved in the extractive sectors. Through the EITI process civil society organisations improve their understanding of the extractive industries, better equipping them to ensure that the revenues from the exploitation and sale of natural resources are being managed in the public interest. The EITI helps to empower civil society by: In Gabon we are seeing how the EITI process is enabling individual citizens, and oversight and advocacy institutions, to access and monitor information on revenues from natural resources which they did not have before. The EITI has been key in creating a forum for the engagement of civil society and establishes their right to engage and building public understanding and awareness of the functioning of the oil, gas and mining sectors. Through this increased participation it has been seen in several EITI countries how the initiative is increasing the capacity of civil society to engage in wider debates regarding public financial management and the management of extractive industries.

REFERENCES

EITI International Secretariat 2010. *Impact of EITI in Africa–Stories from the ground*. Edited by Christopher Eads and Anders Kråkenes, Oslo, p. 16.

For further information visit: www.transparency.org and www.publishwhatyoupay.org

Geological Resources and Good Governance in Sub-Saharan Africa – Runge & Shikwati (eds)
© 2011 Taylor & Francis Group, London, ISBN 978-0-415-58267-4

Investment in extractive industries and sustainable development in Central Africa

Z. Tourere
Political Science and Planning Division, Programming and Planning Unit of MINRESI (Ministère de la Recherche Scientifique et de l'Innovation), Yaoundé, Cameroon

ABSTRACT: The mining industries require enormous investments and generate of important income to the states of Central Africa. The use of this income arouses numerous questioning of public opinions on the wealth distribution and the impact on the economic development and the environment. The rhythm and the type of exploitation of the extractive resources in Central Africa do not seem any more bearable long-term for diverse reasons (ecological, socio-demographic and political). The stakes in the sustainable development are the ones of the survival of the population of Central Africa even that of the humanity in the planet. The possible relations between the mining industries and the sustainable development are the investments and the oil income. In Central Africa, the investments seem certainly limited but the African countries arrange important income received from the sector of the mining industries. The good macroeconomic management and the fair distribution of the recipes of the mining industries, good governance and transparency can assure a sustainable growth and a reduction of poverty and be one of control levers for an effective starting up of the development for the countries of Central Africa.

1 INTRODUCTION

The development of mineral-rich African countries has become a societal issue dependent upon the activity of extractive industries. For almost fifty years African countries rich in oil and mineral resources, enabling them to generate significant financial resources, have failed to develop. Instead of being an asset, mineral resources and oil deposits have become a curse and a source of conflict, fuelling a chain of corruption in Central Africa. They are the principal cause of power struggles and rigged or distorted elections in some countries. This gives rise to the 'paradox of the resources curse', with countries possessing excellent oil revenue but very low levels of human development.

Extractive industries can require significant investment along their entire value chain. The oil industry, for example, can be split into the upstream activities of exploration and production, and the downstream activities of refining and distribution. All these stages demand significant physical investment and a skilled workforce. *What needs to be done in order to convert this potentially considerable investment in extractive industries into sustainable development in African countries?*

2 INVESTMENT IN EXTRACTIVE INDUSTRIES IN CENTRAL AFRICA

Extractive industries extract mineral products present in their natural state, either as solids (coal and minerals), liquids (oil) or gases (natural gas). These products can be extracted in a variety of different ways, such as by underground or open-cast mining, by digging wells or by extracting them from the seabed. Two types of raw material are extracted—hydrocarbons and minerals.

The hydrocarbons extracted include crude oil and gas, involving mining and/or developing oil and gas fields. The activities involved in extracting hydrocarbons may include drilling, constructing and equipping wells, and processing the oil and gas up to the point where it leaves the extraction site.

Mineral extraction involves underground mining, open-cast mining and extraction from the seabed. It may include processes to treat and enrich minerals, such as crushing, grinding, washing, sorting, drying, agglomeration and separation by gravity or flotation.

The oil industry can be split into the upstream activities of exploration and production, and the downstream activities of refining and distribution. The upstream activities include exploration by traditional prospecting methods such as sampling, geological observations and trial drilling. Governments grant some companies rights to search for and extract oil. Downstream activities include the operations necessary to prepare the raw materials for marketing, namely refining and distribution. Refining simply involves distilling oil to separate it into heavier and lighter fractions using a variety of processes. Crude oil is usually transported from production areas to refineries via pipelines or, if long distances are involved, by ship. Rail and road tankers are used primarily for the final distribution of the refined products.

The extractive sector therefore brings together various different structures—private multinational companies, refining companies, national companies and 'independent' research and production companies.

All these activities potentially require a huge amount of investment and generate revenue in the various countries of Central Africa. Public opinion is questioning the way the authorities use revenue from extractive industries due to its impact on wealth distribution, economic development and the environment.

We have for example the discovery of the oil to Doba in 1969 and its economic possibility of development according to the quantities existing in the South of the Chad which were at the origin of the pipeline Chad/Cameroon. Chad being an enclosed country, Cameroon was chosen by the promoters as way of evacuation of the Chadian oil in the export. At first, it is the oil Exxon companies (40%), Shell (40%) and Elf (20%) constituted in a consortium and Chadian and Cameroonian governments which had to finance respectively in 97% and 3% the realisation of this oil pipeline. The socio-politic context of both countries and the impossibility of both states to mobilise in house the required 3% brought these last ones to seek the approval of the World Bank to finance their share. Indeed, these two countries were under structural adjustment and confronted with grave financial difficulties. The participation of the World Bank amounted to 130 million dollars among which 70 million dollars (42 billion FCFA) for Cameroon and 45 million dollars (27 billion FCFA) for the Chad.

The pipeline Chad/Cameroon assures the transport of the Chadian crude oil in Central Africa. It is an oil pipeline 1070 km long between Doba (Chad) and Kribi (Cameroon) with three (03) pumping plants (Doba, Dompta, Belabo), a station of reduction of pressure to Kribi and a submarine oil pipeline binding the Atlantic Coast of Kribi with the floating unity of storage and unloading. The pipeline Chad/Cameroon crosses five (05) ecological zones, namely: the zone of savanna of Doba (Chad) to Deng Deng (Cameroon), the zone of the small forest of Deng Deng to Batschenga, the urban zone of Obala to Ngomedzap, the zone of the dense forest of Ngomedzap to Kribi and the maritime zone of the Atlantic Coast of Kribi in the floating unity of storage and unloading. The exploitation of the pipeline Chad/Cameroon, built from 2001 till 2003, is for a period of 25–30 years.

The American Hydromine Inc company signed on January 13th, 2006 an agreement with the Cameroonian government for the construction of a refinery of alumina for Ngaoundal capable of treating seven (07) million tons of bauxite a year to obtain 2,8 million annual tons of alumina, a méga—oil refinery, the biggest refinery of the crude oil in Africa in the South of Sahara intended for the export. The alumina will be transported by the railroad existing between Edéa and Ngaoundal and the reunification which will be built between Edéa and Kribi.

The American Géovic company holds, since 2003, a licence of industrial undertaking concern of ores: the cobalt, the nickel and the manganese, which will be exploited during 25 years, from 2012. The works have even already begun at the level of Nkamouna, in the region from the east of Cameroon.

The extractive sector in Central Africa requires heavy and expensive investments certain countries of which have no capacities. The countries of Central Africa arrange capacities of refining limited by oil productions, in Cameroon (the SONARA to Limbé), in Congo (a small refinery to Pointe-Noire), etc. ... These capacities limited by refining do not manage to satisfy the local demand. So upstream as downstream to the mining industries, the investments of the countries of Central Africa are limited.

The national companies knew recently difficulties to finance their part of investments in joint-ventures and preferred production sharing contracts today. Furthermore, they are the object of the attention of the financiers in their management for the granting of the loans on States. We have for example: the SNPC in Congo, SNH in Cameroon, etc. ...

The big international companies: ExxonMobil, Shell, ChevronTexaco, Agip, TotalFinaElf, etc. ... are present in Cameroon, in Gabon, in Congo, etc.

These diverse companies express in the African countries their power in term of financial means and technical knowledge. They have a strong capacity of investment and a source of easy credit for States. The division of the profits between these companies and the State is essentially determined by the future receipts, where from the emergence of production sharing contracts.

3 SUSTAINABLE DEVELOPMENT IN CENTRAL AFRICA

Growing population levels and standards of living are ultimately likely to be limited due to the depletion of natural resources in Central Africa. In other words, the type of growth we are currently seeing in Central Africa would appear to be unsustainable in the long term for reasons that are essentially environmental, demographic, social, or indeed even political. Sustainable development[1] is development that meets the needs of present generations without compromising the capacity of future generations to meet their own needs.

The Environment Commission of the 11th Godard and Theys Plan focused the concept of sustainable development on four different interpretations:

- the requirement to meet future human needs;
- the requirement to retain sufficient 'natural capital' such that natural growth is not overtaken by what man extracts;
- the requirement to maintain real per capita income (the neo-classical approach);
- the requirement to balance development with social issues.

The challenges of sustainable development therefore equate to those of mankind's survival on Earth. Three factors emerge from this concept of sustainable development: environmental sustainability, social sustainability and geopolitical sustainability.

The underlying cause of many conflicts and wars in Africa and of global terrorism is oil and mineral resources, which are leading to growing inequality. In 1992, the Rio de Janeiro Summit in Brazil indicated that everything mankind needs to survive is under threat—water, air, soil, underground resources, forests, biodiversity, climate, health, demography and sharing the world's resources among its population. The need to act is therefore urgent, but the distinctive feature of progress is still the slow pace at which international action is being taken. Two possibilities for action[2] are apparent:

- A liberal approach consisting of generalising the market principle so that scarce resources continue to be allocated solely on the basis of private contracts. This approach opens up a new international market in the right to pollute.
- A more interventionist approach consisting of generalising the principles of caution[3] and 'the polluter pays'[4].

However, every possible action has its limitations.

4 POSSIBLE RELATIONSHIPS BETWEEN EXTRACTIVE INDUSTRIES AND SUSTAINABLE DEVELOPMENT IN CENTRAL AFRICA

The oil and mining sectors contribute towards the diversity of economic and industrial activity. They represent a source of revenue and currency, and help to create jobs. In many Central African countries, the various regulations relating to oil, gas and other minerals have been revised to attract foreign investors. Market liberalisation and the introduction of investment-friendly legislation have prompted a revival in the extractive sector, particularly exploration. Central Africa has considerable relative advantages, with a very favourable tax system and mining agreement terms (production sharing agreements and licence agreements).

Public opinion is increasingly questioning the way in which public authorities are using revenue from extractive products, due to its impact on wealth distribution, economic development and the environment.

The exploitation of oil, gas and other mineral resources could leverage sustainable development in countries that have low revenue levels but abundant natural resources. By mining natural resources, these countries have or could have revenue in the form of taxes and duties. Natural gas covers around two thirds of global energy requirements. Oil is a strategic raw material because it meets around 40% of the planet's energy needs. It is virtually the only source of fuel for means of transport such as cars, planes and boats. Without oil, economic activity in today's world would grind to a halt and many sectors would be paralysed, for example, the armed forces.

5 CONCLUSION

Due to the revenue it generates, oil is a resource that is vital for the development of Central African states. It represents an important factor in international trade, with annual production worth 500–750 billion dollars. Good macro-economic management, the equitable distribution of extractive industry revenue, good governance and transparency could provide sustainable growth and a reduction in poverty.

Possible relationships between the extractive industries and sustainable development involve investment and oil revenue. Although investment in Central Africa seems limited, African countries do have access to significant revenue from the extractive sector. Good governance and the transparent management of extractive industry revenue could really jump-start development in Central Africa.

ENDNOTES

[1] The concept of 'sustainable development' first became clearly apparent in 1987 with the Brundtland Report, but its foundations were laid 15 years before that at the Stockholm Conference in 1972 on the relationship between human development and the protection of the environment.

[2] N.B. By refusing to sign the Rio Convention on climate change, the United States has opted for the first approach.

[3] No new activity should be undertaken unless it has been proved to be safe and sufficiently sustainable.

[4] No activity that damages the planet's essential natural capital should be undertaken unless the cost of undertaking it includes the cost of sufficient renewal of this natural capital.

REFERENCES

Etong Oveng, P. 2007. The paradox of the mining industries. *Infoplus Gabon.*

Gendron, C. 2006. *The development sustainable as compromise.* Gendron C. Ed. PUQ.

Guillaud, Y. 2008. *Biodiversity and sustainable development.* UNESCO, 2008.

Harsh, E. 2005. *The investors turn to Africa.* Harsh, E. Ed. Afrique Renouveau.

Kimani, M. 2009. The African mining industry. *Afrique Renouveau.* April 2009.

Ndanga Ndinga, B. 2008. Better make profitable the mining resources of Cameroon. Cameroon *Tribune*, 25 January 2008.

OECD, 2008. *Economic perspectives in Africa.* OECD, African Development Bank, June 2008.

Panitchpakdi, S. 2007. The investment in Africa: the futures challenges. CNUCED.

Theys, J. 2005. *The sustainable development: an underexploited innovation.* Dalloz, Paris.

Regional approaches and activities of the private sector (companies)

Geological Resources and Good Governance in Sub-Saharan Africa – Runge & Shikwati (eds)
© 2011 Taylor & Francis Group, London, ISBN 978-0-415-58267-4

Regional organisations' approach to mining and exploitation in Sub-Saharan Africa

H. Mtegha, C. Musingwini, S.V. Rungan & O. Oshokoya
School of Mining Engineering, University of Witwatersrand, Johannesburg, South Africa

ABSTRACT: Most Sub-Saharan African member states are rich in mineral resources. Member states recognise that the efficient exploitation of these resources could lead to great benefits accruing to their economies. However, most member states do not possess adequate factors needed for mineral development and these include insufficient skills, inappropriate policy and legislative frameworks, inadequate markets, insufficient infrastructure, insufficient technological knowhow amongst others. A regional approach to address these issues provides an opportunity for member states to attract the requisite investment and maximise benefits from mineral exploitation. This paper examines efforts undertaken in mineral resources exploitation in the eight regional economic communities of the African Union in Sub-Saharan Africa. It notes that all Regional Economic Communities (RECs) recognise the need for, cooperate and coordinate natural resources development in general. In particular, three RECs namely, EAC, ECOWAS and SADC have specific programmes in mineral resources development, which range from coordination of effort to development of a common mining code. In general, regional organisations are participating in the minerals sector by focussing on harmonisation of policies, laws and regulations.

1 INTRODUCTION

Regional Economic Communities (RECs) are established throughout Sub-Saharan Africa (SSA), and the Treaty (African Union 1991) establishing the African Economic Community sets a vision of a continental community. Currently, there are eight RECs in Africa (African Union/UNECA, International Study Group 2009) as shown in Table 1 below, of which SSA countries are members in seven of these. Membership to RECs is duplicated in many cases where a member state belongs to more than one grouping.

A regional, hence continental, approach is important to meet Africa's developmental challenges emanating from individual country weaknesses. It is anticipated that such an approach would entail benefits including new trade opportunities, larger markets; increased competitiveness, facilitate larger investments, commit governments to reforms, increase bargaining power, enhance cooperation and improve security (ECA 2004). It is envisaged that the attainment of these benefits would improve the welfare of the peoples on the continent. Progress towards integration is sector-based, for example transport, trade, mining and communications. This starts with cooperation and collaboration through policy harmonisation and alignment at regional level.

The Economic Commission for Africa report (ECA 2004) gives reasons for integration in the minerals sector as:

i. Lack of critical mass to develop the sector by individual member states;
ii. Large markets for downstream value addition;
iii. Exploiting economies of scale, attracting and retaining resources (financial, technical and human).

Cooperation and integration in the minerals sector at the continental level is driven by the African Union guided by the Mining Vision 2050 (UNECA/African Union 2008), which is a product of several initiatives drawn from RECs, amongst others. The Vision aims at utilising mineral resources for industrial development to transform Africa towards modernisation. An integrated African market would be achieved by developing downstream processing and manufacturing; upstream linkages into inputs to the mineral value chain; and side stream to infrastructure including physical (transport, energy, telecommunications) and human skills with the requisite research and development (R&D). A Framework for Action to be done at national, sub-regional and continental levels to implement the vision was developed. To achieve these goals it was recommended that the local private sector, communities and

Table 1. African Regional Economic Communities in SSA (African Union/UNECA, International Study Group 2009).

Regional Economic Community	Geographical spread
Arab Maghreb Union (UMA)	5 members encompass most of North Africa
Common Market for Eastern and Southern Africa (COMESA)	19 members include Egypt in North Africa, all East African countries except Tanzania and seven countries of Southern Africa
Community of Sahel-Saharan States (CEN-SAD)	28 members are in West, Central and Northern Africa
East African Community (EAC)	5 member countries in East Africa
Economic Community of Central African States (ECCAS)	10 members spanning across Central Africa
Economic Community of West African States (ECOWAS)	15 members encompass all of West Africa
Inter-Governmental Authority on Development (IGAD)	12 countries in the Horn of Africa and the northern part of East Africa
Southern African Development Community (SADC)	14 members cover all of Southern Africa

other stakeholders needed to participate fully. In addition, revenues derived from such activities needed pragmatic distribution to benefit all. These actions imply adequate capacities at various levels of government for competent sector management. RECs will play a key role in the realisation of the continental vision through implementation of programmes or projects in key areas.

2 SUB-SAHARAN MINERALS SECTOR

Sub-Saharan Africa (SSA) is greatly endowed with vast amounts of diverse mineral resources, distributed unevenly across its member states. The mineral resources include the following:

i. Metallic substances—platinum, nickel, copper, iron ore, tin;
ii. Non-metallic substances like manganese, salt, phosphate;
iii. Industrial and construction substances—limestone, clay, cement, granite, dimension stones, sand, gravel;
iv. Precious minerals and gemstones—gold, diamond, silver, ruby, sapphire, opal;
v. Semi-precious stones—aquamarine, and other vein gemstones; and
vi. Fuels and petroleum products—crude oil, natural gas, coal.

In numerous cases these minerals do not recognise political borders between the member states. In several instances mineral endowment is similar and common in the member states. What may be different is the level of geological information about the specific mineral resources in particular countries. In terms of mineral development, there are similar problems or obstacles. These problems vary from country to country and their magnitude also differs. These obstacles include the following:

i. Inadequate geological knowledge of mineral resource endowment;
ii. Inadequate capacities and competences to manage the sector, both in the public and private sectors;
iii. Inadequate infrastructure for skills training and the requisite research and development (R&D); and
iv. Inadequate physical infrastructure required for mineral resources development, processing and value added manufacturing.

In the SSA countries, all mineral resources belong to or are vested in the State. This custodianship of the mineral resources places responsibility on the State (government) to ensure that the exploitation of its mineral resources makes a beneficial contribution to the living standards of its citizens. The State, therefore, determines the rate and manner of mineral development through actions, amongst other things, by:

i. Determining necessary procedures for mineral development;
ii. Granting mineral rights to applicants who demonstrate adequate technical, financial and managerial capability to engage in exploitation activities that will benefit the state;
iii. Charging royalties and taxes for any transfer of resources for the purpose of compensating the State for its irreplaceable, non-renewable resources, in order to ensure sustainable investment and development, thereby improving the economic well-being of the nation;
iv. Providing a stable environment in terms of cost of public services and goods, which would be attractive to investors;

- v. Providing state support for research in exploitation techniques;
- vi. Providing effective dissemination of non-confidential, State-held geological information; and
- vii. Reviewing of policies and regulations that constrain downstream development.

It is this responsibility of the state that propels countries to make efforts to address the sector and its issues at the regional level to address bottlenecks with a view to maximising benefits through integration. With similar problems across many member states, common concerted solutions would go a long way towards achievement of national, hence continental goals. These common actions form regional agendas for mineral sector development.

3 ROLE OF THE MINERALS SECTOR

The positive role these vast mineral resource endowments plays, in the member states that are privileged to possess them, cannot be overstated. In the present era of open-trade between countries, the mineral resources of these countries help to foster:

- i. International trade for marketing of mineral products;
- ii. Foreign exchange earnings from exportation of goods;
- iii. Inflow of foreign direct investments in exploration and exploitation;
- iv. Training of local citizens/miners;
- v. Local demand for the promotion of infrastructural development, which increases the economic viability of its metals and industrial minerals deposits;
- vi. Beneficiation (value-adding) activities and refining of minerals.

The resources also potentially promote/attract resource-based industries and clusters of other industries, thereby assisting in diversification of mineral-dependent economies.

The gross domestic product (GDP) has been used over time to ascertain the mining industry's contribution to national economies although it may not be the most appropriate summary of overall economic performance in all countries at all times especially in those countries where largely foreign-operated extractive industries are substantial due to consideration of payments to foreign-owned factors. In most of the Sub-Saharan African countries, the mineral sector is in a fledgling/ underdeveloped stage, with this sector playing a minor role in their national economies (GDP) and exports in countries such as Benin Republic, Somalia, Burkina Faso, Togo, Reunion and Mali. These countries are still largely dependent on export earnings from agriculture and services, with the mining sector taking secondary place in terms of economic activity. Table 2 below is a list of some of these SSA countries in which the mineral sector plays a minor role in the national economy.

While the economies of a considerable part of Sub-Saharan Africa are not mainly dependent on the mining sector, another significant part comprising countries like Equatorial Guinea, South Africa, Gabon, Nigeria, Botswana, and others have the mining sector accounting for a significant part of their Gross Domestic Product. Some of these mining sectors have close to or more than 30% of GDP contribution; contribute to more than 50% of government revenue and over 50% of exports. Table 3 below is a list of some of these SSA countries with significant mineral sectors.

With post-2000 commodity boom, which peaked between 2007 and 2008, several resource-rich, low-income countries displayed high rates of GDP growth (although social indicators did not improve significantly). During the global recession, many of the resource-rich SSA countries with stronger trade ties to the global economy experienced both low global and local demand for commodities. However, as recovery from the recession progresses, improved mining/extraction technology, additional exploration, and the increased international demand for mineral commodities will continue to provide the impetus for the development of the infrastructure necessary to capture these resources and the industry in general. The member states also continue working together more progressively to satisfy their individual needs, thereby, providing a market for the rest of the SSA's mineral products and promoting the development of mineral industries of source countries.

Table 2. Contribution of the minerals sector to GDP in some SSA countries.

Country	Percentage contribution of mineral sector to GDP (economy) and exports
Burkina Faso	As at 2009, the mining sector accounted for 2.8% of GDP although gold accounted for 41% of total exports.
Burundi	As at 2007, manufacturing, mining, and energy accounted for nearly 11% of GDP.
Central African Republic	As at 2004, uncut diamonds made up nearly 60% of export earnings.

Table 3. Economically significant mineral sectors in SSA countries.

Country	Percentage contribution of mineral sector to GDP (economy) and exports
Angola	As at 2004, diamond production accounted for about 95% of non-fuel exports and for about 10% of the non-fuel GDP. Petroleum production accounted for about 75% of government revenue
Botswana	As at 1997, the mining sector accounted for 80% of the value of national exports, about 50% of government revenue, and about 33% of the GDP. As at 2010, the mining sector still accounts for about 33% of GDP.
Equatorial Guinea	As at 2009, hydrocarbons accounted for more than 61% of GDP .
Gabon	As at 2001, the mining sector, which was dominated by crude petroleum production, accounted for about 50% of GDP.
Mauritania	As at 2009, mining (iron, gold and copper) showed satisfactory results, accounting for 36.5% of GDP despite fluctuating world prices.
Nigeria	As at 2010, petroleum accounted for 40% of GDP and 80% of government earnings.
South Africa	As at 2009, agriculture and mining (primary sector) accounted for 7% and industry (secondary sector) accounted for 20%.

Despite the potential of mineral endowments of promoting increased economic performance, a disappointing trend for the most of the mineral-rich SSA countries has been observed over time. In a majority of the Sub-Saharan African countries, it has been observed that these resource-rich countries have not delivered expected economic growth and improved social living standards as compared to other countries in Africa, at similar levels of development. From the data in Table 4, it is observed that as at 2010, the underperformance of most of the mineral-rich Sub-Saharan African countries, as compared with their fellow African countries, is still the case.

Libya, Tunisia, Algeria, Egypt and Morocco, although not classified as Sub-Saharan African countries are equally mineral-rich. However, from Table 4 it can be seen that they are found to fall under High and Medium Human Development categories, whereas their mineral-rich counterparts in Sub-Saharan Africa fall under the low Human Development category with the exception of Gabon, Botswana, South Africa and Namibia, which fall under the Medium Human Development category.

Also, according to Hamilton & Ley (2010), from 1991 to 2008, it was observed that in most SSA countries, there was an improvement in GDP but in light of adjusted Net National Income (αNNI), there was a downward trend implying shrinking national wealth (Fig. 1).

Net National Income (NNI) and adjusted Net National Income (αNNI) are defined as follows (Hamilton & Ley 2010):

$$\textit{Net national income (NNI)} = \textit{GDP} + [\textit{Net foreign factor income}] - [\textit{Depreciation of fixed capital}]$$

$$\textit{Adjusted net national income } (\alpha NNI) = \textit{NNI} - [\textit{Depreciation of natural capital}]$$

From observations from the Human Development categories and the analysis of GDP-αNNI, it can be drawn that the presence of mineral resource endowments does not necessarily pose a threat to economic growth, but points to liquidation and dissipation of the wealth inherent in the resource stocks of some of these countries (Hamilton & Ley 2010) and misuse of these windfalls to build up, broaden and increase the international competitiveness of more stable sectors of the economy (Mainardi 1995). Information from the World Bank indicates that since the 1990s, these underperforming SSA countries have been expending more than their net national incomes, especially during much of the recent resource boom (Hamilton & Ley 2010), apparently due to overconfidence in the potential of mineral resources to pay back loans, under the illusion of an almost infinite inflow of their mineral rents (Fig. 2).

According to Mainardi (1995), the over-spending of SSA countries above their net national incomes has resulted in deficiencies in the allocation of resources giving rise to: i) "an excessive capital-labour ratio" and ii) "a large presence of foreign capital and a significant proportion of export earnings flowing out of the country to service debt obligations, and relatively weaker intersectoral production links".

This mismanagement of the mineral endowments needs to be addressed in order to guarantee a sustainable well-being for both present and future generations. This can be achieved by implementing sustainable development policies that would result

Table 4. Human Development Indices (HDI) for African countries (United Nations Development Programme, 2010).

Country	New HDI estimates for 2010
High Human Development category	
Libya	0.755
Mauritius	0.701
Tunisia	0.683
Algeria	0.677
Medium Human Development category	
Gabon	0.648
Botswana	0.633
Egypt	0.620
Namibia	0.606
South Africa	0.597
Morocco	0.567
Low Human Development category	
Kenya	0.470
Ghana	0.467
Cameroon	0.460
Benin	0.435
Madagascar	0.435
Mauritania	0.433
Togo	0.428
Comoros	0.428
Lesotho	0.427
Nigeria	0.423
Uganda	0.422
Senegal	0.411
Angola	0.403
Djibouti	0.402
Tanzania	0.398
Côte d'Ivoire	0.397
Zambia	0.395
Gambia	0.390
Rwanda	0.385
Malawi	0.385
Guinea	0.340
Ethiopia	0.328
Sierra Leone	0.317
Central African Republic	0.315
Mali	0.309
Burkina Faso	0.305
Liberia	0.300
Chad	0.295
Guinea-Bissau	0.289
Mozambique	0.284
Burundi	0.282
Niger	0.261
Democratic Republic of Congo	0.239
Zimbabwe	0.140

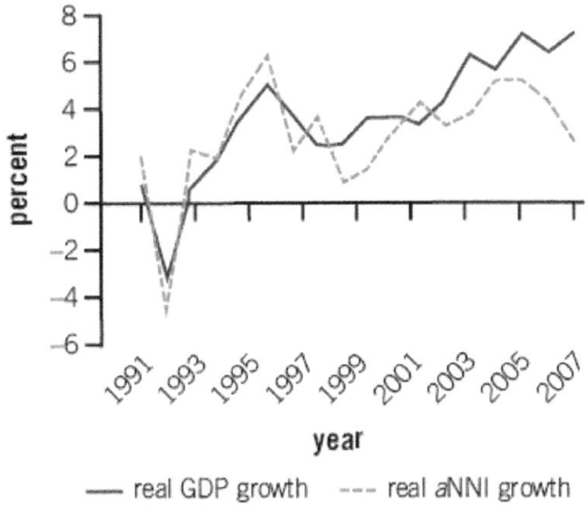

Figure 1. Growth Rates of GDP and *a*NNI in Sub-Saharan Africa, 1991–2008. (Source: Hamilton & Ley 2010).

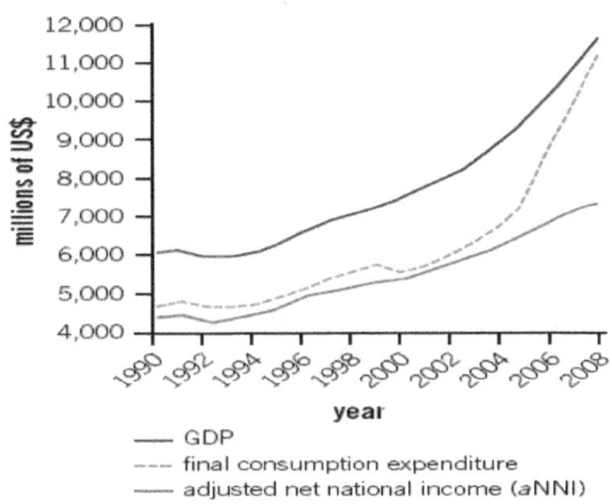

Figure 2. Consumption, GDP, and *a*NNI in Sub-Saharan Africa, 1990–2008 in Constant US$2000. (Source: Hamilton & Ley 2010).

in increasing national wealth. Increasing national wealth requires investment, national savings as well as foreign savings transferred as aid, therefore, proceeds/rents drawn from the mineral resource assets can be invested in the key reproducible inputs to sustain economic growth—produced, human and natural capital, technological know-how, in order to enjoy a constant stream of consumption (Hamilton & Ley 2010).

In all, the contribution of exploitation of mineral commodities to the region, is dependent on long-term (near permanent) and medium-term changes in demand and commodity price changes/trends which are driven by such factors as supply constraints (which can be instigated by factors such as size and quality of reserve, technology needed for exploitation, cartelisation, and available substitutes); a change in international market structure; favourable market conditions; a technological shock, temporary changes around these trends; and different policy responses (Radetzki 1989).

Across the SSA region, there are mutual factors that can be identified as hindrances to mineral resources development and hence their reduced contribution to sustainable development. These factors include weak/inadequate infrastructure—transportation, electricity and telecommunication; weather patterns as relates to their effect on the weak infrastructure; socio-political instability; small markets; lack of reliable/sound educational sector; lack of technological know-how and skilled local expertise. Many of these problems are largely enormous for some individual countries to tackle on their own. In order to promote development and provide mutual assistance for addressing the abovementioned problems, the organisation of countries in the same sub-regions has been fostered. This is in alignment with implementing the Regional Integration in Africa initiative of the African Union's integration and development agenda (Mkwezalamba and Chinyama 2007). These organisations collaborate to brainstorm and proffer effective solutions.

4 REGIONAL ORGANISATIONS APPROACH TO MINING AND EXPLOITATION

A mineral policy represents a vision or aspiration for a country to benefit from its mineral resources. Each country or society may have unique circumstances regarding:

i. Level of geological knowledge or mineral endowment;
ii. Emphasis on mining due to differing contribution of the sector to the economy;
iii. Level of infrastructure development, particularly those that are relevant to mining;
iv. Stages of mineral development, whereby some will have a tradition in mining while others will only be starting; and
v. Each has its own history, politics, tradition and culture, and resources at its disposal for mineral development.

These factors, amongst others, will have a bearing on policy choices by individual countries regarding the mineral sector. That is one of the reasons why mineral policies differ between countries. Otto (1997) contended that a mineral policy is only part of the overall policy framework of a country. The mineral policy should be carefully defined so that the mineral sector and mineral law, which implements policy, do not operate in isolation from other sectors and laws. Otto further added that careful consideration should be given to other policies and practices to ensure compatibility. Therefore, a carefully compiled minerals policy that is able to accommodate the differences between states could be applied at a regional level and could lead to harmonisation in the sector (ECA 2004).

Where several countries are considered, policy instruments within mineral policies could be clustered in themes (Mtegha 2005). Such themes could include mineral administration (such as licensing, social issues, small-scale mining, governance, environmental issues, value addition, R&D, and training). It is necessary to identify common elements and synergies in each theme, carefully considering unique features that are specific to a country and leaving them out of the regional thrust. Issues or concerns for each theme from the countries can be summarised and form part of background information, conclusions derived and policy statements crafted for them. This forms a basis of policy alignment.

The next phase would be the adoption of common standards, which would involve those issues that have accepted levels. These will include, but are not limited to, safety, health, environment, and skills training. Most of these can easily be benchmarked against international or best practices.

A third stage is the adoption of common codes or common elements of legal regimes. Again, a first step would be that adopted by developing a harmonised mineral policy. This is an area probably quite too close to national sovereignty requiring cautious treading. But, as will be seen from cases below, these efforts are already underway in some RECs in Sub-Saharan Africa.

All in all, the overriding impetus of regional effort is to create an enabling environment for entrepreneurs to extract each region's minerals profitably while at the same time affording governments the opportunity to extract maximum benefits arising from mineral exploitation. These considerations take place along the whole extractive value chain from exploration to manufacturing activities. In short, a combined effort addresses individual shortcomings and derives more benefits to all member countries of the region.

The next sections discuss specific SSA regions' approach in mining and exploitation.

4.1 *Common Market for Eastern and Southern Africa (COMESA)*

Article 4 of the Treaty establishing COMESA in Section 6 (h), states that the region will co-operate in the development and management of natural resources, energy and environment. The REC, however, does not have programmes specifically covering the minerals sector. It can be assumed that the intention is there as minerals are part of natural resources. For the moment the REC focuses on the concept of a large and trading unit addressing constraints faced by small developing countries.

4.2 *Community of Sahel-Saharan States (CEN-SAD)*

The members in this REC are within or near the desert. They aim at integrating national plans in identified areas of agriculture, industry, social, cultural and energy. The mineral sector is not specifically pursued. The REC's secretariat is located where it overlaps the Arab Maghreb Union (UMA). This case is not unique as a different agenda may be developed by UMA.

4.3 *East African Economic Community (EAC)*

Article 114 of the Treaty establishing the East African Economic Community (EAC) relates to the management of natural resources. The provisions regarding mineral resources are as follows:

i. Promote joint exploration, efficient exploitation and sustainable utilisation of shared mineral resources;
ii. Pursue creation of an enabling environment for investment in the mining sector;
iii. Promote the establishment of databases, information exchange networks and sharing of experiences in the management and development of the mineral sector using electronic mail, internet and other means for the interactive dissemination of mineral information;
iv. Harmonise mining regulations to ensure environmentally friendly and sound mining practices;
v. Adopt common policies to ensure joint fossil fuel exploration and exploitation along the coast and rift valley; and
vi. Establish a regional seismological network whose primary objective is to monitor seismicity and advice on mitigation measures.

This reflects a cautious approach towards co-operating in mineral resources management. Harmonisation has not yet started and implementation is still in its infancy. So far indications are towards harmonising policies, undertaking joint programmes where they benefit all and adopting common standards in the case of established areas of environment and mining.

4.4 *Economic Community of Central African States (ECCAS)*

The Economic Community of Central African States (ECCAS) has established a protocol on co-operation in natural resources. There is no mention about mineral specific actions. With an increase in the demand for mineral commodities it may only be a matter of time before the sector takes active attention by policy makers.

4.5 *Economic Community of West African States (ECOWAS)*

The ECOWAS issued a Directive (ECOWAS 2009), developed by various stakeholders that included governments, civil society organisations and communities affected by mining. The Directive was agreed on May 27, 2009. The objectives of the Directive are specified in Article 2 as follows, to:

i. Provide harmonisation of guiding principles and policies in the mining sector of Member States to ensure high standards of accountability for mining companies and governments, promoting human rights, transparency and social equity as well as providing protection for local communities and the environment in mining areas within the sub-region;
ii. Provide a mining environment that is responsive to macroeconomic sustainable development and balances the need to provide appropriate incentives to attract investors and to protect the revenue base and resources of Member States;
iii. Improve transparency in mineral policy formulation and implementation processes in mining within the sub region to promote the participation and enhance the capacity of mining communities;
iv. Provide for a harmonised mineral policy and legal framework for Member States; and
v. Ensure that Harmonisation takes into consideration the different stages each Member State is at in relation to mining and how different policies could be enacted and/or developed for different strategies to address the specific needs of each Member State.

Chapter III of the Directive, under Articles 3 to 11, identifies areas that require harmonisation. Each area contains regional intentions to be articulated in eventual implementation strategies. The areas that are covered are:

i. Vesting of Mineral Resources in the State;
ii. Acquisition or Occupation of Land for Mining;
iii. Acquiring Mineral Rights;
iv. Environmental Protection Obligations;
v. Stability Agreement;
vi. Fiscal Framework;
vii. Transferability of Capital;
viii. State Participation in Mining Operations; and
ix. Localisation Policy of Mining Operations.

The main objectives of the Directive are the adoption of the mining policy, and the adoption of a common mining code for ECOWAS (Action Plan 2009). In the development of a Common Mining Policy (CMP), each Member State was studied regarding formulation and implementation of mining policy; assessment, analysis and evaluation of policy tools were undertaken. This then culminated in the compilation of a draft Common Mining Policy Document (CMPD) together with a five years' promotion plan. For effective implementation of this policy, a capacity audit was also planned to be undertaken and capacity-building programme to run concurrently with the promotion plan. To ensure that this Action Plan is implemented, a regional institution was created at the REC level, whose responsibility includes the dynamic acquisition of information for decision-making at the REC level in the development of the CMP.

In the development of a Common Mining Code (CMC), mining legislation in each Member State was studied to examine alignment with the ECOWAS Directive and where divergences occurred, proposals made for convergence. The process of developing the CMC is in a series of steps, including sensitisation of the various stakeholders in the Member States, and expected to culminate in its adoption by the ECOWAS Heads of States and Governments in December 2011. The success of the regional effort in facilitating beneficial mineral development in ECOWAS will depend on the adoption and implementation of the mining policy and mining code as planned by the REC.

4.6 *Inter-Governmental Authority on Development (IGAD)*

The objectives of the Inter-Governmental Authority on Development (IGAD) allude to the sustainable development of natural resources and environmental protection. There seems to be no documented programmes in the minerals sector. Again, these members belong to other RECs where mineral programmes exist, for example Kenya and Uganda belong to IGAD and at the same time to EAC, which has mineral programmes. Member states from REC may undertake different agendas depending upon group consensus.

4.7 *Southern African Development Community (SADC)*

Cooperation in the minerals sector of the Southern African Development Community (SADC), then Southern African Coordination Conference (SADCC), group started in 1984. At that time the REC's overall political agenda was of self-reliance and severance of dependence on the then apartheid South Africa. It was also at a time when most member states were actively participating, in one way or another, in mineral extraction activities. Cooperation was therefore based on actual extraction projects where regional projects were identified and considered for adoption if they benefited two or more countries. For example, a mineral deposit extracted in one country must also benefit a second or third country; and a laboratory set up and promoted in one country must benefit another. One of the very successful projects was the provision of postgraduate geological training and support to Geological Surveys of most SADCC member states by France. This project approach had its own shortfalls, but it helped member states to:

i. Identify common problems and opportunities and hence find common solutions; and
ii. Learn to work together, discuss common sector issues and look for common solutions.

By 1990 it became apparent that the REC's mineral sector needed bold moves to attract private investment into the region. Several initiatives were made including investigating mineral regimes for improvement, hosting REC's mining investment forums (held every four years), and finally started work to harmonise mineral policies and regulatory regimes. One of the developments arising out of this process was the SADC Mining Protocol. The SADC Mining Protocol has six elements:

i. Harmonisation of national policies to the objectives of regional integration;
ii. Improving availability of economic information to the private sector;
iii. Promoting private sector participation in the regional mining industry;
iv. Facilitating the development of human and technological capacities;
v. Promoting responsible small-scale mining; and
vi. Observing internationally accepted standards of health, safety and the environment.

The protocol provided the impetus for subsequent actions that included the development of:

i. A strategic plan to operationalise the mining protocol;
ii. Harmonised framework for mining policies, standards, legislative and regulatory issues; and
iii. An implementation plan of the harmonised framework.

The harmonised framework was benchmarked against international best practice as a basis to attract investment and compete internationally. As a first step, the framework recommended member states to develop their mineral policies according to an agreed template. This would enable application of a uniform thought process and issues that should be articulated. It would thus be easier to develop common threads along the member states on key themes and issues for eventual harmonisation of policy instruments that are in the end translated into legislation. The rest of the framework contains the following themes:

i. Political, economic and social environment, which deal with general macro-economic policy issues and not mining industry specific, but important to it;
ii. General investment regulations, dealing with state participation and exchange controls;
iii. Mining fiscal environment, dealing with international, national and local/regional government tax issues;
iv. Mineral administration and development systems, dealing with rules for licencing;
v. Artisanal and small-scale mining, dealing with appropriate legislation for the promotion and management of the same;
vi. Research and Development, dealing with creation of new knowledge and innovation;
vii. Human Resources and Skills Development, dealing with skills shortages; and
viii. Gender, dealing with women participation in mining.

Each member state would work towards these recommendations as a sovereign state, in which case if all countries are motivated to do the same they would eventually have similar provisions.

The implementation plan (UNECA 2008) of the harmonised framework was for areas that can be done at regional level. These are the areas that once attended to would make the region more investor-friendly, which is, removing bottlenecks or constraints to increased private sector investment and also bringing increased benefits to society. The implementation plan also allows member states to identify common approaches in key issues and work towards integration. The plan is classified into eight themes and these are:

i. Geological and mining information systems, essentially working together to generate and disseminate information to potential investors;
ii. Value Addition, Innovation and R&D, concerning the promotion of these activities for linkages to other sectors of the economy;
iii. Artisanal and Small-Scale Mining, concerning developing common strategies to promote a viable sub-sector;
iv. Safety, Health and Environment, dealing with working through common provisions to adopt similar principles;
v. Human Resource and Institutional Capacities, dealing with strategies to address the skills shortages;
vi. Policy, Regulations and Administration, dealing with adoption of similar provisions;
vii. Social issues and Gender, dealing with common approaches to improve the same; and
viii. Investment promotion, essentially advocating several initiatives as a region and particularly the continuation of SADC Mining Investment Forums.

It is only through working together and developing common approaches that a harmonised mineral regime can emerge in the SADC. The regional mineral regime is slowly emerging on its own.

5 LESSONS LEARNT FROM PRACTICES BY DIFFERENT RECs

Some lessons could be learned from practices by different RECs. These are the following:

i. All RECs recognise the importance of natural resources in general and others have specifically made significant progress towards a group approach to mineral resources management;
ii. All RECs want to or engage in co-operative and integrative arrangements in natural resources;
iii. Most RECs have started harmonising policies and working together in a number of activities;
iv. SADC has taken a cautious step of letting harmonising policies, standards and regulations. It has developed an implementation plan based on common issues;
v. ECOWAS has taken a bold step beyond harmonisation towards development of a common mining code;
vi. RECs are at different stages of participating in the mining and exploitation in the SSA.

On average, the slow pace of harmonisation is understood, taking into consideration that the minerals sector is only part of an economy and it must subscribe to the local agenda of a country, which is invariably different from another. The results of the ECOWAS once implemented will be useful to all other RECs, as it is approaching a unitary regime.

6 CONCLUSION

Regional organisations are participating in the minerals sector by focussing on harmonisation of mineral policies, laws and regulations. The main aim is to make the regions more competitive and attractive to foreign investment, while at the same time obtaining maximum benefits from these extractive activities. This is done through the affirmation of ownership of mineral resources by citizens through the state and deciding the manner in which these resources are developed. RECs aim at removing obstacles faced by individual member states. To achieve this requires political will to work as groups especially where certain actions or decisions call for what would be termed ceding national sovereignty to REC's priorities.

REFERENCES

African Union 1991. *Treaty Establishing the African Economic Community.*

African Union/UNECA 2009. International Study Group, *Africa's Mining Regimes: Framework report* (Draft).

Economic Commission for Africa (ECA) 2004. *Assessing Regional Integration in Africa.*

ECA 2004. *Harmonisation of mining policies, standards and regulatory frameworks in Southern Africa.*

ECA 2008. *Implementation Plan of the Harmonisation Framework.* Draft report.

ECOWAS 2009. Directive C/Dir.3/05/09 *on the harmonisation of guiding principles and policies in the mining sectors.* Sixty Second Ordinary Session of the Council of Ministers, Abuja 26–27 May 2009.

ECOWAS COMMISSION 2009. *Action Plan for the implementation of the ECOWAS Directive on the Harmonisation Principles and Policies in the Mining Sector.*

Hamilton, K. & Ley, E. 2010. Economic Premise: Measuring National income and Growth in Resource-Rich, Income-Poor Countries. *World Bank Report.* http://siteresources.worldbank.org/INTPREMNET/Resources/EP28.pdf

Mainardi, S. 1995. Mineral resources and growth. Towards a long-term convergence? *Resources Policy* 21 (3), 155–168.

Mkwezalamba, M.M. & Chinyama, E.J. 2007. Implementation of Africa's Integration and development Agenda: Challenges and Prospects. *African Union Newsletter.* http://www.africa-union.org/root/ua/Newsletter/EA/Vol.%201,%20No.%201/Mkwezalamba%20&%20Chinyama.pdf

Mtegha, H.D. 2005. Towards a minerals policy for the Southern African Development Community (SADC). PhD Thesis, University of the Witwatersrand, South Africa.

Otto, J. 1977. A national mineral policy as a regulatory tool. *Resources Policy*, 23, ½: 1–7.

Radetzki, M. 1989. Long-run factors in oil price formation. In: Winters, L.A. and Sapsford, D., Editors, 1989. *Primary Commodity Prices: Economic Models and Policy*, Cambridge University Press, Cambridge.

UNECA 2008. Implementation Plan of the Harmonisation Framework. Draft Report.

UNECA/African Union 2008. *African Mining Vision 2050.*

United Nations Development Programme (UNDP) 2010. *Human Development Report 2010–20th Anniversary.* http://hdr.undp.org/en/reports/global/hdr2010

Geological Resources and Good Governance in Sub-Saharan Africa – Runge & Shikwati (eds)
© 2011 Taylor & Francis Group, London, ISBN 978-0-415-58267-4

Instruments and experiences to improve resource governance in a multinational context—cooperation with the Commission of the Economic and Monetary community of Central Africa (CEMAC)

J. Runge
Department of Physical Geography, Goethe-University Frankfurt & CIRA, Centre for Interdisciplinary Research on Africa, Frankfurt am Main, Germany

ABSTRACT: Implementation of international transparency standards like EITI and KCS is crucial also for African regional organisations. Good resource governance depends on ownership and political will that is expressed by professional political leaders. This process is studied and highlighted by the example of the Commission of the Economic and Monetary community of Central Africa (CEMAC). Its six member states are rich in geological resources, however, the majority of the population within CEMAC is still living in severe poverty. Joint measures of World Bank, EU and German Technical Cooperation (GTZ) together with multi-stakeholder participation and integration prepare the ground for more transparency and the sustainable fighting against poverty. The diamond sector in the Central African Republic (CAR) serves as a showcase how EITI can be implemented even if economic framework conditions are dominated by informal mining structures.

1 INTRODUCTION

The Commission of the Economic and Monetary community of Central Africa (*Communauté Economique et Monetaire de l'Afrique Centrale*—CEMAC, www.comcemac.org) is an African regional organisation that was established in 1994 as a successor institution of the former Central African customs union founded in 1964 (*Union Douanière et Economique des Etats de l'Afrique Centrale*—UDEAC). CEMAC headquarters are located in Bangui, the capital of the Central African Republic (CAR). CEMAC has several executive bodies like the conference of the head of states and the Central Bank BEAC (*Banque des Etats de l'Afrique Centrale*) based in Yaoundé, Cameroon.

Six countries with a total population of about 35 million inhabitants of the western Central African region are belonging to CEMAC: Equatorial Guinea, Gabon, Cameroon, Republic of Congo, Chad and Central African Republic (CAR). Actually Sao Tomé and Principe have an observer status within CEMAC. With the exception of Equatorial Guinea (Spanish) the Central African Economic and Monetary community with the common Francs CFA currency (€1 = FCFA655.957) belongs to the francophone area.

The CEMAC region is quite heterogeneously composed regarding its geographical, demographic and economic features (e.g. size of state territory, landlocked or with direct access to the ocean and world markets, occurrence of natural resources, number of inhabitants, infrastructure, average incomes and state of development). According to the Human Development-Index of UNDP (2008) the landlocked, wide expanse exclaves of CAR (position 171) and Chad (position 170) represent the CEMAC's economic and developmental bottom. The average per capita income in 2007 came up to US$740 in CAR and US$21,230 in Equatorial Guinea due the oil boom (World Bank 2008). On the macroeconomic level CEMAC showed an average economic growth of 5.8% in 2008 (EU 2009).

The CEMAC region is rich in geological resources. Recently oil and gas make up almost 80% of the total exports. Being the smallest CEMAC country—only 28,051 km² in size and a population of 616,459 inhabitants—Equatorial Guinea is within a Sub-Saharan context the most important producer of crude oil following Angola and Nigeria. In 2007 production in crude oil reached 17.5 million of tons, followed by Gabon (12.1), Republic of Congo (11.0), Chad (7.3) and Cameroon (4.3). However, none of these countries are a member of the OPEC. These five CEMAC states contribute 11% of the total African oil production (EU 2009). In addition to hydrocarbon based resources there

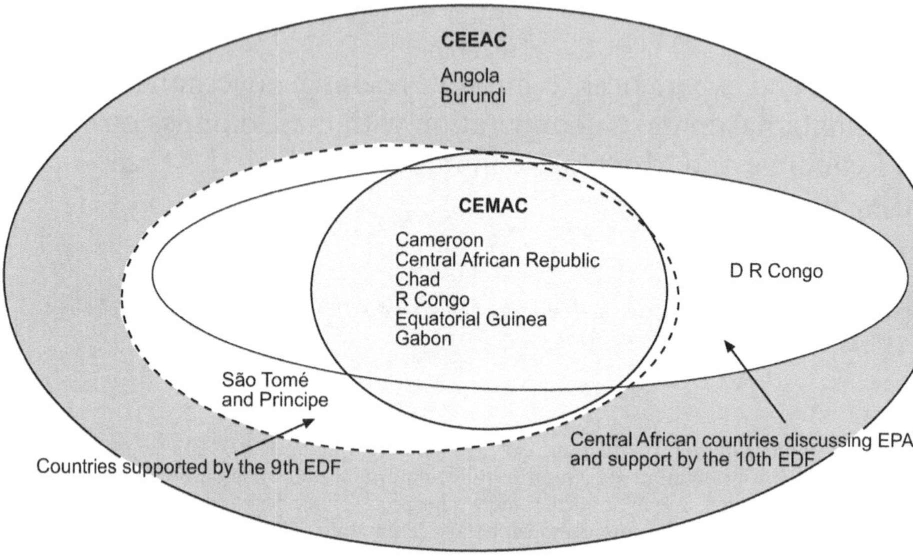

Figure 1. Regional country configuration of CEMAC and CEEAC member states in Central Africa and application of the 9th and 10th European Development Fund (EDF) programme and the Economic Partnership Agreements (EPA) (redrawn and modified after EU 2009, p. 23).

are still a lot of not yet exploited deposits in the sub region like iron ore, cobalt, nickel, bauxite, gold, uranium and others that offer a huge future potential for extractive industries.

In spite of everything, the majority of the population within CEMAC is still living in severe poverty. In September 2004 the CEMAC Ministers of Finances decided in Paris to make an effort to support the implementation of the Extractive Industries Transparency Initiative (EITI) for strengthening good governance and reducing poverty in its member states. Since November 2007 the German Technical Cooperation (GTZ) has been supporting CEMAC by setting up the sub regional, so-called 'Afrika NA' project, on 'strengthening resource governance in the extractive sector'.

It has to be underlined that so far CEMAC has not been recognised by the African Union (AU) as the official regional organisation for western Central Africa. This lead has been given to the larger (11 countries) sister organisation CEEAC (*Communauté Economique des Etats de l'Afrique Centrale*), established in 1983 with the main office in Libreville, Gabon (Fig. 1).

2 EITI IMPLEMENTATION IN CEMAC

EITI implementation is mainly an individual nation-state based process. It depends on the political will of a government to join the EITI. Therefore, the influence of a regional organisation as CEMAC is to a certain way limited and restricted. It can be mainly understood as a contribution to enhance the sensitivity of the topic and to trigger the process by supporting the dialogue between stakeholders like governments, companies and civil society.

When EITI came into existence in 2002 on the Johannesburg UN world summit, the instrument still was an initiative that required a totally voluntary participation of potential candidates. This short period of a 'free' decision for EITI changed later on when World Bank started to insist on EITI as a transparency condition and standard to meet further requirements for collaboration and grants for loans. These days, five of six CEMAC countries are already candidate countries according to the EITI rules (www.transparency.org). Only in Chad—where World Bank recently (2009) revoked collaboration with the government due to disturbances in relation with the Doba-Kribi pipeline—the EITI processes have slowed down and no candidate status for Chad is nowadays (2009) in sight.[1]

The GTZ CEMAC project REMAP (*Renforcement de la Gouvernance dans le secteur des matières premières en Afrique Centrale*) supports EITI implementation in the CEMAC sub region by strengthening the multi-stakeholder approach on all kinds of transparency in relation to extractive industries. Joint workshops and conferences with the national EITI secretariats, the exchange of best practices on the implementation process and already made experiences are exchanged. In CAR the Kimberley Certification Scheme for Diamonds is another objective of the project. A CEMAC based data collection (Geographical Information System, GIS) on geological resources was started in 2008 that serves—in collaboration with a similar unit at the Central Bank BEAC (SITIE) in Cameroon—as a reference for interested stakeholders from civil society as well as from professional mining companies.

As part of GTZ's HIV workplace mainstreaming policy the informal mining sector in CAR is another objective of REMAP. More than 3000 free HIV tests have been offered to miners in the so called artisanal Gold- and Diamond sector (see contribution by Runge & Ngakola in this volume) in collaboration with local hospitals and mining societies (e.g. AXMIN Aurafrique Ltd.).

3 MULTI STAKEHOLDER DIALOGUES—INTERVIEW WITH CEMAC'S COMMISSIONAIRE DR ZOBA

Ownership and political will of CEMAC's leaders and national governments are crucial and most important to put into action the above outlined objectives. Dr. Bernard Zoba is CEMAC's commissioner for infrastructure and sustainable development. He is responsible for natural resources including mining in the sub-region. He exchanged ideas with the author who is a geoscientist and who had advised CEMAC from 2007–2010 on issues of governance in the resource sector on behalf of the German Federal Ministry for Economic Cooperation and Development (BMZ). In September 2009 during an international conference in Yaoundé (www.yaounde2009.net) they spoke about their points of view, their insights and their expectations.

Central Africa is a region that is very rich in resources. But very little of this wealth reaches the people. How can this 'resource curse' be dealt with?

Zoba: The resource wealth in itself is not the problem. After all there are countries—such as Botswana—where resources contribute a great deal to affluence and sustainable development. But if mineral deposits are not properly managed and corruption comes into play, things go wrong. Nevertheless, this is not an inescapable curse but a complex management problem. This conference is one of the means by which we are trying to change the situation.

During the conference many companies and international organisations, as well as government representatives, have voiced their opinions. How are responsibilities divided in the resource sector?

Runge: In a globalised resource market, it is more important than ever not to consider individual countries' geological resources in isolation. If we do nothing to address the political and economic instability of many resource countries, resource customers all over the world will remain involuntary accomplices in poverty and violence. The situation is made worse by the fact that, alongside the 'usual' stakeholders, aspiring economic powers such as China that are extremely hungry for resources are investing in Africa. It is therefore important that the international community pulls together to establish mandatory standards.

One term crops up again and again in the working groups: participation. Why is this so important?

Runge: The key to solving the problems in the resource sector lies in general improvement of democratic and participatory structures. The first step is to identify the needed structure through democratic dialogue. The resulting information then provides a planning basis for sustainable development in the mining sector.

Zoba: Participation requires the integration of people who in the past have had little or no voice. For example, small-scale miners must be better integrated; their rights must be better respected and acknowledged, but there must also be more obligations and rules, such as transparency standards. Through a fair dialogue we can work together to tackle the numerous social and environmental problems that still exist in this area. This process also prevents conflict and diminishes the attraction of smuggling and fraud. This encourages economic development and promotes peace and security within the country.

Policymakers, NGOs and advisers are constantly working to improve resource management in the countries of Africa and throughout the world. What approaches are most conducive to progress?

Runge: The most promising approach to the establishment of international transparency standards is without doubt the Extractive Industries Transparency Initiative. The Kimberley certification process for diamonds has also met with some success. Both these mechanisms have strengths and weaknesses. In the EITI, the process is very formal and bureaucratic and ignores figures relating to the shadow economy that continues to exist. Under the Kimberley Process it is relatively easy to manipulate the certificates, and it is clear that outside North America and Europe there are still markets where non-certified diamonds are bought and sold.

Where do you see the greatest need for action at present?

Zoba: We need mechanisms for imposing sanctions on operators who fail to comply with the EITI principles and the Kimberley Process. Legislation at national level would be helpful in disciplining the local resource industry. Internationally, though, offences are hard to prove. There is as yet no effective means of punishing countries. It is therefore important for governments themselves to support this process one hundred per cent. If they do not, they are merely making political declarations.

If the initiatives are so hard to implement, is there any chance at all of greater transparency in the resource sector?

Runge: Yes, there is, but an initiative is not a magic wand. In my view what is often lacking is what we in development cooperation call 'local ownership'—the assumption of direct responsibility by the governments with which we work. And awareness of the problem is also an issue: people in politics and industry sometimes think that because the EITI is being implemented, everything is going fine. But the EITI alone will not change these countries. For example, transparency regarding public revenue from the extractive sector is useless if no one utilises the information that is provided. Civil society and parliaments must therefore be strengthened so that governments can be held to account.

Zoba: It is therefore a priority to get governments and businesses to understand that transparency and fairness are not a bad thing for them. On the contrary: they are a means to an end, to boost our prospects of political, economic and social development.

In other words, an improved perception of the positive aspects of good governance, Mr. Runge, what can development cooperation do here?

Runge: One of the aims of German development cooperation is close coordination with the partner governments and other donors. This comprehensive approach operates primarily through the use of existing initiatives: for example, in this way it has been possible to help the Central African Republic acquire EITI candidate status. This means that five of the six CEMAC countries are now EITI candidates. To promote dialogue and understanding, the German advisory programme in Bangui regularly organises seminars with the national EITI working groups.

Dr. Zoba, how does CEMAC set about achieving progress in resource management?

Zoba: CEMAC can support the EITI process and other initiatives and strengthen the national decision-making processes of member countries. Regional economic integration and freedom of movement for all CEMAC citizens are important planning goals for us in this context. The economy has a strong interest in the guaranteeing of the legally normative framework for long-term investment in the resource sector. Another area in which CEMAC will be becoming more involved in future is that of cross-border infrastructure schemes that can be financed from resource revenues—especially roads, railways and the planned airline AIR CEMAC.

Looking to the future, how can value creation from resources contribute to sustainable development?

Zoba: The resource sector is one of the most important sectors of the economy. The majority of the Millennium Development Goals could be met through revenue from geological resources. They could contribute enough money to national budgets to permanently reduce or completely eradicate poverty. Financial development aid would soon be virtually superfluous. That presupposes good governance …

Runge: Even if we sometimes seem to have a long way to go, there are some encouraging signs. To quote one example, in the Central African Republic the PWYP initiative had not caught on at all. Under the influence of the Yaoundé conference things got moving and soon Publish What You Pay (PWYP) will form part of the transparency and governance mechanism in the CAR. It is also noticeable that the large mining companies are becoming more open to such initiatives: they want to be seen as responsible members of society. Nowadays a poor reputation for social and environmental standards can have serious consequences for a company on the stock exchange. These are steps in the right direction. We just need to keep going.[2]

(Source: CEMAC 2010a, pp. 16–19).

4 CEMAC COUNTRY CASE STUDY: CENTRAL AFRICAN REPUBLIC (CAR)

Diamonds in CAR are alluvial diamonds. They are linked to secondary detrital formation in extended sandstone plateaus in the west and in the north of the country. To date no trace of kimberlitic primary deposits has been found.

Prospecting is carried out by informal small or medium sized undertakings organised in miner unions. Diamonds are mainly found in the southwest around the cities of Carnot and Berberati and in the north-east around Bria. Official total production indicated by Central African statistics is around 500,000 carats per year corresponding to an export value of around US$60 mio. taking into account fraudulent exports (smuggling), which are estimated at 30% of the total the real production may reach about 650,000 carats a year (KCS 2003). New GIS based estimations by the REMAP project suggest that even more than 70% of all diamonds are illegally exported. Also some discrepancies between the official production statistics for CAR diamonds and the statistics of importing countries for Central African diamond imports are significant.

By support of World Bank and REMAP the Central African Republic became the 5th EITI candidate country in the CEMAC region in November 2008. Validation has to be successfully fulfilled till November 2010 (the final decision on CAR is expected to take place on the 5th International EITI Conference in March 2011 in Paris, France).

The first national EITI report for CAR based on data for the fiscal year 2006 showed a shortfall of FCFA 463,582,771 or approximately €708,000. As the official government data seems to depict only a small portion of the real—still unknown—diamond and gold production in CAR, these data give a first, still quite superficial insight of what is going on within the hardly ever controlled informal CAR mining sector.

5 CONCLUSION AND OUTLOOK

Within five of six CEMAC countries EITI implementation is progressing. Efforts to increase transparency in the different sectors of extractive industries is slowly advancing. However, the availability of non aggregated data is still often limited and the overall data quality is occasionally to be called into question.

Governmental influence on national EITI secretariats and staff is sometimes (too) high (e.g. Cameroon, Republic of Congo) and the independence of the process and of its representatives are obviously hindered. NGOs and civil society agents are in some cases (e.g. Gabon, CAR) not sufficiently taken into consideration and the multi-stakeholder mechanism slows down. Exaggerated per diems payments to EITI officials, for example, have partly scandalised the process in the public perception like in Cameroon.

Nonetheless, on the CEMAC level the EITI process is no longer observed as a kind of 'geological spying' as it was mentioned early in 2008 by a high CEMAC official when looking at the GTZ REMAP's objectives. Nowadays CEMAC seems to be interested in proposing a sub-regional contribution on resource governance by the harmonisation of exploration and exploitation bills which will support efforts for a better regional and economic integration. This becomes especially true by including the EPA and EDF programmes (Fig. 1) in the EU into the overall resource governance discussion (CEMAC 2010b).

ENDNOTES

[1] In February 2010 Chad was—after some earlier separation from the EITI process—nominated as an EITI candidate.

[2] The interview was led by Christina Höschele.

REFERENCES

CEMAC 2010a. *Geological Resources and Good Governance in Central Africa, Conference report, 24–25 September 2009*, www.yaounde2009.net, Bangui, 35 p.

CEMAC 2010b. *CEMAC 2025: vers une économie régionale intégrée et émergente: Programme Economique Régional 2009–2015*, Bangui: 30., www.comcemac.org

EU 2009: *Afrique Centrale—Communauté Européenne—Document de stratégie régionale et Programme indicatif régional (PIR) pour la période 2008–2013, Bruxelles.*

KCS, Kimberley Certification Scheme 2003. Report of the review mission of the Kimberley process to the Central African Republic, 8–15 June 2003.

UNDP 2008. *Human Development Report 2007/2008. Fighting climate change: human solidarity in a divided world. http://hdr.undp.org/en/media/HDR_20072008_EN_Complete .pdf*: 399

World Bank 2008: *Equatorial Guinea at a glance, http://devdata.worldbank.org/AAG/gng_aag.pdf, and Central African Republic at a glance / http://devdata.worldbank.org/AAG/caf_aag.pdf*

Geological Resources and Good Governance in Sub-Saharan Africa – Runge & Shikwati (eds)
© 2011 Taylor & Francis Group, London, ISBN 978-0-415-58267-4

Corporate Social Responsibility (CSR) in Cameroon

E. Dibeu
Canadian High Commission, Yaoundé, Cameroon

ABSTRACT: Corporate Social Responsibility (CSR) is the way in which a corporation interacts with stakeholders in society to address complex social issues. It is implemented through some international guidelines consisting of a framework of voluntary standards and principles for responsible business conduct. The Government of Canada (GOC) and Canadian companies play a key role in the promotion of CSR internationally. A case study of a CSR initiative program is provided by Alucam, Rio Tinto Alcan's subsidiary in Cameroon.

1 WHAT IS CORPORATE SOCIAL RESPONSIBILITY?

Corporate Social Responsibility (CSR) is defined as the way in which a corporation interacts with stakeholders in society to address complex issues such as labour rights, environmental protection, bribery and corruption, and human rights.

CSR is pursued by business to balance economic, environmental and social objectives while addressing stakeholders' expectations and enhancing shareholder value.

CSR initiatives are broad-ranging and include activities by individual companies and industry sectors as well as international standards and norms endorsed and promoted by government.

2 INTERNATIONAL GUIDELINES

The Guidelines are a government-endorsed framework of voluntary standards and principles for responsible business conduct, providing general and specific recommendations in various areas such as environment, employment and industrial relations, consumer interest, taxation, etc.

2.1 *The Organisation for Economic Cooperation and Development (OECD) guidelines for multinational enterprises*

The Guidelines (www.oecd.org/daf/investment/guidelines) constitute a set of voluntary recommendations to multinational enterprises in all the major areas of business ethics, including employment and industrial relations, human rights, environment, information disclosure, combating bribery, consumer interests, science and technology, competition, and taxation. Adhering governments have committed to promote them among multinational enterprises operating in or from their territories.

The instrument's distinctive implementation mechanisms include the operations of National Contact Points (NCP), which are government offices charged with promoting the Guidelines and handling enquiries in the national context.

2.2 *IFC performance standards on social and environmental sustainability*

The International Finance Corporation (IFC) applies its Performance Standards to manage social and environmental risks and impacts and to enhance development opportunities in its private sector financing in its member countries eligible for financing. The Performance Standards may also be applied by other financial institutions electing to apply them to projects in emerging markets. Together, the eight Performance Standards establish standards that companies are to meet throughout the life of an investment by IFC or other relevant financial institution:

- Performance Standard 1: Social and Environmental Assessment and Management System;
- Performance Standard 2: Labor and Working Conditions;
- Performance Standard 3: Pollution Prevention and Abatement;
- Performance Standard 4: Community Health, Safety and Security;
- Performance Standard 5: Land Acquisition and Involuntary Resettlement;
- Performance Standard 6: Biodiversity Conservation and Sustainable Natural Resource Management;

- Performance Standard 7: Indigenous Peoples;
- Performance Standard 8: Cultural Heritage.

Performance Standard 1 establishes the importance of: (i) integrated assessment to identify the social and environmental impacts, risks, and opportunities of projects; (ii) effective community engagement through disclosure of project-related information and consultation with local communities on matters that directly affect them; and (iii) the client's management of social and environmental performance throughout the life of the project. Performance Standards 2 through 8 establish requirements to avoid, reduce, mitigate or compensate for impacts on people and the environment, and to improve conditions where appropriate. While all relevant social and environmental risks and potential impacts should be considered as part of the assessment, Performance Standards 2 through 8 describe potential social and environmental impacts that require particular attention in emerging markets. Where social or environmental impacts are anticipated, the client is required to manage them through its Social and Environmental Management System consistent with Performance Standard 1.

2.3 *Extractive industries transparency initiative*

The Extractive Industries Transparency Initiative (EITI) supports improved governance in resource-rich countries through the verification and full publication of company payments and government revenues from oil, gas and mining.

2.4 *The Equator Principles*

The Equator Principles are an adaptation of the IFC standards that have been adopted by financial institutions. The Equator Principles Financial Institutions (EPFIs) have adopted those Principles in order to ensure that the projects they finance are developed in a manner that is socially responsible and reflect sound environmental management practices. By doing so, negative impacts on project-affected ecosystems and communities should be avoided where possible, and if these impacts are unavoidable, they should be reduced, mitigated and/or compensated for appropriately.

The Principles are intended to serve as a common baseline and framework for the implementation by each EPFI of its own internal social and environmental policies, procedures and standards related to its project financing activities. The EPFIs will not provide loans to projects where the borrower will not or is unable to comply with their respective social and environmental policies and procedures that implement the Equator Principles.

2.5 *The global reporting initiative*

The Global Reporting Initiative (GRI) is a network-based organisation that has pioneered the development of the world's most widely used sustainability reporting framework and is committed to its continuous improvement and application worldwide. In order to ensure the highest degree of technical quality, credibility, and relevance, the reporting framework is developed through a consensus-seeking process with participants drawn globally from business, civil society, labour, and professional institutions. This framework sets out the principles and indicators that organisations can use to measure and report their economic, environmental, and social performance.

The cornerstone of the framework is the Sustainability Reporting Guidelines (www.globalreporting.org/ReportingFramework/G3Guidelines/). The third version of the Guidelines—known as the G3 Guidelines—was published in 2006, and is a free public good.

2.6 *The voluntary principles on security and human rights*

Established in 2000, the Voluntary Principles on Security and Human Rights—an initiative by governments, NGOs, and companies—provides guidance to extractives companies on maintaining the safety and security of their operations within an operating framework that ensures respect for human rights and fundamental freedoms. The Voluntary Principles (VPs) are the only human rights guidelines designed specifically for oil, gas, and mining companies.

The VPs are nonbinding and offer an operational approach to help companies function effectively. The VPs can help a company:

- Conduct a comprehensive assessment of human rights risks associated with security, with a particular focus on complicity.
- Engage appropriately with public and private security in conflict prone areas.
- Institute proactive human rights screenings of and trainings for public and private security forces.
- Ensure that the use of force is proportional and lawful.
- Develop systems for reporting and investigating allegations of human rights abuses.

2.7 *United Nations global compact*

Launched in July 2000, the UN Global Compact is both a policy platform and a practical framework for companies that are committed to sustainability and responsible business practices. As a leadership initiative endorsed by chief executives, it seeks to align business operations and strategies everywhere with ten universally accepted principles in the areas of human rights, labour, environment and anti-corruption. Through a wide spectrum of specialised work streams, management tools, resources, and topical programs, the UN Global Compact aims to advance two complementary objectives:

- Mainstream the ten principles in business activities around the world;
- Catalyse actions in support of broader UN goals, including the Millennium Development Goals (MDGs).

By doing so, business, as the primary agent driving globalisation, can help ensure that markets, commerce, technology and finance advance in ways that benefit economies and societies everywhere and contribute to a more sustainable and inclusive global economy.

The UN Global Compact is not a regulatory instrument, but rather a voluntary initiative that relies on public accountability, transparency and disclosure to complement regulation and to provide a space for innovation.

3 CANADA'S APPROACH TO CSR

Canada is a member of the OECD and a signatory to the Guidelines for Multinational Enterprises. The Government of Canada encourages and expects all Canadian companies working internationally to respect all applicable laws and international standards, to operate transparently and in consultation with the host government and local communities, and to develop and implement CSR best practices. Internationally-agreed voluntary principles are a key part of Canada's CSR approach.

Canada encourages countries to effectively manage the various sectors with an emphasis on economic and social development, environmental stewardship, and respect for human and labour rights, including the transparent development of legislation which strike a balance between host government responsibilities and corporate activity.

3.1 *Implementing the guidelines in Canada—the Canadian National contact point*

As part of its mandate, the Trade Commissioner Service hosts Canada's National Contact Point (NCP) for the OECD Guidelines for Multinational Enterprises. The Canadian NCP is a government entity led by the Department of Foreign Affairs and International Trade with the participation of several departments, including Industry, Human Resources Development, Finance, Environment, Natural Resources and the Canadian International Development Agency.

The National Contact Point NCP (www.ncp.gc.ca) promotes awareness and effective implementation of the OECD Guidelines, in addition to assisting in the resolution of any complaints raised regarding non-compliance with the Guidelines in any specific instances.

3.2 *Current Canadian CSR initiatives and context*

The Government of Canada (GOC) and Canadian companies play a key role in the promotion of Corporate Social Responsibility (CSR) internationally.

While CSR practices have been employed by Canadian companies in many sectors over many decades, Canada's extractive sector has become a focus of interest for Canadians and foreign governments given the very significant size of Canadian Direct Investment Abroad in this sector. Given this emphasis, the GOC hosted four Roundtables on CSR and the Canadian Extractive Sector in 2006. The independent Advisory Group for the Roundtables released a consensus report and recommendations in March 2007.

The Government of Canada has responded to the recommendations of the Roundtables Advisory Group, announcing its new CSR Strategy for the Canadian extractive sector operating abroad: Building the Canadian Advantage: A Corporate Social Responsibility (CSR) Strategy for the Canadian International Extractive Sector (www.international.gc.ca/trade-agreements-accords-commerciaux/ds/csr-strategy-rse-stategie.aspx).

3.3 *Canada's CSR strategy*

The Government of Canada announced its CSR Strategy for the Canadian extractive sector operating abroad in March 2009. The Building the Canadian Advantage Strategy includes four main pillars:

- Continuing Canadian International Development Agency assistance for foreign governments to develop their capacity to manage natural resource development in a sustainable and responsible manner;
- Endorsement and promotion of widely-recognised and voluntary international CSR guidelines for corporate social responsibility performance and reporting;
- Support for the development of a CSR Centre of Excellence to be established outside government as a one-stop shop to provide information

for companies, non-governmental organisations and others; and

- The creation of a new Office of the Extractive Sector CSR Counsellor to assist in resolving social and environmental issues relating to Canadian companies operating abroad in this field. A competency-based selection process will be launched shortly to identify qualified candidates for this position.

Building the Canadian Advantage will improve the competitive advantage of Canadian extractive sector companies operating abroad by enhancing their ability to manage social and environmental risks. It recognises that, while most Canadian companies are committed to the highest ethical, environmental and social standards, those that lack this commitment can cause harm to communities abroad and undermine the competitive position of other Canadian companies. Through its CSR policy, the government will:

- Support initiatives to enhance the capacities of developing countries to manage the development of minerals and oil and gas, and to benefit from these resources to reduce poverty.
- Promote, primarily through the Department of Foreign Affairs and International Trade and Natural Resources Canada, in addition to the OECD Guidelines the following widely-recognised international CSR performance guidelines with Canadian extractive companies operating abroad:
  - International Finance Corporation Performance Standards on Social & Environmental Sustainability for extractive projects with potential adverse social or environmental impacts;
  - Voluntary Principles on Security and Human Rights for projects involving private or public security forces; and,
  - Global Reporting Initiative for CSR reporting by the extractive sector to enhance transparency and encourage market-based rewards for good CSR performance.
- Set up the Office of the Extractive Sector CSR Counsellor to assist stakeholders in the resolution of CSR issues pertaining to the activities of Canadian extractive sector companies abroad. Dr. Marketa Evans was appointed as the first CSR Counsellor for the Canadian international extractive sector on October 2nd, 2009.
- Support the development of a CSR Centre of Excellence within the Canadian Institute of Mining, Metallurgy and Petroleum to encourage the Canadian international extractive sector to implement these voluntary performance guidelines by developing and disseminating high-quality CSR information, training and tools.

3.4 *Canada's corruption of foreign public officials act*

3.4.1 *Highlights*

Canada's legislation to implement the OECD Convention, the Corruption of Foreign Public Officials Act, S.C. 1998, c.34 came into force on February 14, 1999. The new law makes it a criminal offence to bribe a foreign public official in the course of business. Businesses convicted under the Corruption of Foreign Public Officials Act could face heavy fines, and individuals could be sentenced to a maximum of five years in jail. For more information, please see the Department of Justice's Guide to the Act (http://www.justice.gc.ca/eng/dept-min/pub/cfpoa-lcape/index.html).

3.4.2 *Bribing a foreign public official*

Section 3 of the new Act prohibits the bribery of a foreign public official (representing a foreign state or public international organisation) to obtain or retain an advantage in the course of business.

This offence is punishable on indictment. The law sets no maximum limit for fines which judges could order corporations to pay. Individuals could be sentenced to a maximum of five years imprisonment. It is an extraditable offence.

It is also a crime to conspire or attempt to bribe a foreign public official. Corporations and individuals could also be prosecuted for aiding and abetting in committing the offence.

Under the new law, "facilitation payments", or payments made to expedite routine service provided by a foreign public official, are not bribes and therefore are not subject to prosecution. It is not an offence if the benefit that was given is lawful in the foreign public official's country or public international organisation.

Reasonable expenses, incurred in good faith and directly related to the promotion, demonstration or explanation of products and services or to carrying out a contract with the foreign state, are also exempt.

3.4.3 *Possession and laundering*

Section 4 of the new Act makes it illegal to possess property or proceeds of property obtained or derived from the offence of bribing a foreign public official or from laundering that property or those proceeds.

Section 5 makes it illegal to launder the property or proceeds obtained or derived from the offence of bribing a foreign public official.

If prosecuted by indictment, the maximum penalty for these offences is ten years imprisonment. Corporations would face fines with no maximum limit set out in law. If prosecuted by summary conviction, the maximum fine would be not more than

US$50,000 and individuals found guilty would be sentenced to a term not more than six months imprisonment, or both.

Illegally obtained property and proceeds could be seised, restrained or forfeited. The new Act incorporates the proceeds of crime provisions in the Criminal Code for use in prosecutions of the new offences.

3.4.4 *Other federal laws*

The Corruption of Foreign Public Officials Act is part of a government Bill (Bill S-21) that will also strengthen other federal laws to combat corruption such as:

- Income Tax Act: the offence of bribing a foreign public official is added to the list of offences found in section 67.5 of the Income Tax Act to deny claiming bribe payments as a deduction.
- Criminal Code: all the offences in the Corruption of Foreign Public Officials Act are added to the corruption-related offences defined as an "enterprise crime offence" in the Criminal Code. The following sections of the Criminal Code are also added: section 123, municipal corruption; section 124, selling or purchasing office; and section 125, influencing or negotiating appointments or dealing in offices.

The new offences in the Corruption of Foreign Public Officials Act are added to section 183 of the Criminal Code, making it lawful for police to use wiretap and other electronic surveillance to collect evidence in the investigation of these cases.

4 CASE STUDY: RIO TINTO ALCAN'S CSR INVOLVMENTS IN CAMEROON

4.1 *Overview*

Rio Tinto Alcan (RTA) is a worldwide leader in the aluminium business, and one of the world's largest producers of bauxite, alumina and aluminium. The company bases its actions and decisions on sustainable development principles, health, safety and environmental excellence being its top priorities.

In Cameroon, RTA operates through Alucam JV, a 46.7% interest, created in 1957. Alucam's other stakeholders are Government of Cameroon (46.7%), Agence française de développement (5.6%) and other investors (1.0%).

Alucam is a world-class smelter with a capacity increased from 58 kilo-tonnes per annum (ktpa) to 100 ktpa in 1981 following a modernisation program:

- Environment: ISO 14001 certification obtained in 2004;
- Solid waste disposal: world class facility (2005),
- Quality: OHSAS 18001 certification obtained in 2005, first in Sub Saharan region;
- Safety: Among best in class within RTA.

Production of aluminium facilitated the development of downstream industries:

- Socatral (sheet and plate);
- Alubassa (kitchen utensils).

4.2 *Alucam's contribution to Cameroon*

4.2.1 *Economic contributions*

- Contribution to Cameroon's GDP estimated at 3%;
- Cameroon's 3rd largest exporter in the country;
- 1000 direct and 1270 indirect jobs created;
- 290 local suppliers;
- Significant contributor to the State budget (3% or FCFA 53 billions in 2007).

In middle term period, RTA's contribution to the country will be more important, with the launching of its industrial projects (Alucam extension project and RTA global project estimated at 9 billion $US).

4.2.2 *Social contribution*

- Efficient HIV treatment program recognised by the WHO and the EU;
- Medical clinic available for employees and the local population;
- Infrastructure building program including housing units and fresh water wells;
- Support for the local education system.

4.3 *RTA is pursuing its role of Cameroon development catalyst, supported by a thorough community development plan*

Community Development is an integral component of RTA engagement for sustainable development; it encompasses in particular specific aspirations expressed by local communities in the areas of RTA's presence:

- Aspirations for economic development with an active participation of Cameroonian companies;
- Aspirations for social development with the improvement of people daily life in the villages (fight against the endemic diseases, better access to drinking water, electricity, education, medical care, professional training).

RTA has thus launched several programs in the framework of Alucam extension project and RTA global project. These programs are implemented with local NGO's and Government partnership.

4.3.1 *Alucam's program*

RTA contributions are supporting development objectives of Cameroon and are an example of successful partnership already in place.

INITIATIVES	ACHIEVEMENTS
Governance	• Implement RTA Code of conduct with reinforcement to our current suppliers, • Coordinate "Forum of the Ethical Companies of Cameroon" with objectives: – To share best local practices in order to promote governance at all levels, – To influence public sector and government.
Water access	• With the partnership of local communities and NGO's, sponsorship of construction, maintenance and management of wells (Edéa, Song Mbengué areas).
HIV program	• Significative impact of our current program in Edéa area (awarded by UE and WHO), • Benchmark to National HIV initiative, • Edéa is recording lowest rate of infection of the country.
«Partnership of Enterprises for Maritime Sanaga Development (PED)»	• Partnership of companies with objective to implement important socio-economic projects for "Sanaga Maritime" Communities in particular on: – Health (HIV/AIDS, Oncho-cercoses), – Environment and Safety, – Education, • These Projects are implemented by NGO's for partners and local Communities.
RTA Regional Industrial Development (RID)	• RTA comprehensive economic development program for communities covering in particular, Upgrading program of local SME's, Development of structuring projects and valorisation of the Aluminium global industry through Alucam.

4.3.2 *Some achievements in 2008*

INITIATIVES	ACHIEVEMENTS
Water Access	Implementation of 15 drinking water supply points in the villages and 11 drinking water supply points in 11 public schools.
« PED » Health	Free medical treatment extended to poor people with HIV.
« PED » Environment	• Construction of a public dump in Edéa • Partial revamping of Edéa public lighting • Revamping to standards of Edéa stadium.
« PED » Education	• Installation of toilets in 11 public schools • Donation of learning material to public schools.
RID	• Preliminary starting up of a pilot SME's Upgrading program under Alucam extension project with the objective in the long term to transfer this model to the Government • Development of structuring projects (feasibility studies planned): – Biomass: Valorisation of local sawdust presently unexploited – Micro Hydroelectric power plants in the view to develop countryside electrification.

5 CONCLUSION AND OUTLOOK

Governments and companies are increasingly interested in measures of Corporate Social Responsibility (CSR). The outlined good experiences of development partnerships between Canadian and Cameroonian businesses, institutions and civil society groups in the extractive industries sector showed and underlined that it is worth paying greater attention to CSR activities in the future which contribute to sustainable development. This includes measures that promote compliance with internationally ecological and socio-economic standards which could also trigger the enhancement of the competitiveness of local enterprises and companies. Nowadays, almost every CSR activities are taken as economic result-oriented measures and not as selfless philanthropy, quoting former UN secretary Kofi Annan who expressed this fact by saying 'We do not request companies to do things different from their normal business, but to conduct normal business in different ways' (cited from GTZ 2010).

REFERENCES

Conference Board of Canada 2011. *Governance and Corporate Social Responsibility*. http://www.conferenceboard.ca/topics/GCSR/default.aspx

Department of Justice Canada 1999. *The corruption of foreign public officials act a guide*. http://www.justice.gc.ca/eng/dept-min/pub/cfpoa-lcape/guide.pdf

Equator Principles (EPs) 2006. *A financial industry benchmark for determining, assessing and managing*

social & environmental risk in project financing. http://www.equator-principles.com/principles.shtml
Foreign Affairs and International Trade Canada 2005. *Mining in Developing Countries–Corporate Social Responsibility. The Government's Response to the Report of the Standing Committee on Foreign Affairs and International Trade*. http://www.international.gc.ca/trade-agreements-accords-commerciaux/assets/pdfs/scfait-response-en.pdf
Foreign Affairs and International Trade Canada 2009. *Building the canadian advantage: a Corporative Social Responsibility (CSR) strategy for the Canadian international extractive sector*. http://www.international.gc.ca/trade-agreements-accords-commerciaux/assets/pdfs/CSR-March2009.pdf
Foreign affairs and international trade Canada 2011. *Canada's National Contact Point (NCP) for the OECD Guidelines for Multinational Enterprises (MNEs)*. www.ncp.gc.ca
Global Reporting Initiative (GRI) 2000–2006. Suistanability Reporting Guidelines. *Sustainability*. http://www.globalreporting.org/Home
GTZ, Deutsche Gesellschaft für Technische Zusammenarbeit 2010. Center for Cooperation with the Private Sector: *Corporate Social Responsibility and Development Cooperation, GTZ's contribution*, Berlin, Eschborn, p. 19.
International Council on Mining and Metals (ICMM). www.icmm.com
Natural Resources Canada 2009. *Mining Information Kit for Aboriginal Communities*. http://www.nrcan-rncan.gc.ca/mms-smm/abor-auto/min-min-eng.htm
Organisation for economic co-operation and development (OECD) 2008. *Guidelines for Multinational Enterprises*. http://www.oecd.org/dataoecd/56/36/1922428.pdf
Organisation for economic co-operation and development (OECD) 2010. *e3 Plus A Framework for Responsible Exploration*. http://www.pdac.ca/e3plus/index.aspx
Prospectors and Developers Association of Canada (PDAC) 2010. *Corporate Social Responsibility in the Mineral Industry*. www.pdac.ca/pdac/advocacy/csr/index.html

Geological Resources and Good Governance in Sub-Saharan Africa – Runge & Shikwati (eds)
 ISBN 978-0-415-58267-4

Rating the social and environmental quality of commodities

R.D. Häßler
Product & Market Development, oekom research AG, München, Germany

ABSTRACT: Gold mining is invariably associated with large-scale interventions in nature. The widespread use of cyanide to separate gold from ore for example has an impact not only on ecosystems but also on the health of people living in the areas surrounding gold mines. At gold mines in developing and newly-industrialised countries, employment and social standards often do not meet international standards. There are consequently frequent conflicts over land use in the areas around the gold mines. Hundreds of workers die in gold mines each year due to inadequate workplace safety measures and uncontrollable risks. Almost all mining companies operating internationally have begun to draw up social and environmental standards for their mining activities. Nonetheless, there are flaws in the way in which these standards are implemented, as the numerous breaches which have been substantiated in the overwhelming majority of the companies examined here show. Overall, there are large differences in the social and environmental risks associated with gold mining, depending on the country the gold comes from and the company which mined it. It can generally be stated that the social and environmental standards of large mining corporations are better than those of smaller local mine operators. However, this is not true for human rights standards.

1 FACTS AND FIGURES

In 2009, approximately 2575 tonnes of gold were mined worldwide. Its principal uses are in the manufacture of jewellery, in the electrical engineering industry and for optical products, as well as in (dental) medicine. More than 50% of gold is destined for the jewellery industry, with only around 11% being used in industry and medicine. Furthermore, gold is seen as a long-term capital investment and is of great significance particularly in politically and economically unstable times. It is this last function of gold that is reflected in the current trends in the price of gold. The graph in Figure 1 shows the movement in the price of gold over the last five years. In the wake of the current financial and economic crisis and burgeoning anxiety about inflation, gold is currently gaining importance as an investment.

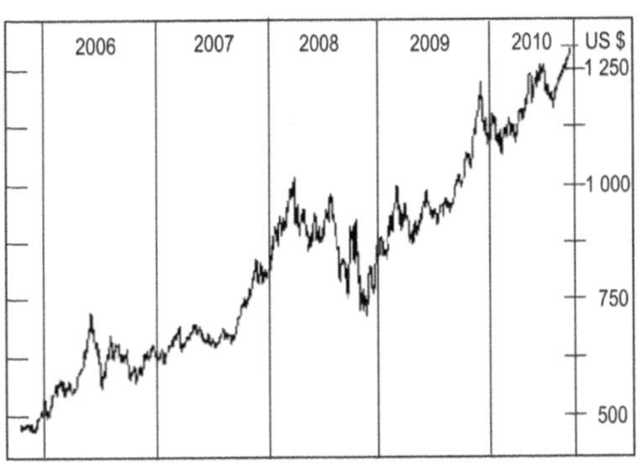

Figure 1. Trends in gold price (US$) 2006–2010. (Source: www.comdirect.com).

2 ENVIRONMENTAL, SOCIAL AND GOVERNANCE (ESG) ISSUES CONNECTED WITH THE EXPLORATION AND MINING OF GOLD

2.1 *Human Rights (HR)*

Open-cast mining in particular requires large areas of land, determined by the extent of the deposits, and this often leads to large-scale clearing of forests and to the—sometimes compulsory—resettlement of the local inhabitants. However, the traditional way of life of the indigenous population is usually closely linked to the land, the natural water resources and the local flora and fauna. In many places, the reserves of raw materials are already being exploited by artisanal small-scale mining "enterprises" using traditional methods.

As a result of large-scale industrial mining activities, it may no longer be possible for local inhabitants to carry on with their traditional way of life and methods of obtaining food. This can lead to protests and frequently also to trespassing on mining sites, which occasionally gives rise to violent reprisals by security personnel, local military or police forces.

The most common initial response to conflicts over land use is to offer compensation payments. In order to prevent violence, human rights standards, programmes and training courses for security forces have been developed. However, the primary concern of both governments and mining companies should be to safeguard for the long term the human rights and livelihoods of the population in the neighbourhoods of all mining sites and to enable the local inhabitants to maintain their traditional way of life. The question of whether or not the indigenous population has a legal claim to the land under the law of the country concerned should not play a role here.

2.2 *Labour Rights (LR)*

The protection of health and safety at work is a key issue where standards for employees are concerned. In this sector, due to the processes and/or sites involved, accident risk levels are high at numerous workplaces. Regulatory and industry codes have been developed and companies have implemented comprehensive preventive measures, the success of which is evidenced by falling accident rates. However, accidents at work resulting in fatalities are still a major problem in the industry and are a cause of social discontent in mining countries. In South Africa's underground gold mines alone, 67 people died in accidents between January and August 2010.

Working conditions in the gold mines, which are often operated by small local entrepreneurs, are atrocious. There are generally no occupational health and safety measures in place, and child labour is widely used. According to data from UNICEF, there are between 60,000 and 200,000 children working in gold mines in Burkina Faso, for example. Approximately 70% of these are still under the age of 15.

2.3 *Environmental Impact (EI)*

Alongside air pollution from climate-changing gases, dust and sulphur dioxide, the disposal of waste and of processing residues, some of which, e.g. cyanide compounds, acids, heavy metals and sometimes even radioactive substances, are highly toxic, presents a major challenge to mining countries and mine operators. The practice, which continues even now in some places, of disposing of processing sludge containing heavy metals, toxins and acids in rivers or coastal waters (riverine or marine tailings disposal) has catastrophic consequences for aquatic ecosystems and thus also for local drinking water supplies and fishing. Where strata occur in deposits in a reduced form (e.g. sulphides), these may, through contact with oxygen in the air and with rain water, form large quantities of acids, which leads to the acidification of soil and water and can also mobilise toxic heavy metals.

Mining companies often employ a process known as cyanide heap leaching. With the aid of cyanide leaching, they can separate out even minute quantities of gold from the pulverised rock. During this process, highly toxic residues are formed. The water that has been contaminated with cyanide is collected in large containment reservoirs. These lakes of poison often overflow and pollute entire regions. The use of mercury, on the other hand, to extract gold (primarily from river sand) is no longer as important as it was for the large mining corporations. However, this process is still employed by small local enterprises. The mercury attracts the gold dust and is then evaporated. During this process, poisonous gases are released unfiltered into the air and rivers. It is estimated that approximately 100 metric tons of mercury end up in the Amazon each year.

While new projects generally incorporate sound precautionary measures, abandoned plants and contaminated sites pose enormous threats to the environment. In some cases, companies seek legal indemnity for any long-term environmental impact through one-off payments to governments, which are often far below required clean-up and rehabilitation costs. Poor environmental regulation and weak legal enforcement in some countries aggravate the environmental impacts of mining.

2.4 *Corruption (CO)*

Corruption remains widespread in the mining sector (Tab. 1). Almost 5% of the mining companies in the oekom research universe have violated standards in relation to corruption. However, this figure is only the tip of the iceberg; experts put the figure for unreported cases of corruption at over 80%.

Table 1. Corruption perceptions index. Scale from zero (high perceived corruption) to ten (low perceived corruption).

Country	Corruption perceptions index
Uzbekistan	1.7
Russia	2.2
Indonesia	2.8
China	3.6
Peru	3.7
Ghana	3.9
South Africa	4.7
United States	7.5
Australia	8.7
Canada	8.7

Source: Transparency International (2010).

Some gold mines are in countries in which, according to the anti-corruption initiative Transparency International, corruption is rife. This is particularly true of Russia and Indonesia, but corruption is also commonplace in China, Peru and South Africa.

Many of the companies operating internationally in the raw materials industries are affiliated to the Extractive Industry Transparency Initiative (EITI). The EITI process is intended to help ensure that money from these industries reaches the relevant governmental budgets in a way that is traceable and that it does not just trickle away under corrupt authorities or governments.

2.5 *Countries of origin*

Gold deposits and their extraction are concentrated in approximately 20 countries worldwide, the eight largest mining countries accounting for more than 60% the total output. China currently extracts the most gold, followed by the United States, South Africa and Australia (Fig. 2). The largest gold reserves, according to current estimates, lie in South Africa. Gold mining is of particular economic importance in developing and newly industrialised countries, where it makes a significant contribution to government revenues foreign exchange earnings.

Social and environmental conditions in the countries of origin vary extremely widely. In many of the countries there are low human rights and labour rights standards in some places corruption is rife (Tab. 1). Environmental legislation and its implementation, as well as the exposure and vulnerability of the existing ecosystems, also differ greatly from country to country.

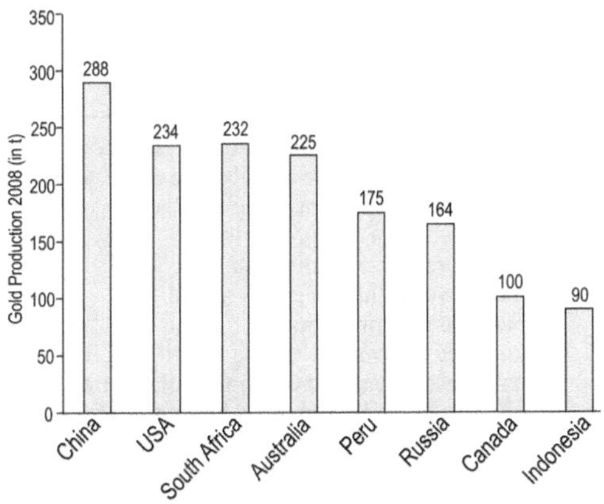

Figure 2. Gold production in selected countries 2008. (Source: United States Geological Survey 2008).

3 COMPANIES INVOLVED IN EXPLORATION AND MINING OF GOLD

Gold-mining rights are held by (semi-)state-owned and private companies, a number of which are listed on the stock markets. The companies from oekom research's universe described below cover approximately 30% of global gold mining in 2009. Two types of companies can be distinguished here: firstly those such as Barrick Gold and Goldcorp, which specialise in gold mining, and secondly companies which principally mine other metals, for example BHP Billiton and Rio Tinto, and for which gold is a "by-product" of the mining of other ore deposits. Gold is, for example, frequently found in association with copper deposits.

In metals production, the use of secondary raw materials (scrap metal) entails substantial economisation of energy, as energy intense processes, such as the open-cast or underground mining of ores, its transport, processing and beneficiation are omitted when using recycled metals. In addition, the use of toxic chemicals and generation of hazardous waste are less intense and land use and related problems do not apply. Thus, metals recycling has considerable environmental advantages. The large international diversified mining groups and companies specialised in gold mining, however, have not (yet) made tangible efforts to use this potential as the mining of metal ores has been and remains their main activity.

Besides these, particularly in developing and newly industrialised countries, there is a plethora of small local mine operators. Artisanal and Small Scale Gold Mining (ASGM) is currently estimated to be responsible for 12% of the worlds gold production or approximately 330 tons per year. In addition to the 13 to 20 million small-scale miners directly involved in the industry, ASGM supports the livelihood of over 100 million people in 70 countries. ASGM generally has significantly lower environmental and social standards than the global players. However, the large corporations are just as likely as the local mine operators to display shortcomings in adhering to human rights standards.

3.1 *ESG company profiles*

The following country (Tab. 2) and company (Tab. 3) profiles are extracted from the comprehensive corporate ratings oekom research conducts of currently more than 1400 companies worldwide using more than 100 social and environmental criteria. We have selected assessment fields for these profiles that are of high relevance in this context: human rights, health and safety, environmental standards in mining and transparency on payments to governments.

Table 2. ESG country profiles (for methodology and grey scale explanation see Figs. 3–4).

<table>
<tr><td rowspan="3">Australia</td><td>Human rights:</td><td>2</td><td>Labour rights:</td><td>1</td><td>Environmental impact:</td><td>2</td><td>Corruption:</td><td>1</td></tr>
<tr><td colspan="4">National specifics relating to gold mining</td><td colspan="4">Strengths & weaknesses</td></tr>
<tr><td colspan="4">Large parts of Australia are desert or semi-arid lands, and local communities depend on the use of groundwater. An increase in water consumption due to mining activities leads to lowering of the groundwater table, alters the hydrological cycle and has a negative impact on local Aboriginal communities, their farming activities and the environment. Moreover, several mines in Australia are sited close to natural conservation areas such as the wilderness area in western Tasmania, which is listed as a World Heritage Area, and Lake Cowal, and thus pose a threat to the habitat of flora and fauna.</td><td colspan="4">+ human rights largely observed
+ the government respects the right of peaceful assembly or freedom of association
+ good conditions of employment and especially good health and safety conditions
– environmentally unfavourable energy mix</td></tr>
<tr><td rowspan="3">Canada</td><td>Human rights:</td><td>1</td><td>Labour rights:</td><td>2</td><td>Environmental impact:</td><td>2</td><td>Corruption:</td><td>1</td></tr>
<tr><td colspan="4">National specifics relating to gold mining</td><td colspan="4">Strengths & weaknesses</td></tr>
<tr><td colspan="4">Canada has abundant freshwater resources, with 9% of the world's renewable water supply. In addition, freshwater lakes provide an ideal habitat for several fish and bird species. In recent years, many lakes in Canada have been reclassified as tailings dumps and are used by mining companies as waste dump sites. The Metal Mining Effluent Regulations impose limits on releases of heavy metals and prohibit the discharge of effluent that is lethal to fish. The companies have to implement Habitat Compensation Plans, to offset the loss of fish habitat as a result of the use of these water bodies as tailings impoundment areas. Nevertheless, these activities endanger natural water bodies and ecosystems. Furthermore, mining activities affect local indigenous communities by changing the environment of sacred sites. In Nunavut, mining activities endanger the Inuit way of life.</td><td colspan="4">+ human rights largely observed
+ the government respects the right of peaceful assembly or freedom of association
+ good conditions of employment and especially good health and safety conditions
+ high level of renewable fresh water resources per capita
– high per capita water and energy consumption</td></tr>
<tr><td rowspan="3">China</td><td>Human rights:</td><td>5</td><td>Labour rights:</td><td>4</td><td>Environmental impact:</td><td>3</td><td>Corruption:</td><td>4</td></tr>
<tr><td colspan="4">National specifics relating to gold mining</td><td colspan="4">Strengths & weaknesses</td></tr>
<tr><td colspan="4">China is the world's major producer and the second biggest consumer of gold. Small-scale mining, in particular, is expanding rapidly and employs large numbers of people, often in dangerous conditions. Small-scale gold miners use mercury, and activities in some mountainous areas of China have already caused a series of mercury pollution problems. The central government has enacted increasingly tough laws and measures on occupational safety, but it lacks the professional bureaucracy, institutional will and resources to enforce them. Furthermore, in China most local government officials hold shares in mines. Therefore, their</td><td colspan="4">+ rich in biodiversity and natural resources
– authoritarian regime and absence of free elections
– restricted diversity of trade unions or obstruction of union activities
– widespread use of child labour
– poor overall human rights record
– widespread corruption
– inadequate overall working conditions; forced and uncompensated overtime, unlawful wage deductions, violations of minimum wage regulations and unhealthy working conditions
– environmental pollution is very serious and ecological destruction is increasing</td></tr>
</table>

(*Continued*)

Table 2. (*Continued*).

	interest in keeping the mines open to generate profits typically overrides any concerns about worker safety or environmental protection. Although there are official inspectors to control the health and safety and environmental conditions at mining sites, widespread corruption and state repression mean that there is no guarantee that grievances will be reported.	
Indonesia	Human rights: 3 | Labour rights: 4	Environmental impact: 4 | Corruption: 4
	National specifics relating to gold mining	Strengths & weaknesses
	Although in Indonesia the use of mercury in gold mining is illegal, due to its toxicity to both human health and the environment, large amounts of mercury are still used in small-scale gold mining. The political will or capacity to prevent the use of this toxic metal is insufficient. Due to Indonesia's tropical climate, heavy rainfalls cause frequently landslides and floods in open pits. Although flooded open pits show the highest mercury levels, the local populations use them for many purposes such as fishing and bathing. Illegal mining is a growing problem in Indonesia and illegal miners often operate in poor conditions with few environmental precautions. Illegal miners dump mercury, cyanide, oil, garbage and tailings along the riverbanks. Furthermore, Indonesian coastal waters are highly polluted because toxic mining waste is often dumped into the ocean by mining companies.	+ rich in biodiversity and natural resources – widespread use of child labour – poor overall human rights record – widespread corruption – inadequate overall working conditions
Peru	Human rights: 3 | Labour rights: 4	Environmental impact: 3 | Corruption: 4
	National specifics relating to gold mining	Strengths & weaknesses
	In Peru, many indigenous people live near mining areas and depend on the land for subsistence crops, farm animals and medicine. Mining activities lead to the loss of agricultural land and interfere with sustainable ways of life. Water from the tailings often contains high concentrations of metals, cyanide and other toxic non-metals and contaminates water courses, causing problems such as the disappearance of fish, illnesses among cattle and the loss of medicinal plants. A high illiteracy rate and low levels of education among indigenous people in Peru hamper their participation in decision-making processes. Although mining companies create jobs and economic growth in the areas around the mines, working conditions are poor and small-scale mines feature the very worst forms of child labour. The poverty rate is still very high in mining areas, and the local population claims not to benefit from mining operations. Mineral and other subsoil rights belong to the state, a situation that often causes conflict between mining interests and indigenous communities.	+ the law provides for freedom of association, and the government generally respects this right in practice – widespread corruption – inadequate overall working conditions; – an undetermined number of children work as forced labourers – language barriers and inadequate infrastructure in indigenous communities

(*Continued*)

Table 2. (*Continued*).

Russia	Human rights:	4	Labour rights:	4	Environmental impact:	3	Corruption:	4
	National specifics relating to gold mining				Strengths & weaknesses			
	In Russia, many mining companies are partly state-owned and have a monopoly status. Environmental standards and health and safety standards are often not fully implemented and it cannot always be guaranteed that grievances are reported. Two-thirds of Russia's gold reserves lie in Siberia and in the far east of Russia. Most mining, especially in these areas, affects the living conditions of the indigenous peoples. Most indigenous people live mainly from fishing, livestock-rearing, hunting and herding domestic reindeer. However, the animals change their migration paths when they are disturbed by gold mining activities, and cyanide, used for the separation of gold, contaminates the water. Native peoples living in the vicinity of the mines are often not adequately informed of the risks to the environment and the danger to their traditional economy nor are they given the opportunity to participate in decision-making on mining projects.				+ rich in biodiversity and natural resources – authoritarian regime – widespread use of forced labour – poor overall human rights record – widespread corruption – inadequate overall working conditions – generally high emissions of air pollutants relative to GDP			
South Africa	Human rights:	3	Labour rights:	3	Environmental impact:	3	Corruption:	3
	National specifics relating to gold mining				Strengths & weaknesses			
	The rate at which the temperature rises with depth is very low in South Africa compared to the world average, at just 9°C per km depth. This makes it possible to mine to depths of 3000–4000 m. The nation's gold mines include two of the deepest mines in the world. In deep mines, working conditions are difficult and dangerous, and mining accidents are frequent. The deeper miners go, the greater the risk of seismic activity. Tremors cause rock falls that are responsible for most of the mine deaths. In addition, mining companies must displace groundwater for the duration of the mining operations by pumping it out, thereby carrying naturally present heavy metals to the surface and contaminating water courses. Furthermore, deep gold mines need more energy than open pits for lighting and air-conditioning to make working conditions in the mines bearable.				+ the law provides for freedom of association, and the government generally respects this right in practice – widespread use of child labour – widespread corruption – low level of renewable fresh water resources per capita			
USA	Human rights:	2	Labour rights:	2	Environmental impact:	2	Corruption:	2
	National specifics relating to gold mining				Strengths & weaknesses			
	Most gold produced in the US comes from large open-pit mines in the state of Nevada. Nevada is an arid area with scarce precipitation during the year. Mining activities with a huge consumption of water have a negative impact on the groundwater level and therefore on the environment. Furthermore, several mining sites are of spiritual and cultural importance for indigenous peoples (like the Shoshones in Nevada) and conflicts of interest regarding land rights are common.				+ human rights largely observed + the government respects the right of peaceful assembly or freedom of association + good conditions of employment and especially good health and safety conditions – lack of participation in international environmental agreements e.g. on climate protection – high per capita water and energy consumption – environmentally unfavourable energy mix			

Table 3. ESG company profiles (for methodology and grey scale explanation see Figures 3–4).

BHP Billiton	Human rights:	C+	Health & safety:	C–	Mining standards:	C	Transparency:	C–
Overall-Rating PRIME	Controversies relating to gold mining				Strengths & weaknesses			
	• ongoing contamination in Papua New Guinea due to the Ok Tedi legacy				+ comprehensive and detailed mine closure standards in place at all active mining sites + reasonable measures to minimise the impact of mining on biodiversity + group-wide implementation of a strategy for addressing climate change and related sector-specific risks + comprehensive environmental standard for suppliers of raw materials – unresolved environmental legacies with regard to ongoing water pollution in Papua New Guinea – mining operations in several protected areas			
Gold Fields	Human rights:	B+	Health & safety:	C–	Mining standards:	B+	Transparency:	A
Overall-Rating PRIME	Controversies relating to gold mining				Strengths & weaknesses			
	• alleged cyanide and heavy metal pollution of streams in the context of gold mining in Ghana • general allegations concerning human rights violations by mining firms and security staff in Ghana • investigation into mine accident due to rope failure causing nine fatalities in South Deep mine, South Africa				+ pension schemes and health benefits offered to all employees in low and middle income countries + transparency regarding the breakdown of relevant payments to governments by country – high energy and carbon intensity of gold mining operations in South Africa – occurrence of several fatal accidents in recent years			
Xstrata	Human rights:	B+	Health & safety:	B–	Mining standards:	C	Transparency:	B–
Overall-Rating PRIME	Controversies relating to gold mining				Strengths & weaknesses			
					+ comprehensive standards for mine closure and rehabilitation + adequate procedures to ensure facility safety and emergency preparedness + implementation of various measures taken to enhance community awareness and outreach – no material activities to promote transparency and certification of origin in ore and metal trade chains – involvement in projects negatively affecting human rights of communities, e.g. in Colombia			
Rio Tinto	Human rights:	D+	Health & safety:	B–	Mining standards:	B–	Transparency:	D+
Overall-Rating NOT PRIME	Controversies relating to gold mining				Strengths & weaknesses			
	• major violations of human rights with regard to the Grasberg Mine joint venture in Indonesia between 2003 and 2006				+ mine closure and rehabilitation programmes in place for all operated projects			

(Continued)

Table 3. (*Continued*).

	• environmentally controversial mining activities, e.g. in Indonesia and Madagascar				+ comprehensive measures to analyse and reduce potentially negative social impact of company activities + comprehensive strategy implemented on protecting biodiversity – application of riverine tailings disposal at least at on site – involvement in projects criticised for low human rights records – occurrence of several fatal accidents in recent years			
Barrick Gold	Human rights:	C+	Health & safety:	C–	Mining standards:	B–	Transparency:	C
Overall-Rating NOT PRIME	Controversies relating to gold mining				Strengths & weaknesses			
	• human rights violations in connection with mining operations in Papua New Guinea and Peru • ongoing contamination due to riverine tailings disposal and environmental legacies in Papua New Guinea and on the Philippine island of Marinduque • fatal accidents and alleged inadequate safety management at Nevada underground mines				+ mine closure and rehabilitation programmes in place for all operated projects + comprehensive measures taken to enhance community awareness + reporting includes description of controversial aspects/projects – occurrence of several fatal accidents in recent years – involvement in exploration projects criticised for human rights violations – lack of transparency regarding a company policy on mining in protected areas			
Newmont Mining	Human rights:	D	Health & safety:	C–	Mining standards:	C–	Transparency:	B+
Overall-Rating NOT PRIME	Controversies relating to gold mining				Strengths & weaknesses			
	• various allegations regarding pollution or mining activities in sensitive areas, e.g. in Romania and Ghana • majority stakeholder in mining project criticised for human rights violations in Peru • highly controversial activities, e.g. allegedly inadequate compensation, at Ahafo and Akyem sites in Ghana				+ transparent disclosure of the breakdown of relevant payments to governments by country + group-wide implementation of safety management systems and downward trend of accident rate – only limited information with regard to measures to reduce the environmental impact of tailings – lack of transparency regarding the energy and carbon intensity of metals production processes – operation of projects that significantly affect internationally protected areas			
AngloGold Ashanti	Human rights:	D	Health & safety:	D+	Mining standards:	D+	Transparency:	A–
Overall-Rating NOT PRIME	Controversies relating to gold mining				Strengths & weaknesses			
	• allegations regarding severe human rights violations at mining operations in Ghana in 2006 • allegations concerning heavy pollution of rivers in Ghana caused by cyanide spills in 2007				+ mine closure and rehabilitation programmes in place for all projects operated + transparent disclosure of payments to governments by country – occurrence of several fatal accidents in recent years – no information provided on consideration of seismic, geologic and/or climatical conditions when planning for a proposed site			

(*Continued*)

Table 3. (*Continued*).

				– lack of transparency regarding a management framework for tailings storage or disposal – lack of transparency with respect to the application of low-impact exploration methods			
Goldcorp	Human rights:	C	Health & safety:	C	Mining standards:	C–	Transparency: D+
Overall-Rating NOT PRIME	Controversies relating to gold mining			Strengths & weaknesses			
	• severe human rights issues at the Marlin gold and silver mine in Guatemala • fined HNL 1 m (approximately EUR 35,000) for long-term arsenic pollution of water in Honduras in 2007 • alleged water pollution in Honduras in 2009			+ regularly updated mine closure plans in place for all active mining projects + group-wide implementation of safety management systems and downward trend of accident rate – major environmental controversies and allegations of major human rights violations in recent years in Guatemala and Honduras – no detailed information provided on the management of tailings storage and disposal facilities – lack of transparency with regard to the application of low-impact exploration methods – no information with regard to a policy regarding mining activities in protected areas			

The overall analysis is graded on a twelve-point scale from A+: The company performs exceptionally well to D–: The company shows little engagement. Oekom research awards prime status to those companies which according to the oekom Corporate Rating are among the leaders in their industry and which meet industry-specific minimum requirements. Approx. 100 criteria are used for the rating of companies. The following categories with a strong link to the risks connected with the extraction of gold have been selected for this profile: Human rights policy and management; Health and safety management and accidents; Environmental standards in mining, e.g. application of low-impact exploration methods and hazardous substances management; Transparency of payments to governments.

	AU	CA	CN	ID	PE	RU	US	ZA	Others
BHP Billiton	X								
Gold Fields	X					X		X	Ghana
Xstrata	X	X							Argentina
Rio Tinto	X	X		X			X		Chile
Barrick Gold	X	X				X	X	X	Argentina, Chile, Papua New Guinea, Tanzania
Newmont Mining	X			X		X	X		Bolivia, Ghana, Mexico, New Zealand, Romania, Suriname
AngloGold Ashanti	X						X	X	Argentina, Brazil, Ghana, Guinea, Mali, Namibia, Tanzania, others
Goldcorp		X					X		Argentina, Chile, Guatemala, Honduras, Mexico

Figure 3. Gold mining operations of the companies rated by oekom research. Source: oekom research (2010) (for methodology and grey scale explanation see Figure 4).

The "overall rating" refers to the results of best-in-class rating based on the mentioned 100 criteria. oekom research awards "Prime" status to those companies which according to the oekom Corporate Rating are among the leaders in their industry and which meet industry-specific minimum requirements.

3.2 *Companies activities in relevant countries*

Table 3 provides an overview on which of the companies evaluated by oekom research operate gold mines in which of the countries examined:

- Companies which, like BHP Billiton and Rio Tinto, for example, mine other metals as well as

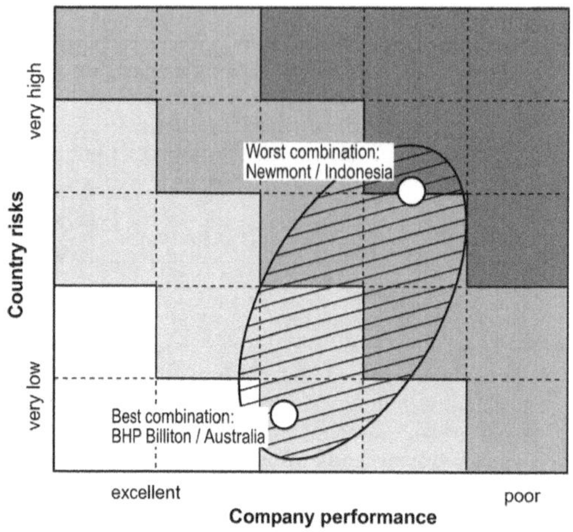

Figure 4. Risk-performance classification for gold: The overall assessment shown in this matrix is based on the rating of the environmental and social rating of the companies involved in gold mining and of the major countries of origin. (Source: oekom research, 2010).

gold generally operate gold mines in only a few countries. For these companies, gold is generally a by-product of the extraction of other raw materials, notably copper. Companies specialising in gold mining (such as Barrick Gold and Goldcorp), on the other hand, are as a rule involved in gold mines in a whole range of countries.

- The table shows that the companies specialising in gold mining—with the exception of Gold Fields—routinely receive lower evaluations in oekom research's ratings than the diversified mining companies. This is due to the particular social and environmental problems associated with gold mining as compared with those encountered in the extraction of other raw materials.
- The ESG-related quality of the gold concerned can be derived from a combination of company performance and country risk. When doing this, it is important to note that gold mining fundamentally entails large-scale impacts on the natural environment and that aspects such as working conditions, safety at work and corruption are an issue in many mines. In this respect there is no gold that would meet strict ESG criteria.
- The ESG-related quality of the gold concerned can be derived from a combination of company performance and country risk. When doing this, it is important to note that gold mining fundamentally entails large-scale impacts on the natural environment and that aspects such as working conditions, safety at work and corruption are an issue in many mines. In this respect there is no gold that would meet strict ESG criteria.
- At the same time, however, there are clear differences in the regulatory frameworks governing gold mining in different countries and in the standards implemented by companies. While gold mined by BHP Billiton, Rio Tinto and Xstrata in Australia, for example, can be rated relatively highly from the ESG point of view, gold mined by Newmont Mining in Peru is associated with huge social and environmental problems, as neither the company nor the country involved meet basic standards in the areas of environment, human rights, labour rights and corruption.

4 METHODOLOGY

4.1 *Country Risk Assessment*

The Country Risk Assessment documents and evaluates the social and environmental situation, especially in NICs and developing countries, using valid and acknowledged international sources. It is based on the methodology of oekom's Emerging Markets Risk Assessment (EMRA®), which contains two modules: as part of the risk classification process, countries are classified according to their social and environmental risks. As a basis for the evaluation, oekom research consults a variety of internationally recognised indices. A country ranking and a visual representation of the country classifications on a world map enable comparison of the results.

The individual country profiles contain a detailed risk assessment covering six thematic areas and give details of key social and environmental risks such as labour and human rights, civil rights and freedom of press, child and forced labour, corruption, bribery and money laundering and environment and natural catastrophes. The analysis is based on internationally acknowledged sources such as the UN, the World Bank, the International Labour Organisation (ILO), Transparency International and the WWF.

4.2 *Corporate rating*

Oekom currently covers more than 3,000 companies from more than 50 sectors and countries. The assessment of the social and environmental performance of a company as part of the Corporate Rating is carried out on the basis of over 100 social and environmental criteria, selected specifically for each industry. For OCES the analysis focuses on stock-market-listed shares from the oekom universe.

Oekom research awards prime status to those companies which according to the oekom Corporate Rating are among the leaders in their industry and which meet industry-specific minimum requirements.

In addition oekom research carries out comprehensive negative screening covering a great number of ethically controversial business fields and practices for each company, e.g. child labour, controversial environmental practices, human rights violations and labour rights violations.

REFERENCES

Business & Human Rights Resource Centre 2010. *www.business-humanrights.org*

Comdirect 2010. *www.comdirect.com*

Mines and Communities 2010. *www.minesandcommunities.org*

MiningWatch Canada 2010. *www.miningwatch.ca*

OECDwatch 2010. *www.oecdwatch.org*

Tranparency International 2010. *Corruption Perceptions Index (CPI). www.transparency.org*

United States Geological Survey 2008. *www.usgs.gov*

Artisanal mining, gender and HIV/AIDS

Geological Resources and Good Governance in Sub-Saharan Africa – Runge & Shikwati (eds)
© 2011 Taylor & Francis Group, London, ISBN 978-0-415-58267-4

Artisanal mining activity—a benefit or a burden for sustainable development in Central Africa?

A. Bomba Fouda
Extractive industries data management, SCTIIE-CEMAC, Yaoundé, Cameroon

ABSTRACT: Artisanal mining is an activity that has been traditionally practised by some ethnic groups in Central Africa for centuries. Since the end of the 20th century, it has expanded rapidly. The phenomenon is, to a great extent, the result of the impoverishment of the population and/or economic crises. It has boomed within a context of economic liberalisation and globalisation. In Africa, the annual value of artisanal production is estimated to be over a billion Euros. The activity has therefore become a determining factor in the fragile national economies of Central Africa. The main issue here is to examine the conditions necessary for a productive co-existence or even a partnership between the artisanal mining sector and major investors.

1 DEFINITIONS AND BACKGROUND OF ARTISANAL MINING

According to Cameroon's mining code, mining is defined as any form of extracting solid, liquid or gaseous mineral substances, irrespective of the procedure or method used, from on or beneath the ground in order to extract useful substances. It therefore includes all processes that are directly or indirectly necessary for, or are related to, this activity. Artisanal mining is just one type of mining. Cameroon's mining code defines artisanal mining as any exploitation where the activities consist of extracting and concentrating mineral substances by methods and procedures that involve little or no mechanisation. Artisanal mining involves processes carried out by individuals or small groups, often family groups. Artisanal mining takes place without any planning, using methods and tools that are often rudimentary and ancestral to extract and process resources about which little is known. It is not characterised by the quantities or timescales involved, but takes place along with other activities (agriculture) within a context of opportunistic gathering.

Three different ways of organising and carrying out artisanal mining can be seen in Central Africa:

1. An activity that complements traditional activities such as agriculture, to obtain a modest income. Artisanal mining represents one of many activities within village communities, supplementing subsistence farming to provide a minimal income. It is a calendar-based activity and conforms to the principles of traditional social organisation.
2. Artisanal miners specialising for a certain period of time in exploiting a mineral resource that is not necessarily located close to a settlement from which they originate.
3. People of disparate origins flocking to newly discovered deposits, attracted by the prospect of making their fortune. When technical constraints become too great, the miners leave the location and head elsewhere.

The harmful effects of artisanal mining are of a technical, economic, environmental, social and health-related sanitary nature. Artisanal mining is characterised by rudimentary methods and tools, technical deficiencies that result in unstructured exploitation, and below-average extraction levels of the resource in question. Artisanal production of a resource at a given location is incomplete, because the deeper layers of the deposit are not extracted. Artisanal mining often only involves the upper layers of the deposit. This exploitation is not only incomplete, but also unstructured.

Artisanal mining activity can lead to environmental damage such as unproductive agricultural land, damage to forests and water pollution caused by sediments. It also has a mechanical impact on the environment. Extensive use of chemicals (for example, mercury used for amalgamating gold) in the processing of mineral resources can have major impacts on health, such as poisoning. Artisanal mining also raises the issue of challenges and conflicts relating to land usage and control. Mining areas remain very sensitive because controlling and limiting them depends on their geological and physical context rather than on administrative action. This is often a source of conflict between communities and/or states.

The spontaneous migration resulting from people rushing to mining sites leads to overpopulation and promiscuity. Communities of individuals of diverse ethnic origins with no cultural points of reference tend to reject the cultural constraints of nearby communities, and sometimes also national legislation. This refusal or rejection leads to ethnic conflict, which occurs frequently between artisanal miners, local communities and sometimes the administrative authorities. Bringing together various ethnic groups creates a new type of social organisation that is artificial and short-lived compared with customary local practices. It is also a source of promiscuity, leading to the spread of sexually transmitted diseases (including AIDS), low levels of school attendance, excessive consumption of alcohol and drugs, etc. Working conditions at mining sites are often very hazardous and difficult. They result in increased numbers of deaths from many illnesses caused by vibration, chemicals, noise, asphyxiation, rock falls, etc.

Artisanal mining sites are often temporary camps located in remote areas far away from any educational and health infrastructure. The lack of sanitation causes endemic disease such as malaria, yellow fever, cholera and tuberculosis.

Artisanal mining has long been seen as a social scourge or burden; in other words, an activity with no significant economic consequences. Its harmful effects have frequently been cited by national and international development institutions. But despite this negative image, artisanal mining is seen by some Africans as an activity that generates wealth and helps people survive; in other words it promotes the economic and social development of poor people and deprived areas.

In Central Africa, artisanal mining relies on an abundant workforce with little or no training, thus reducing unemployment. It provides inactive people with jobs and an income. It restricts the rural exodus by providing work in isolated areas. Artisanal mining makes it possible to work small deposits that large companies would not be in a position to extract. For some African countries, artisanal mining is therefore a social and economic development priority.

Artisanal mining is very often likened to small-scale mining. Small-scale mining is the term used to describe any mining activity carried out on a non-industrial scale. It covers companies with a minimum of fixed facilities and machinery that extract moderate-sized deposits in a planned manner using semi-industrial procedures. According to the United Nations (UNCTAD 1997), a small-scale mine produces less than 50,000 tonnes per year with an investment of less than one million euros. For materials, this production figure is between 100,000 and 200,000 tonnes per year. It has an annual turnover of less than €1.5 million, employs fewer than 40 employees and has a lifespan of less than five years.

2 POSSIBILITIES FOR A PRODUCTIVE COEXISTENCE OR PARTNERSHIP BETWEEN ARTISANAL MINING ACTIVITY AND MAJOR INVESTORS

If it were better managed, artisanal mining could be seen as an important factor for economic and social development in the fragile economies of Central Africa. There are ways in which artisanal mining activity and major investors could coexist in a productive manner. Possible ways of direct and indirect coexistence include:

- development aid for artisanal mining
- the intermediate solution of small-scale mining.

Assistance in modernising artisanal mining cannot be the subject of a single model because of the specific nature of the associated communities, the various issues related to artisanal mining, and the numerous social criteria involved. These issues include controlling poverty and conserving biodiversity. The numerous social criteria (geographical, historical, political, economic) mean that a case-by-case approach is needed. To be effective, any assistance must address the many different facets of artisanal mining and be of a long-term nature. Modernising artisanal mining will require long-term action in the fields of education, awareness-raising, training, technical assistance and access to finance for future integration by major investors.

Small-scale mining is seen as a small-scale private business that is driving economic development. It is more flexible and adaptable, and could act as a platform for designing, using and disseminating innovative procedures. Within the reach of local entrepreneurs and well integrated, small-scale mining could make artisanal mining more effective. However, it should be noted that small-scale mining is still not very compatible with the profit-driven approaches of western industry, when considered in its capacity as a halfway house to artisanal mining. A lack of financial resources also limits the investment capacity of small-scale mining to gradually transform artisanal mining.

3 CONCLUSION AND OUTLOOK

Extracting mineral resources, which are essentially non-renewable, may appear to go against the principles of sustainable development. Given the amount of capital generated by industrial mining, some economists believe there is considerable potential for reinvestment in sustainable local development. This hypothesis assumes that the mining sector adopts an approach based on social corporate responsibility. International institutions have developed the concept of "corporate social responsibilities" (see contribution of E. Dibeu in this volume).

The problem of developing artisanal mining is complex because at a global level, the activity occurs in a wide variety of different contexts depending on the resources, the agents involved, the geographical areas, and the political, administrative, economic, cultural and social situations. In other words, artisanal mining is an issue involving many different criteria and agents. There are different approaches to its development in different countries.

The profitability approach has often been put forward by national and international institutions when looking at rationalising and formalising the activity. Cooperation agreements have been used to mobilise human and financial resources for technical and institutional support initiatives. Artisanal mining often follows a subsistence-based approach rather than focusing on profitability.

Small-scale mining must avoid competing with artisanal mining and adopt a complementary role, acting as an intermediary between major investors and artisanal mining. It needs to focus on new technical resources that are beyond the reach of artisanal activity.

The impact of artisanal mining on sustainable development varies from country to country. It depends on the challenges faced by the mining sector across the three pillars of sustainable development, namely the economic, social and environmental challenges. Artisanal mining is often seen as temporary—the challenge is to make it sustainable and ongoing. The two possibilities proposed here entail risks regarding support for local communities, which may be restricted to selective humanitarian assistance or small-scale support that could be limited to granting funding to a few opportunistic actors.

Preliminary socio-economic studies into identifying suitable case-by-case solutions for the profitable coexistence of major investors and artisanal mining are to be encouraged. Moreover, integrating artisanal mining into the various mining codes of Central African states, which should be standardised, has become essential for sustainable development in these countries.

REFERENCES

BIT 1999. *Problèmes sociaux et de travail dans les petites exploitations minières*. Bureau International du Travail (BIT), Genève, 1999.

Ernst et al. 2007. *Rapport sur les revenus pétroliers et miniers de la République Gabonaise pour l'année 2005*. ITIE—Gabon, 2007.

Ernst et al. 2008. *Rapport sur les revenus pétroliers et miniers de la République Gabonaise pour l'année 2006*. ITIE—Gabon, 2007.

Foute, R.-J. 2008. Exploitation minière—le Cameroon attire. *Cameroon Tribune*, 07.02.2008.

Gendron, C. 2006. Le développement durable comme compromis. *PUQ*, 2006.

Harsh, E. 2005. Les investisseurs se tournent vers l'Afrique. *Afrique Renouveau*, 2005.

Mazars/Hart Group 2006. *Rapport de conciliation des chiffres et des volumes dans le cadre de l'ITIE au Cameroon pour les années 2001–2004*. ITIE—Cameroon, 2006.

Mazars/Hart Group 2007. *Rapport de conciliation des chiffres et des volumes dans le cadre de l'ITIE au Cameroon pour l'année 2005*. ITIE—Cameroon, 2007.

MINIMIDT 2001. *Code minier du Cameroon. Loi N° 001 du 16 avril 2001 portant code minier*. Ministère de l'Industrie, des Mines et du Développement Technologique (MINIMIDT), 16.04.2001.

Ndanga Ndinga, B. 2008. Mieux rentabiliser les ressources minières du Cameroon. *Cameroon Tribune*, 25.01.2008.

Theys, J. 2005. *Le développement durable: une innovation sous-exploitée*. Dalloz, Paris 2005.

Touna, R. 2008. Production de cobalt dès 2010 au Cameroon. *Jeune Afrique*, 30.06.2008.

UNCTAD 1997. *Environmental management of small-scale and artisanal mining sites in developing countries*. L.R. Blinker. UNCTAD Consultancy Report.

Geological Resources and Good Governance in Sub-Saharan Africa – Runge & Shikwati (eds)
ISBN 978-0-415-58267-4

Institutional aspects of artisanal mining in forest landscapes, western Congo Basin

J. Schure & V. Ingram
Centre for International Forestry Research (CIFOR), Central Africa Regional Office, Yaoundé, Cameroon

J.C. Tieguhong
Technical Training and Research Centre for Development (TTRECED), Yaoundé, Cameroon

C. Ndikumagenge
International Union for the Conservation of Nature (IUCN-PACO), Central and West African Office, Yaoundé, Cameroon

ABSTRACT: This contribution examines the multiple impacts of artisanal mining in the high-biodiversity forest of the Congo Basin's Sangha Tri-National Landscape (TNS), and proposes measures for improving livelihoods in the area. It was concluded from a literature review, interviews and site visits that: diamonds and gold are an important but highly variable income source for at least 5% of the area's population; environmental impacts are temporary and limited, mainly caused by mining inside the parks; overlaps between artisanal small-scale mining (ASM), large-scale mining (LSM), timber concessions, and national and trans-boundary protected areas have intensified competition for land resources; and despite the existence of legal frameworks, ASM is largely informal. Cross-boundary agreements concerning the TNS do not address mining, albeit a regional approach of mining policies is recommendable to reinforce beneficial outcomes for the landscape and the area's population.

1 INTRODUCTION

1.1 *Artisanal mining in the Congo Basin*

Small-scale mining employs over 13 million people in developing countries; and an additional 100 million people are dependent indirectly on this sector for their livelihoods (CASM 2009, Danielsen 2000). Artisanal and small-scale mining (ASM) refers to mining by individuals, groups, families, or cooperatives with minimal or no mechanisation, often informally and/or illegally (Hentschel *et al.* 2002). Of over 40 different minerals that are exploited by artisanal miners, gold and diamonds account for more than half of the number of people involved. Sub-Saharan Africa produces over 60% of world's artisanally mined diamonds. In general, ASM provides cash income for poor people, requires few financial and technical inputs and is labour intensive, with modest levels of production and efficiency. Miners are generally simultaneously engaged in other activities, such as agriculture (CASM 2009).

Gold and diamond deposits have been found throughout the Congo Basin and are the two major minerals exploited in the Sangha Tri-National Landscape (TNS) in Cameroon, Central African Republic (CAR) and the Republic of Congo (RoC). The mineral sector of CAR and Cameroon is relatively small compared to major producers such as Angola and the Democratic Republic of Congo (DRC), and is confined to mainly artisanal, small-scale production (Sale 2006). The 700 km border area between Cameroon and CAR is diamond rich (Gweth 2006). Gold mining started in Cameroon in 1933 and between 1934 and 1984, output was in the range of 20 tons, equating to an annual production average of 400 kg, which has a value of two billion FCFA (US$4.3 million) (Lang 2007). In CAR, mining also started in the 1930s with alluvial discoveries, with gold production peaking in the early 1980s at 521 kg and diamonds at 609,360 carats in 1968. Recent data are shown in Table 1. Since 2003, investors have been increasingly interested in industrial mineral exploitation, and a number of companies have been granted exploration licenses in Cameroon and CAR (e.g. for Cameroon: African Aura Resources Ltd. of the United Kingdom (gold), Geovic Cameroon Plc of Canada (cobalt-manganese-nickel), Hydromine Inc. of the United States (iron ore), Mega Uranium Ltd. of Canada (uranium), and Sundance Resources Ltd. of Australia (iron ore); for CAR

Table 1. Key data on the artisanal mining sector in Cameroon and Central African Republic.

	Cameroon	CAR
Main minerals	Bauxite, cobalt, iron ore, nickel, gold, sand, and uranium (1, 5)	Clay, diamond, gold, limestone, and sand and gravel (2)
Producing regions	East, Centre, Sud, Nord (1, 5)	Berberati, Haute-Kotto, and Haute-Sangha (2)
Quantity diamonds produced (2008)	12,000 carats (1)	377,209 carats (3)
Value diamonds produced (2008)	US$1,519,124 (7)	US$47,752,282 (3)
Quantity gold produced (2008)	1200–1800 kg (4, 1)	10 kg (2)
Value gold produced (2008)	US$99,783,761–149,675,641	US$831,531
Number of artisanal miners	20,000–30,000 (5)	50,000 (6)

Sources: Newman (2009)(1); Bermúdez-Lugo (2009)(2); Kimberley Process (2009)(3); Sale (2006)(4); CAPAM (2006)(5); Dietrich (2003)(6); estimation (7).

diamonds: Energem Resources Inc. of Canada, Gem Diamonds Ltd. of the United Kingdom, and Pangea Diamond Fields Plc of the British Isles) (Newman 2009, Bermúdez-Lugo 2009).

1.2 *Artisanal mining impacts*

ASM is labour intensive and requires few capital and technical investments, which means it typically provides a substantial contribution to local development in the form of employment and cash income. Its importance to livelihoods has become increasingly acknowledged after decades of criminalisation, informality and this sector of the economy being overlooked in the majority of developing countries.

However, besides the potential of the sector to contribute further towards poverty reduction, recent studies have underlined a range of negative issues related to its operations. These are primarily associated with social and environmental impacts, with frequently raised concerns about:

- Miner's health and safety (Banchirigah 2006, Walle & Jennings 2001);
- Informality and lack of legal status (Siegel & Veiga, 2009, Sinding 2005). Criminalisation of informal mining activities causes exclusion of miners to participate in decision making that concerns their lives (Tschakert 2009);
- Child labour (Hilson 2008, ILO n.d.);
- Poverty traps and dependence, particularly in rural communities with a heavy reliance on mining as the sole economic engine, reinforced when miners cannot reinvest, middlemen control finances and (specific groups of) miners have little bargaining power (Hilson & Pardie 2006, Fisher 2007, Sinding 2005);
- Conflicts over land and resettlement, mainly between government, small-scale miners, large-scale operations and the local population who practice agriculture at the same site (Hilson & Potter 2005, Hilson *et al.* 2007) and;
- Environmental impacts: water and air pollution (notably from metals and chemicals used, such as mercury for gold amalgamation) (Shandro *et al.* 2009, Babut *et al.* 2003, Limbong *et al.* 2003), river and dam siltation, unrecovered open pits (which can trap animals, cause accidents and create a breeding ground for mosquitoes) and loss of biodiversity (deforestation, over-fishing and poaching) (Hentschel *et al.* 2002, Labonne & Gilman 1999, UNESC 2003).

These concerns about the impacts have arisen in response to an increase in artisanal mining activity, especially in areas characterised by high biodiversity, sensitive landscapes and where people are poor (World Rainforest Movement 2002).

Poverty and environmental conditions can be closely linked, as Labonne & Gilman (1999) highlight: poverty results from a denial of choices and opportunities and implies living on marginal and vulnerable environment, as well as further exhausting this environment. Most rural poverty is exacerbated by a lack of access to sufficient resources such as land, productive soils and water. Sustainable livelihood strategies imply that the economic needs of individuals and communities are integrated into the maintenance and improvement of the environment.

1.3 *Institutionalisation of artisanal mining*

A second body of work on ASM has dealt with the institutionalisation and organisation of the sector. To counter negative impacts, an array of technology, support-related and poverty-reduction projects focussing on ASM have been implemented over the past 15 years. Whilst difficult to generalise the outcomes of these interventions, because every mining site and country has its very own specific characteristics, many efforts have been criticised for having inadequately addressed miners' needs (Sinding 2005, Hilson &

Banchirigah 2009). Common lessons that can be drawn include:

- Albeit an often observed flaw of external intervention projects, ASM support projects have been rebuked for focussing too much on technical assistance or providing alternative livelihoods without taking into consideration the actual needs of miners (Hilson & Banchirigah 2009, Tschakert 2009). Childs (2008) contends that 'Fair Trade' initiatives could help in demonstrating alternative approaches to manage the sector by supporting participation and partnerships.
- Power relations between and among households and local organisations need to be taken into account and miners need a voice to avoid further exclusion of already vulnerable groups while institutionalising mining activities (Fisher 2007). The role of women and children in contributing to miners' household incomes is also often not acknowledged. Hilson (2005) argues that a poor understanding of the demographics of target populations has led to negative outcomes. Improved policy and assistance in the sector should be on the basis of accurate data on the number of people operating in ASM regions, their origins and ethnic backgrounds, ages, and educational levels.
- Formalising ASM may have become more difficult over the last decade, as Hilson (2009, p. 3) estimates, because 'foreign-controlled large-scale mining economy is now firmly entrenched in sub-Saharan Africa, and shadow networks long in place in diamond and gemstone production chains have become even more rooted'. It is not always clear whether legalising mining activities have actually helped the poor. Assistance programs often start by scaling-up activities and applying techniques which in the end require less local labour. A better understanding of the conditions for 'effective formalisation' is needed to know, how, and, under what conditions, this can contribute to economic and social development (Siegel & Veiga 2009).
- ASM reforms should encompass good sector governance, although implementing these policies on the ground requires institutional capacity and complex and sometimes expensive, monitoring mechanisms (Sinding 2005). Often the dynamics behind past sector reforms and how these have led to the expansion of ASM are poorly understood. Focus on large-scale mining (LSM) and privatisation have caused job losses and land evictions that pushed people in ASM (Banchirigah 2006). A major challenge is when artisanal mining takes place in poor or post-conflict countries where capacity for organising the sector is lacking (Maconachie 2009, Vlassenroot & van Bockstael 2008).

Governments can play an important role in redistribution revenues from mining (from royalties, taxes and license fees) to finance government services. Prerequisites for this are 'good governance' and appropriate policies, regulations, and fiscal regimes in the country together with mechanisms and capacity for implementation. A legal framework defining the rights of miners and communities, setting standards for environmental impact assessment and mitigating measures, and requiring financial and social obligations for new mining operations is also crucial. For artisanal mining, the challenge is to improve organisation and techniques and channel a proportion of resulting revenues to promote alternative, sustainable activities, given the finite nature of the resources. All of these aspects co-determine the eventual outcomes of mining for development and the environment.

Enabling positive governance outcomes at different levels entails understanding 'institutions'. Institutions can be defined as 'regularised patterns of behaviour that emerge from underlying structures or sets of rules in use' (Leach *et al.* 1999, p. 266). Institutions can be found on the international, national, regional and local levels. Wiersum (2009, p. 4) distinguishes '1. Formal institutions (or bureaucratic) based on official rules or even established by law. [and:] 2. Informal institutions (or socially embedded) in the form of unwritten codes and rules' and describes institutions being part of an ongoing dynamic process of transformation manifested in local and external institutional arrangements. The latter 'informal institutions' also seem especially relevant in the ASM sector that is often characterised by customary law and practices.

1.4 *Objective*

The Center for International Forestry Research (CIFOR) was requested by the International Union for the Conservation of Nature, Central and West African Office (IUCN-PACO) to research the impacts of artisanal gold and diamond mining on livelihoods and the environment in the TNS area. This study was part of the multi-partnership "Landscape and Livelihoods Strategy" (LLS) initiative, where the consequences of mining were of particular concern in the Sangha Tri-National Landscape (TNS).

The economic activities of the 191,000 people living in the vicinity of the TNS include logging, hunting, fishing, collection and sale of non-timber forest products, slash-and-burn-agriculture and artisanal gold and diamond mining (Tieguhong &

Ndoye 2007). All of these activities are critical to sustain people's life, but at the same time impose potential threats to the valuable landscape.

As information about management of natural resources in the Congo Basin is limited and the scale to which activities affect the landscape and lives of people is poorly understood, this poses difficulties for implementing appropriate policies (Tieguhong & Zwolinski 2008).

The aim of the study conducted in 2008 and 2009 was to provide information on the problems, functioning and prospects of artisanal mining and to provide recommendations for supporting a future small-scale mining sector that provides equitable access to the resource for the miners in the most sustainable manner.

This paper provides an overview of the impacts of ASM in the TNS Landscape and looks at

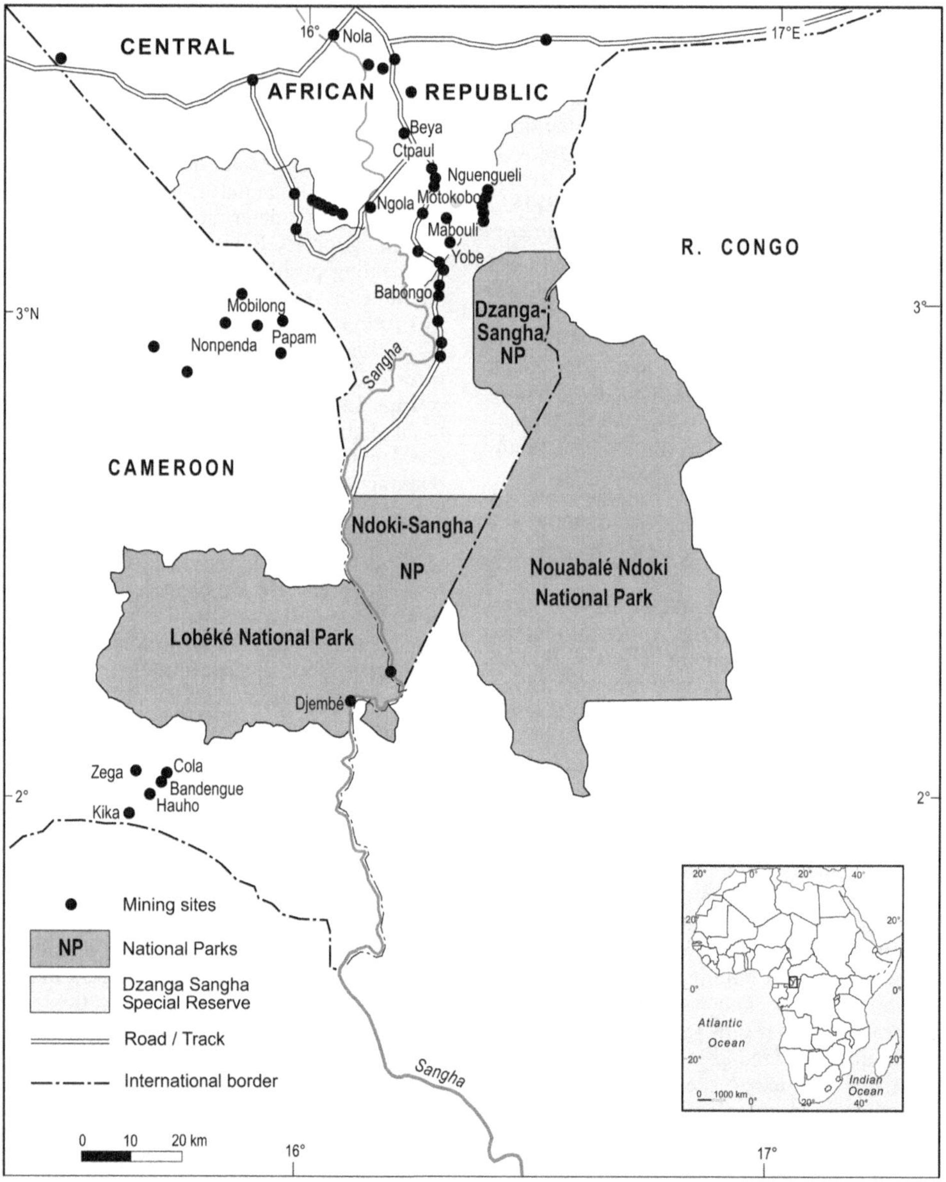

Figure 1. Artisanal mining sites in the Sangha Tri-National Landscape.

the institutional arrangements for mitigating those impacts and promoting sustainable livelihoods. First, it presents the main findings about the livelihoods and environmental impacts of artisanal mining. Detailed findings about these socio-environmental impacts of ASM in the TNS areas are presented in a full report (Tieguhong *et al.* 2009). Secondly, the institutional framework at local, national and regional levels is analysed to respond to the question of whether the current institutional set up is adequate to deal with issues around artisanal mining in the TNS Landscape.

2 MATERIALS AND METHODS

2.1 *Study area*

The study area is the Sangha Tri-National Landscape (TNS) shared by three countries: Cameroon, the Central African Republic (CAR) and the Republic of Congo (RoC) (located at 2°00′–3°32′N; 15°28′26″–17°34′8″E see Fig. 1). The landscape covers an area of 36,236 km^2 and has an elevation of 330–700 m asl. The Congolese section of the Landscape extends over the administrative departments of Sangha and Likouala. It covers 21,470 km^2 and includes Nouabalé-Ndoki National Park plus five forest management units (FMU), which cover an overall area of 17,280 km^2 and form the buffer zone of the national park. In the north, the area is delimited by the FMU of Mokabi; in the south by the FMUs of Pokola and Toukoulaka; in the east by the FMU of Loundoungou and in the west by that of Kabo. In the west, Nouabalé-Ndoki National Park borders on Dzanga-Ndoki National Park and Dzanga-Sangha Special Reserve in CAR. The CAR section covers 4644 km^2 and includes Dzanga-Ndoki National Park and Dzanga-Sangha Special Reserve. The Cameroonian section is centred on the Lobéké National Park.

The TNS contains a rich variety of flora and fauna. It is comprised of 93% dense rainforest, 5.6% mixed swamp forest and 0.3% of the area is grassy clearings. Less than 0.6% of the area is forest cultivation mosaic. The forest is rich in tree species with commercial value such as *Terminalia superba* (Limba), Sterculiaceae, in particular *Triplochyton scleroxylon* (Ayous), and Ulmaceae; and, is a sanctuary for some endemic or vulnerable tree species such as: *Autranellacongolensis, Pericopsis elata* (Afrormosia), *Diospyros crassifl ora* (Ebony) and *Swartzia fistuloides* (Pao rosa or African tulip wood) (on the IUCN Red List of Threatened Species) and *Entandrophragma* and *Khaya, Aningeria altissima* (Anigre), *Mansonia altissima, Pausinystalia macroveras* (Tsanya) and *Gambeya pulpuchra* (Longhi). In the CAR sector, 105 species of land mammals have been identified, among which are key protected species such as the gorilla *Gorilla gorilla*, hippopotamus *Hippopotamus amphibius*, and the African forest elephant *Loxodonta africana cyclotis.* Especially in Bayanga (CAR) and Nouabalé-Ndoki (RoC), elephant and gorilla populations offer important income-earning opportunities through ecotourism. The forests of the Landscape, with protected areas covering 21.5% of the entire area (752,000 ha), have been recognised as one of the priority areas for forest conservation in the northwest Congolese forests ecoregion. In 2000, the three countries signed a cross-border cooperation agreement with a view to improve conservation of the protected areas. The governments and international organisations, such as World Wildlife Fund (WWF), Deutsche Gesellschaft für Technische Zusammenarbeit (GTZ), Wildlife Conservation Society (WCS), International Tropical Timber Organisation (IITO), International Union for the Conservation of Nature (IUCN) and United Nations Educational, Scientific and Cultural Organisation (UNESCO), actively engage in the conservation of the protected areas and work towards sustainable management of the buffer zones.

The TNS has about 191,000 inhabitants with a very low average density of 0.7 inhabitants/km^2. Population growth in the South East Technical Operation Unit (part of which lies in the TNS), has been estimated at 1.88% (Sandker *et al.* 2009). Miners in the TNS have diverse ethnic backgrounds. In the Cameroonian part, indigenous people (Baka and Bangando, also known as Bagongo) comprised 27% of the artisanal mining population. The largest immigrant ethnic group were the Mpiemo of CAR, Foulbe or Haoussa, Kako and Mvongmvong. In CAR, 71% of the miners were indigenous groups (Ba'aka, Bosongo and Sangha-Sangha) and the Bilo, Bossangoa and Ngondi constituted the main immigrants.

The major economic activities around the landscape include: logging, mining, hunting, fishing, agriculture, livestock breeding, gathering, conservation, tourism and trade (in small commodities such as soap, cigarettes, palm oil, salt and palm oil) (CBFP 2006, Tieguhong *et al.* 2009). Direct threats to the landscape emerge from many of these activities, in combination with improved access to forests due to logging roads and a population increase. Household surveys conducted in 2008 in the South East Technical Operational Unit indicate an average income of US$250 per capita per year, which is substantially lower than the average Cameroonian per capita income of 1010 US$ (Sandker *et al.* 2009).

2.2 *Methods*

The methodologies used to collect the data presented in this study included:

- Field visits to 17 mining sites between August and December 2008 (13 in Cameroon and four in CAR) located within 50 km of the TNS for observation, mapping and photographic documentation. The Republic of Congo's nearest mining sites (Boloko, Golana and Pandama) were not within the 50 km distance of the TNS and therefore not part of this study.
- Interviews with 131 (63 gold and 68 diamond) miners, and complementary semi-structured interviews with key actors (such as the park conservators, government representatives in charge of mines, forest and the environment, and representatives of international non-governmental organisations). Topics covered in the interviews included: general perceptions on mining activities and other income earning activities, and perceptions on environmental degradation, benefit flows, characteristics of mining and governance of the sector.
- A literature review of scientific studies, reports of ministries and support organisations, national laws and regulations, and mining permits.

Data were analysed using twelve equations in the calculation of annual quantities of minerals, costs of production, gross and net revenues, as well as the aggregation of observed values. Field data entry was done in the CPros version 3.0 typing mask and transferred using Stat-Transfer version 5.0 into SPSS Program version 12.0 for analysis. STATA version 8.0 was used for the logistic regression analysis. Miners were dichotomised into those who were highly dependent on artisanal mining and those with a lower dependence on mining. The Gini coefficient was used to test inequality associated with dependence on mining income. Information provided by partners and field organisations was recorded as their perceptions and later used to cross-check data provided by artisanal miners.

3 RESULTS

3.1 *Socio-economic impacts*

Most of the miners interviewed had permanent or temporary residence in Zega (CAR), Mboya (CAR), Nguenguili (Cameroon) and Ngola (Cameroon). The majority of mining activity took place year-round but activities (especially the alluvial mining in the Sangha River) reduced in the rainy season when flooded conditions impeded the work. Most mining camps had agricultural crops planted, and in some cases livestock (fowls, goats and sheep) were reared. The livestock was particularly important for the ritual of making sacrifices to the god of diamonds.

Two types of artisanal miners were observed during this study: divers and diggers. Diggers are those who dig pits in the soil to extract economic minerals, and divers are those who dive into the Sangha River to scoop sand and soil (from the riverbed) to get their diamonds. Although miners come from diverse ethnic groups, the vast majority (95% in Cameroon and 87% in CAR) are permanently based in nearby towns and villages. The average age of miners is 36 years old, and miners generally have many years of experience in the ASM sector (17 years of experience in CAR and 9.5 years of experience in Cameroon). In CAR, no women lead mining activities, whereas in Cameroon 13% of the head miners are women. However, in both countries, most miners are married, and assisted by (family) labour, involving many women and children. Education among the miners is generally low, with over 70% of the miners with only primary schooling level or having had no formal education at all (Tieguhong *et al.* 2009).

The main finding related to socio-economic outcomes was that the livelihoods of at least 5% of the area's population, totalling 4600 people (517 miners, their dependents, and labourers), are based primarily on artisanal mining. Although the income generally pays for important household needs, the range of income gained among miners is highly disparate. All minerals extracted in the region on an artisanal scale are sold unprocessed. The mean annual net revenues from gold and diamonds—after reduction of the costs of production materials, labour and transport—were 575,338 FCFA (std = 461,913) (US$1130) and 812,644 FCFA (std = 676,487) (US$1596), respectively, in Cameroon. This represents more than four times the income of the average 'non-miner' in the region (of about US$250) (Sandker *et al.* 2009). On the CAR side of the TNS, diamond miners profited less with an average annual net income of 368,084 FCFA (std = 904,427) (US$723).

Ethnicity was a major factor in explaining a higher income from artisanal mining. As expected, migrant miners with skills, capital, better education and more experience could earn more from artisanal mining than indigenous Bangandos and Ba'aka/Baka pygmies.

In both Cameroon and CAR, over 90% of miners used their income to meet four to six basic needs, including food, education, healthcare, radio/TV (information tools), clothing and housing construction. Meeting these needs fits some of the most important Millennium Development Goals (MDGs) of reducing poverty.

The calculated Gini coefficient (from a scale of 0 indicating complete equality to 1 showing inequality) for absolute income was 0.50, and for absolute non-mining income, was 0.43. This means that there was a higher income disparity between miners than between the other people in the region. With a standard deviation of up to three times the average income, actual revenues ranged from considerable profit to significant losses. Net annual losses of up to 1,032,450 FCFA (US$2028) were found in CAR and 400,000 FCFA (US$786) in Cameroon. The risks of losing income were of special concern because most miners did not save any of their income. Additionally, the results of the logistic regression analysis indicate a significant dependence on mining incomes ($p < 0.05$). Ethnic groups, educational levels and time spent mining were the main variables that determined level of mining dependency.

Mining was the principal activity for 79% and 88% of the artisanal miners in Cameroon and CAR, respectively, although mining is often part of a multiple income generation strategy, combined with between one to six other activities. Agriculture was the second important source of income, followed by non-timber forest product gathering in Cameroon and fishing in CAR.

Small-scale miners in the TNS offered different views on production trends and problems encountered. In Cameroon, miners believed production was increasing and related that this was mainly a result of increasing prices, hiring labour and purchasing of new tools. On the contrary, miners in CAR said that production was decreasing, blaming over-exploitation and a lack of external support. Miners reported a variety of problems related to their daily activities, the most pressing of which are: lack of food and medicine, harassment by conservation officials, dishonesty of their sponsors, low production, harsh government laws and actions, an inability to detect materials and minerals, price fluctuations and lack of capital. In Cameroon, the three major problems (stated by 76% of interviewees) were: a lack of materials for detecting and exploiting minerals, a lack of food and medicines at the sites and low production. In CAR, the two main problems identified, expressed by 65% of interviewees, were: low production and lack of materials for detecting and exploiting minerals.

The fact that over three-quarter (90% in Cameroon and 77% in CAR) of the overall income of small-scale miners originates from natural resources—not only minerals, but also NTFP gathering, hunting and fishing—shows the enormous contribution of the natural environment to miners' livelihoods. This raises the question of how environmentally-sustainable these activities are, and how artisanal mining activities impact upon the very same natural environment that miners depend on?

3.2 *Environmental impacts*

Local stakeholders interviewed indicated that miners, albeit a small percentage, were operating within the interior of the reserve, especially in the northern section of the Dzanga-Ndoki National park in CAR and the southern part of Lobéké National Park in Cameroon. Several related environmental risks and impacts were noted, such as: water and soil pollution, disturbing of fish breeding grounds, an increase in infrastructure in the forest environment, un-recovered exploited mining pits and poaching of wild animals. The research findings confirmed that some miners operate within the parks. In Cameroon, 20% of the miners indicated that they (also) mined inside the reserve, in contrast with only 1 (out of 32 miners) in CAR. However, the overall scale and conduct of artisanal mining in the TNS Landscape was found not to drastically threaten environmental values.

Environmental impacts from artisanal mining in the TNS Landscape appear to be of limited scale and duration. The majority of mining took place along streams, causing direct but insignificant impacts such as diversions, siltation and sedimentation of water sources. Only a limited felling of trees or land clearance was observed. Small surfaces were cleared during the period of mining, which was often seasonal, with only limited farming activities taking place. The indirect effects of working in the forest areas included timber and non-timber forest product exploitation, including bush meat and medicinal herbs, by 21% of Cameroonian and 28% of CAR respondents, who indicated that such activities provided alternative sources of income. No miner reported using mercury or cyanide for gold extraction, nor had these polluting activities been observed by any of the stakeholders.

This conclusion of small-scale and temporal environmental impacts due to artisanal mining could, however, change in the future. An expected increase of the number of artisanal miners (migration and an overall estimated population growth in the region of around 1.88% per year) and the influx of large-scale mining operators are expected to place increasing pressure on the landscape. This, combined with a lack of environmental awareness among the miners, could cause destruction. In both countries, over 53% of the artisanal miners stated that gold and diamond are infinite resources, and 67% believed that mining had no negative environmental impacts whatsoever.

Another concern linked to future impacts, raised by some stakeholders, was the fact that the

Table 2. Types of government support of ASM in TNS.

Government support or program	Frequency	Percentage by country (%)	Overall percentage in TNS (%)
Cameroon			
Education	1	1	1
Technical training	28	28	21
Exploration equipment	4	4	3
No support	66	67	50
Total	99	100	76
CAR			
Education	9	28	7
Technical training	4	13	3
Exploration equipment	2	6	2
No support	17	53	13
Total	32	100	24
Total	131	100	100

buffer zone of the Lobéké National Park has been attributed to mining operators under research permit titles. A WWF map shows the overlaps of mining and timber concessions with national parks in Southeast Cameroon (WWF 2008). These multiple allocations of land for different uses, with no coordination between the ministries issuing permits or land use rights, demonstrate the lack of coordination between the responsible authorities and the potential for conflicting interest, not just between timber and mining operations and protected area managers, but also between LSM and the activities of the local population. One example of the latter is the operation of C&K Mining, a joint Cameroon-Korean company that has explored potentially huge diamond deposits (estimated volume of 740 million carats) at Mobilong in the East Province (Gweth 2008). Local miners in this area, none of whom possess official mining titles, are wary of the presence of the company, fearing they could be evicted if the company starts mining operations.

These socio-economic and environmental impacts of artisanal mining can be placed in the institutional context of local, national and regional forms of organisation that co-determine the managing of risks and opportunities and the eventual outcomes for the population and the landscape.

3.3 *Institutional framework at the local, national and regional levels*

How formal and informal forms of organisation are perceived and structured at the local level is first presented, followed by an analysis of the key institutions at a national and international level.

3.3.1 *Local institutions*

Of the miners interviewed, over 70% and 63% in Cameroon and CAR, respectively, worked for themselves. Mining in groups or cooperatives was unusual, except for the recently introduced project by the Cameroonian Support and Promotion Framework of Mining Activities Organisation (CAPAM). About 29% in Cameroon and 37% in CAR were working for sponsors, who purchase materials, food and medicine for their workers. Mining sites are commonly headed by a site chief (*chef de chantier*), usually the oldest or the most experienced person, who retains special mining rights and exercises leadership at the camp. For example, the chief of a diamond mining site in Ngola (CAR) was informally entitled to 25% of sales. Formally, the chief is obliged to pay an annual government tax of 30,050 FCFA (US$59).

There was correspondingly no organised sale of products in the TNS; each miner sells winnings individually to buyers. The buyers, generally known as collectors, travel to buy the minerals at pre-determined prices and as the miners have little bargaining power, they are largely price takers. Price variability, especially for diamonds, is large and miners lack knowledge and tools to determine their value and prices themselves. Small-scale miners who work for sponsors generally do not even know the unit price of their product. About 32% of Cameroonian and about 47% of CAR miners obtained financial support for their operations from either a village sponsor or an external sponsor. Sponsored miners were unaware of how much business capital had been provided to them, claiming they were basically fed and paid to perform their job, indicating how individual buyers control miners through sponsorship. The miners, in turn, are expected to be loyal and sell their products exclusively to their sponsor; harassment results when a miner has been discovered selling to another buyer. Miners related that transport to market usually was problem free, explained by the fact that the small volume—high value of the product makes it largely undetectable by government officials.

When asked about government support for small-scale mining in the TNS, 67% and 53% of miners in Cameroon and CAR, respectively, indicated that it was completely absent. However, 29% in Cameroon mentioned education and technical training, specifically making reference to the diamond miners in the Mboy region, where the government agency CAPAM has been providing equipment and technical assistance to small-scale miners since 2006. At Mobilong, CAPAM gave motorised pumps to artisanal miners free of charge, which otherwise are rented at 5000 FCFA (US$10) per day. The miners of this site attributed the fact that they have accumulated capital to the support they received from CAPAM. In CAR, education and technical training accounted for 40% of the government-support identified by miners (Tab. 2).

When artisanal miners were asked if they had experienced problems during operations, 91% and 66% in Cameroon and CAR, respectively, stated they had no disturbances of any kind. Further questioning on the nature of problems faced, however, revealed that government agents, conservation officials and individual buyers as control agents caused harassment. Government agents are the major source of harassment in CAR, but conservation agents were the leading source in Cameroon. Strategies for dealing with such problems in Cameroon were to run away (22%), bribe the controller (11%), speak angrily (33%) or stay quiet (33%).

In CAR, artisanal miners react either by speaking angrily (64%) or show their papers to prove they are operating legally (36%). This underscores not only the weaker legal position of miners in Cameroon but also the general level of corruption in both countries.

Artisanal mining in the TNS region is informal and largely illegal. This is particularly the case on the Cameroonian side, where none of the interviewees confirmed having paid tax of any kind or having a legal mining permit. The delegate of mines, for the Boumba and Ngoko division in Cameroon, indicated that miners do not want to follow the rules and regulation of the sector and refuse to pay the yearly 5000 FCFA (US$10). However, in the CAR, a little over 56% of the artisanal miners pay the labourers' tax and possess an annual mining card, miners pay an annual labourers' tax of 2000 FCFA (US$4) and the site head pays an annual tax of 30,050 FCFA (US$59). Collectors or buyers normally pay an annual tax of up to 1,100,000 FCFA (US$2160). The higher level of legality in CAR demonstrates that regulation is possible. Yet, about 62% of the artisanal miners in Cameroon and 81% in CAR reported not being aware of the respective country's mining code and allegedly could not comment on its enforcement. Local stakeholders expressed the need for sensitising miners on the code and legalising of their operations. Most miners (64% and 67% in Cameroon and CAR respectively), however, said it was not easy to obtain legal mining papers. Paradoxically, most of those who said it was easy to obtain legal papers did not actually have them at the time this research was carried out. The miners who had legal papers in CAR mentioned that the benefit of having the papers was basically the freedom to exploit and sell minerals. Despite the fact that 'harassment by conservation agents' and 'harsh government laws and actions' are mentioned as problems, miners still see the government as the main actor with power to assist. The question that remains to be answered is whether interventions directed at these problems can break the vicious cycle of informality and dependency to enable small-scale miners to earn more profits from mining. Miners cited a number of opportunities for resolving such problems, which are detailed in Table 3.

Over two thirds of the miners in both Cameroon and CAR believed that the government should help them obtain materials such as tools or ensure sensitisation of the mining code, Artisanal miners called for greater transparency in sales offices, reduction in the costs of obtaining legal papers and stabilisation of prices. Stakeholders in the TNS are often

Table 3. Opportunities suggested by artisanal miners in TNS for resolving problems.

Opportunity/ Request	Frequency	Percentage by country (%)	Overall percentage in TNS (%)
Cameroon			
Assistance to obtain working materials	55	49	38
Assistance to obtain legal papers	26	23	18
Open a sales agency in village	10	9	7
Stabilise prices	6	5	4
Create a cooperative	9	8	6
Receive training in modern mining techniques	7	6	5
Total	113	100	77
CAR			
Assistance to obtain working materials	12	36	8
Assistance to obtain legal papers	10	30	7
Open a sales agency in village	4	12	3
Stabilise prices	1	3	1
Create a cooperative	6	18	4
Total	33	100	23
Total	146	200	100

advised to organise the miners and formalise their activities as a first step to solve some of the issues. Groups could become trained on better technologies and environmental impacts.

3.3.2 *National institutions*

Cameroon and the Central African Republic, both have a foundation for good governance in the sector with a legal framework structured by a Mining Code, Tax Code, Land Tenure Code, and regulations on the environmental and protected areas.

In Cameroon, the main bodies dealing with geology and the mining sector are: the Ministry of Industries, Mines and Technological Development (MINIMIDT) which also has the responsibility for the national geological survey through the Direction of Mines and Geology, and; the Ministry of Scientific and Technical Research, an agency that oversees a variety of research institutions in the areas of geology and geophysics, hydrology, and energy. Mining activities are regulated by a legal framework comprising the following: the Mining Code under MINIMIDT; the Tax Code including customs, labour and investment codes under the Ministry of Finance; and the Environmental Code under the Ministry of Environment and Nature Protection.

The legal framework for Cameroon's mining sector follows French law. The Mining Code consists of a law (1964) which regulates mineral substances, and another law (1978) which defines taxes, including royalties and mining taxes. The latter was supposed to define the fiscal framework for mining, but this did not happen until 2001 when the new Mining Code (Law No. 1 of April 2001) was promulgated with the assistance of the World Bank. It comprises the fiscal laws necessary for the regulation of the sector with provisions for investors to negotiate on a case-by-case basis for the establishment of mining companies. According to the code, all mineral resources belong to the state. Prospecting, exploration and mining activities for any mineral deposit are regulated by permits, which are awarded for quarrying, prospecting/research, exploration, exploitation, and mining concessions (Republic of Cameroon 2001). This legal framework has reduced administrative burdens and put the authorities in a better position to evaluate investment opportunities (which often require rapid decisions). One perceived advantage of the new code is a reduction in the role of the state in mining operations as well as its discretionary powers. On the other hand, there is an increase in the state's role as a supervisor and regulator of the mining sector.

The creation of the Support and Promotion Framework of Mining Activities Organisation (CAPAM), in 2003, followed a new provision of the 2001 Mining Code enabling the setting up of an autonomous unit to facilitate, assist and promote small-scale mining and aid up-scaling to large-scale mining operations. In 2006, CAPAM channelled 50 kg of gold and 300 carats of diamonds in its market facilitation structure. The revenues are being used to invest in materials, pay tax (3% for gold, 8% for diamonds) and 15% is given to the local council, 10% to local population and 25% to the monitoring and control organ. One of the advantages for the miners would be greater certainty about prices for their production with help of sales according to an approved price list. This approach appeared difficult to implement; miners were not well informed about the price list and still had no capacity to determine the diamonds' quality and value category. In 2008, the purchase of diamonds from artisanal miners by CAPAM was suspended by the Minister.

The 2001 Mining Code of Cameroon differentiates between ASM and LSM but at the same time, gives provisions for the two to operate at the same site, recognising the importance of the livelihoods of local people and referring to the fact that the former goes to less profound depths than the latter. A related challenge recognised by provisions of the 2001 Mining Code is to mitigate problems caused by both ASM and LSM operations in the same area. However, there have not been any practical tools developed to date to deal with the situation of ASM versus LSM.

The mining sector in the CAR falls under the responsibility of the Ministry of Mines and Energy. The General Directorate of Mineral Resources implements the law and policies regarding mining permits. The Mining Code of 1961 was revised in 2004 in order to provide more flexible and attractive policies to stimulate investments in the sector, including fiscal incentives (e.g. tax exemption on equipment during exploration) and establishing an organisation responsible for geological exploration and prospecting claims (CAR 2004). By law, all mineral resources in the ground or at the surface are property of the State and access can be granted by means of permits. The new Mining Code defines six categories of permits, including artisanal mining. The permits serve different purposes, subjected to various surface areas, validity and delivery authorities, and provide for exclusive rights on the defined property; when a deposit is discovered, the right to mine is guaranteed to the owner of the exploration permit, which can be sold or transferred with the authorisation of the Ministry of Mines (CAR 2004).

The government maintains statistics concerning diamond production and trading through the *Bureau d'Évaluation et de Contrôle de Diamant et d'Or* (BECDOR). BECDOR was established in 1982 to oversee the internal diamond market and to valuate official exports. It also maintains a database concerning all diamond production in the country. BECDOR estimates that there are approximately 50,000 licensed diamond diggers, or 'creuseurs', in the CAR. Labour taxes are collected from 56% of miners. The artisanal miners sell their production to about 160 certified collecting agents who, in turn, sell this production to two purchasing offices located in Bangui.

3.3.3 *International and regional institutions*

At the international level, artisanal mining and its potential for development have received increasing attention over the past decade. Focussing on the most relevant international and regional initiatives and the involvement of Cameroon and CAR, four are key to the ASM sector.

The Communities Artisanal and Small-scale Mining (CASM) is a global networking and coordination facility with a stated mission "to reduce poverty by improving the environmental, social and economic performance of artisanal and small-scale mining in developing countries". CASM is currently chaired by the UK's Department for International Development, and is housed at the World Bank headquarters in the United States. CASM Africa has its secretariat based in South Africa (CASM 2010). Central African countries can benefit from its facilities but until now, they have not really taken advantage of them.

The International Labour Organisation (ILO)'s Convention on Safety and Health in Mines, 1995 (No. 176), covers all mines and provides the minimum safety standards against which all mine operations should be measured. The Convention sets out procedures for reporting and investigating accidents and dangerous occurrences in mines. Governments that ratify it undertake to adopt legislation for its implementation, including the designation of the competent authority to monitor and regulate various aspects of safety and health in mines (Walle & Jennings 2001). So far, none of the Central African countries have ratified the convention (ILO 2010).

The Kimberley Process was initiated by African diamond-producing countries in May 2000 to develop an international certification scheme for rough diamonds in order to prevent conflict diamonds from entering legitimate markets (Kimberly Process 2004). This process, supported by the World Diamond Council and the United Nations, and implemented by a United Nations vote in 2003, requires the certification of all diamonds mined and upon every transfer of ownership of the rough diamonds. The Process brings together a broad range of international stakeholders in the diamond trade, including government officials, industry representatives and non-governmental organisations. At present, Côte d'Ivoire is the only country under embargo by the United Nations for the export of conflict diamonds; this occurred in December 2005. The CAR, DRC, Gabon and Republic of Congo are currently member countries. In 2007 Cameroon affirmed its intention to join the Kimberley Process but this has not been put into practice yet. Cameroon joining the Kimberley Process will be important for improving monitoring mechanisms and monitoring cross-border trade.

The Extractive Industries Transparency Initiative (EITI) followed a 2003 conference to improve transparency of oil, gas and mining payments by companies to governments. The general idea is that revenues from oil and mining would become public information, and would be spent in a more transparent manner to benefit the country's population. At a country level, a first step is the foundation of a multi-stakeholders group with representatives of government, companies and civil society, which works towards an action plan with rules for disclosure and monitoring. Cameroon is currently a candidate country of EITI and has until March 2010 to undertake validation. Hitherto, Cameroon has only reported oil revenues in its 2006 and 2007 reports, covering the period 2001–2005. The country has promised to start reporting revenues from mining exploitation in its next report that would cover 2006–2009. Until now, ASM has not been part of the EITI discussion (Valéry Nodem, Coordinator *Réseau de Lutte contre la Faim*, member-organisation of Publish What You Pay (PWYP) Cameroon, pers. comm.). CAR also has the status of EITI candidate country with until November 2010 to validate its work plan. CAR has diamond mining companies as partners in the stakeholder forum. Other countries in the sub-region that are EITI candidate include Gabon, the Republic of Congo and the Democratic Republic of Congo (EITI 2010).

At a regional level, there are no specific initiatives on ASM although the similarity of the sectors throughout various countries of the Congo Basin suggests the usefulness of having such a forum for exchange on technical aspects. Moreover some cross-border issues ask for regional solutions. Cross-border smuggling can result from different institutional contexts or differences in mineral taxes between countries. For the case of Cameroon and CAR, the difference in institutional setting is the fact that Cameroon is no applicant of the Kimberley Process and has less visible institutional structures when it comes to ASM in the TNS region. Also there is a different official tax base for Cameroon with a diamond tax of 8% and CAR with a higher 12% taxes (Barthélémy *et al.* 2008).

The Yaoundé Declaration signed in 1999 by the governments of six Central African countries; Chad, Cameroon, CAR, Republic of Congo, Equatorial Guinea and Gabon, set the foundation of the Central African Forest Commission (COMIFAC) and became an important agenda for cross-regional conservation and development goals. It resulted in cross-boundary agreements, among which the agreement between Cameroon, CAR and Republic of Congo about conservation of the TNS Landscape. The 2005 Brazzaville Treaty and the adoption of the Convergence Plan (*Plan de Convergence*) by ten African countries—Burundi, Cameroon, CAR, Chad, Democratic Republic of Congo, Equatorial Guinea, Gabon, Republic of Congo, Rwanda and, Sao Tome and Principe—created the provision of free cross-border movement of park guards in the TNS (Carroll 2008). However, at no point do the agreements which followed the Yaoundé Declaration involve ASM as integral part of discussing conservation and development in forest landscapes.

4 DISCUSSION

In the Sangha Tri-National Landscape, artisanal mining activities have had both negative and positive impacts. It is questionable, however, whether the current institutional arrangements are capable of addressing the industry's negative aspects. Adequate institutional mechanisms not only refer to the laws and regulations of the countries but involve also the 'informal institutions' at various levels and their ability to deal with institutional challenges around ASM. As identified in the introduction, main challenges are: knowing the limits and opportunities of external interventions; understanding power relations between actors; and finding out what 'effective formalisation' and 'good sector governance' should entail in the specific TNS context.

The study results indicate that mining in the TNS landscape contributes significantly to livelihoods, as diamonds and gold provide on average higher levels of income than traditional alternative activities for over 4600 miners, labourers and dependents in the landscape. Nevertheless, the huge income range of miners shows the high financial risk some of the miners carry and overall incomes are still too low to lift households out of poverty. The challenge to educate miners on how to avoid wide income swings and losses, and improve incomes is not sufficiently embedded in current policies or activities of the government, development organisations or private sector in either country.

Armstrong's (2008) study of the artisanal diamond sector, lists some recommendations for improving revenues, including improving knowledge about pricing and technology; improving access to lower-risk credit; making accessible and competitive licensed buyers or buying offices; joining fair trade initiatives; and value-added activities and local processing of the minerals. For knowledge transfer to be effective, programs should accommodate local needs and show its use in terms of profitability, simplicity and efficiency to increase benefits with minimum health and environmental risks (Hinton *et al.* 2003).

An expected outcome of making ASM more profitable through support initiatives is that more people will be attracted to the sector. This is even the purpose of the Cameroonian support agency CAPAM, which aims to increase the number of small-scale miners from the estimated 20,000–30,000 to a workforce of 60,000 (CAPAM, 2006). This carries the risk of increasing dependency of many on this finite source when it is not embedded in wider development strategies (Vlassenroot & van Bockstael 2008).

Diversification of livelihoods, often recommended as way to improve the sustainability of ASM (Hinton *et al.* 2003), would clearly be nothing new to the miners in TNS, who, although largely dependent on mining incomes, conduct between one to six additional income earning activities. Deliberate government or external support programs on diversification of income opportunities should build upon these activities that miners already practice.

A first step of outside intervention programs is often the organisation of miners in groups. The logic behind grouping miners is to have them share investment costs and benefits and at the same time to have an entity with whom the government or development agency can work. This often implies the transfer of ASM into larger-scale operations (Tarras-Wahlberg *et al.* 2000). Blore (2008), who studied miner cooperatives in Latin America, contends that the groups suffer serious constraints when it comes to sharing investments and revenues; however, they can benefit from working together in other areas such as sharing land and costs of mining claims and administrative procedures, exchanging information and working together on community development initiatives. The problem with organising miners is that it often builds upon a false assumption that the sector is 'chaotic and unorganised, when in fact it is highly organised' (Hilson 2008, p. 223). The picture drawn in this study confirms this, as it shows miners who have been in the business for a long time and although not 'formally', are already highly organised in mining camps. This level of organisation could offer a good entry point for supporting wider issues, such as the by the miners expressed wish for obtaining tools and legal papers. A noticeable difference of perspectives on solutions between actors was that none of the miners expressed the wish for better organisation of miner groups while most of the external stakeholders (local NGOs and officials) mention this as one of the key priorities. This difference needs to be further explored before imposing types of organisation that might not fit the working structure of miners. Unfortunately, the lessons of PASAD, a project for technical support to ASM in CAR funded by *Caisse Française de Développement* (1996–1998), have never been evaluated (Barthélémy *et al.* 2008). However, current experiences of the activities of CAPAM in Cameroon and the USAID financed pilot project 'Property Rights and Artisanal Diamond Development Program' in CAR (Vlassenroot & van Bockstael 2008) will hopefully provide valuable context specific lessons.

A key point of institutional attention for any support program includes addressing power relations, between and among the actors, in order to avoid further inequality and marginalisation of certain groups. Logistical support must not only exist but need also to be considered accessible by the miners (Sinding 2005). Especially the process of issuing mining titles is prone to exclusion and can eventually lead to abuse of vulnerable groups by other miners and mining companies (Fisher 2007). The study findings show the average lower miner revenues of the indigenous Bangandos and Ba'aka/Baka pygmies compared to other miners. Currently there is no institutional arrangement to disseminate knowledge from the migrant miners with better skills, capital and educations to indigenous groups.

An aspect of power relations are the so called 'vested interests' in the valuable mineral sector, which should be identified before any reform takes place (Hilson 2009; Maconachie 2009). The actors with vested interests in mining in TNS are the miners, the site chiefs, middlemen and sponsors. Miners much depend on the site leader and sponsor for their revenues. The site chiefs, who receive a percentage of miners' income based on informal arrangements, are not likely to support a permit system when this would endanger their revenues. On the other hand, awareness-raising about benefits of mining titles for miners could entice them to become part of a formalisation process and become less dependent on their current sponsors. Another potential source of future conflict clearly signalled in Cameroon is the overlap of ASM and LSM concessions, timber concessions and National Parks in TNS.

Formalisation is often considered as the way forward for integrating ASM into the national economy and building a mechanism to look into sustainability and livelihood issues. With one fifth of miners already operating within protected areas of TNS and an expected increase of ASM and LSM activities in the area, maintaining the current low level of environmental impacts will only be possible if policy measures are enacted. Balancing improvements and formalisation in the sector with measures to discourage mining inside the protected areas are key priorities. Three elements for ASM formalisation are: 'A legal and regulatory framework ensuring security of tenure and property rights, acknowledging the necessary participation of local authorities and backed-up by a sound geological survey and cadastral system; The delineation and creation of artisanal mining zones; The use of miners' identity cards' (Armstrong *et al.* 2008, p. 110). The first two elements are not sufficiently embedded in current ASM practises in CAR and certainly not in Cameroon where the sector is largely informal. Livelihoods are hazarded by the lack of coherence in the legal framework (Mining Code, Tax Code and Environmental Code) and the low or non-existent level of tax payments do not enable government agencies to channel revenues from the sector back to support it. Miners at the Cameroonian side of TNS do not carry mining permits. In CAR, 56% of the miners interviewed do possess a license and pay the annual tax.

For miners to voluntary pay their licence, they first need to believe that the tenure and tax system will provide them with some benefits. This asks for real benefit transfer within the system and awareness-raising about these advantages. Benefits may come from tenure rights that offer miners security and a longer term perspective (Sinding 2005). Property rights are the 'basis of legal redress when rights are violated by government or company as they frequently are' (Siegel & Veiga 2009, p. 52).

However, the transaction costs of putting in place a formal system for ASM can be out of proportion with potential revenues it will create, as might be the case in Cameroon where revenues are modest and the number of miners relatively large and difficult to monitor. Sindings (2005) suggested the possibility of leaving out the smallest operators from the formalisation process for this very reason. But leaving them out of formalisation also means that they stay out of sight and remain excluded from accessing rights and support. A rather light structure for these type of miners combined with easy accessible logistical services seems more appropriate.

Another element brought to the forth by this study is the element of trust. Trust between miners and local officials and vice versa is needed for any of the more formal mechanisms to function. Miners in TNS report harassments of government agents (mainly in CAR) and conservation agents (mainly in Cameroon). One advantage of having legal papers shows in CAR, where a third of the miners solve their confrontation with agents by proving that they operate legally. Other encouraging signs of a positive atmosphere for formalisation are that miners do propose 'having access to legal titles' as one of the solutions for improving their livelihoods.

The main challenge is to construct a flexible type of formalisation that accommodates the needs of miners, takes into account power structures and can be the operational arm of broader good sector governance.

Good sector governance in the context of TNS should first of all encompass more than one sector. Integrated resource management on a landscape and regional, cross boundary level, can help to balance the positive and negative impacts of ASM. The current competing and contradicting stakes and authorities in the TNS landscape increases the risk of degradation of the environmental values of the landscape and enhances the exclusion of less-represented groups who are the most dependent on these resources: poor, rural ethnic groups such as the Baka pygmies.

Frost *et al.* (2006) proposed eight guidelines for successful resource management, based upon experiences from the TNS landscape. These include a focus on multi-scale analysis and intervention; developing partnerships and engaging in action research; facilitating change rather than dictating it; promoting visioning and the development of scenarios; recognising the importance of local knowledge; fostering social learning and adaptive management; concentrating on both people and their natural resources, including biodiversity; and embracing complexity. Using these guidelines can help particularly in empowering local stakeholders to be more articulate advocates and active participants in their own development and conservation efforts.

Incentives, knowledge, and property rights are key point of attention for supporting outcomes of formalisation and 'good resource governance'. Property rights will stimulate investment in knowledge to minimise environmental risks and 'expanding existing mining areas creates incentives for miners to remain settled in their current locations, potentially limiting the ecological impacts of ASM to a particular zone' (Siegel & Veiga, 2009: 55). Property right systems should be well communicated made clear and understandable, enforceable and supported by a cadastral system that also includes prospecting of potential ASM and discloses information about mineral deposits (Armstrong *et al.* 2008).

Although ASM has received more attention in global policy debates over the past decade and possible positive outcomes in terms of development have become increasingly recognised, there are few international frameworks in which the rights of ASM are being discussed. CASM is the only platform that specifically deals with ASM, but again, Cameroon and CAR are not yet active participants. The EITI can enhance overall transparency of the sector but focuses on large-scale mining and is still in a planning phase for both countries. The Kimberley Process is interesting when it comes to the diamond sector monitoring in CAR but also implies a backward position for Cameroon, which is not a member (yet) and therefore unable to access the international diamond market of the KP member countries (Barthélémy *et al.* 2008).

At the regional level, there are cross-boundary agreements about the natural services of the park. However, these TNS agreements are mainly conservation and forest exploitation based and they do not address mining activities. A regional integrated approach towards current issues in both the forest and mining sector would be highly recommendable and reinforce the outcomes of the TNS agreements. Regional collaboration around mining issues is especially relevant when it comes to monitoring shared mining deposits, such as the 'border-rivers with mineral rich alluvions: Mboumbé between Cameroon and the CAR; Sangha between the CAR, Cameroon and the Republic of Congo' (Barthélémy *et al.* 2008, p. 32) and mitigating smuggling of high value minerals. Formalising regional agreements that include harmonisation of taxes is seen as one of the solutions to minimise cross-border trafficking of high value minerals (Armstrong *et al.* 2008). Furthermore, countries could join forces in prospecting of geological deposits in trans-border zones, in the monitoring and organisation of commodity trade and offering support to artisanal miners. The Republic of Congo and CAR already signed an agreement in 1998 about common management of the diamondiferous zones in their border region but as this was never enforced, these plans would need to become revitalised first (Vlassenroot & van Bockstael 2008). Barthélemy *et al.* (2008, p. 39), in their study about trans-border issues around mining in Central Africa, advise the governments of the TNS countries, Cameroon, CAR and Republic of Congo to 'undertake together an ambitious project that could be called 'Control and Development of the Exploitation of the Diamond Resources of Central Africa' and use for this the existing structures of CEMAC (Economic and Monetary Community of Central Africa) and CEEAC (Economic Community of Central African States). Such a proposed structure, preferably not only on diamonds but also on gold and other valuable natural resources, seems of great value, especially if it adopts an integrated natural resource management angle and could build upon the benefits of the existing trans-boundary TNS agreements.

5 CONCLUSIONS AND RECOMMENDATIONS

Key recommendations to regional governments, ministries, non-governmental organisations, private enterprises and development agencies are to improve the coherence of strategies across the mining and forestry sectors in order to enhance livelihoods and minimise environmental impacts. Special attention should focus on the mitigation of conflicting interests: between small-scale and large-scale mining activities; and with regard to mining activities in timber concessions and/or in protected areas. Mining companies should state explicitly how they will interact with local communities and artisanal miners during their daily operations and as part of their overall social responsibility.

By harmonising mining policies and resource governance strategies in the Congo Basin and in the three countries (Cameroon, the Republic of Congo and the Central African Republic) in the TNS in particular, a coherent approach to mining in the TNS area would enable the negative issues of artisanal and small-scale mining and trans-boundary trafficking to be more efficiently tackled. The COMIFAC countries are recommended to integrate the mining sector in their Convergence Plan (*Plan de Convergence*). The outcomes are expected to strengthen existing Park-related trans-boundary agreements on sustainable management that follow the 1999 Yaoundé declaration signed by Central African governments.

Empowering miners by informing and sensitising them about their rights under the national mining laws and how to access mining titles and obtain legal permits should increase their bargaining power with both buyers and when confronted by corruption and lack of knowledge by officials. Finally, miners' livelihoods can be improved by:

- Transferring knowledge about sustainable techniques, tools, valuation and prices.
- Helping them to further organise themselves, for example by creating forums for information exchange and sharing experiences about production, processing, financial management and marketing.
- Supporting profitable diversification of livelihood with alternative income activities such as gathering NTFPs, farming and raising livestock.

ACKNOWLEDGEMENTS

This study is based upon research recommended by LLS partners in TNS, and realised with the support and assistance of a team of enumerators, numerous local organisations and individuals from CEFAID-Cameroon, GTZ, WWF, the Ministry of Forests and Wildlife, the Federation of Miners, GIC/Mineur Boumba and local chiefs. We are grateful to all those involved, for devoting time and sharing expertise and experience.

We would like to especially thank Gavin Hilson and Kirsten Hund for their valuable comments on an earlier draft of this work.

All mistakes in this contribution are the sole responsibility of the authors and reactions remain welcome.

REFERENCES

Armstrong, W., D'Souza, K. & Pooter de, E. 2008. Monitoring, formalisation and control of the artisanal alluvial diamond mining sector. In K. Vlassenroot & S. van Bockstael (eds), *Artisanal diamond mining: perspectives and challenges*: 93–125. Gent: Academia Press.

Babut, M., Sekyi, R., Rambaud, A., Potin-Gautier, A., Tellier, S., Bannerman, W. & Beinhoff, C. 2003. Improving the environmental management of small-scale gold mining in Ghana: a case study of Dumasi. *Journal of Cleaner Production* 11(2): 215–221.

Banchirigah, S.M. 2006. How have reforms fuelled the expansion of artisanal mining? Evidence from sub-Saharan Africa. *Resources Policy* 31(3): 165–171.

Barthélémy, F., Eberlé, J.M. & Maldan, F. 2008. Trans-border artisanal and small-scale mining zones in Central Africa: Some factors for promoting and supporting diamond mining. In K. Vlassenroot & S. van Bockstael (eds), *Artisanal diamond mining: perspectives and challenges*: 20–40. Gent: Academia Press.

Bermúdez-Lugo, O. 2009. The mineral industries of Central African Republic, Côte d'Ivoire and Togo. *USGS 2008 Minerals Yearbook.* Retrieved 19 Dec. 2009, Available at: http://minerals.usgs.gov/minerals/pubs/country/africa.html

Blore, S. 2008. Artisanal diamond miners' cooperatives: What are they good for? In K. Vlassenroot & S. van Bockstael (eds), *Artisanal diamond mining: perspectives and challenges*: 159–189. Gent: Academia Press.

CAPAM. 2006. *Creation du Capam.* Retrieved 10 Jan. 2010, Available at: www.capam.site.voila.fr

CAR. 2004. *Ordonnance No.04.001 Portant Code Minier de la République Centrafricaine.* Bangui: République Centrafricaine, Présidence de la République.

Carroll, R. 2008. *The Congo Basin: Large-Scale Conservation in the Heart of Africa.* Washington: WWF.

CASM. 2009. *ASM Workers.* Retrieved 10 Oct. 2009, Available at: http://www.artisanalmining.org

CASM. 2010. *Who is CASM.* Retrieved 2 Jan. 2010, Available at: http://www.artisanalmining.org

CBFP. 2006. *The Forests of the Congo Basin, State of the Forest 2006*. Congo Basin Forest Partnership.
Childs, J. 2008. Reforming small-scale mining in sub-Saharan Africa: Political and ideological challenges to a Fair Trade gold initiative. *Resources Policy* 33(4): 203–209.
Danielsen, F., Balete, D.S., Poulsen, M.K., Enghoff, M. & Nozawa, C.M. 2000. A simple system for monitoring biodiversity in protected areas of a developing country. *Biodiversity and Conservation* 9(12): 1671–1705.
Dietrich, C. 2003. *Diamonds in the Central African Republic: Trading, Valuing and Laundering. Occasional Paper 8*. Smillie, I. (ed.), Partnership Africa Canada, International Peace Information Service, Network Movement for Justice and Development.
EITI. 2010. *EITI countries*. Retrieved 3 Jan. 2010, Available at: http://eitransparency.org/countries
Fisher, E. 2007. Occupying the margins: Labour integration and social exclusion in artisanal mining in Tanzania. *Development and Change* 38(4): 735–760.
Frost, P., Campbell, B., Medina, G. & Usongo, L. 2006. Landscape scale approaches for integrated natural resource management in tropical forest landscapes. *Ecology and Society* 11(2): 30.
Gweth, P.N. 2006. *Fundamental reasons at the base of the creation of CAPAM. Information Review for the Promotion of Mining Activities in Cameroon. Information Review for the Promotion of Mining Activities in Cameroon*. Yaounde: 5–11.
Gweth, P.N. 2008. *Le pays multiplie les efforts pour faire connaitre ses richesses minières*. Cameroon. Retrieved 28 Nov. 2008, Available at: www.jeuneafrique.com
Hentschel, T., Hruschka, F. & Priester, M. 2002. *Global report on artisanal & small-scale mining*, IIED, WBCSD, MMSD.
Hilson, G. 2008. A load too heavy: Critical reflections on the child labor problem in Africa's small-scale mining sector. *Children and Youth Services Review* 30(11): 1233–1245.
Hilson, G. 2009. Small-scale mining, poverty and economic development in sub-Saharan Africa: An overview. *Resources Policy* 34(1–2): 1–5.
Hilson, G. & Potter, C. 2005. Structural adjustment and subsistence industry: Artisanal gold mining in Ghana. *Development and Change* 36(1): 103–131.
Hilson, G. & Pardie, S. 2006. Mercury: An agent of poverty in Ghana's small-scale gold-mining sector? *Resources Policy*, 31(2): 106–116.
Hilson, G. & Banchirigah, S.M. 2009. Are Alternative Livelihood Projects Alleviating Poverty in Mining Communities? Experiences from Ghana. *Journal of Development Studies* 45(2): 172–196.
Hilson, G., Yakovleva, N. & Banchirigah, M.S. 2007. To move or not to move: Reflections on the resettlement of artisanal miners in the western region of Ghana. *African Affairs* 106(424): 413–436.
Hinton, J.J., Veiga, M.M. & Veiga, A.T.C. 2003. Clean artisanal gold mining: a utopian approach? *Journal of Cleaner Production* 11(2): 99–115.
ILO. 2010. Convention No. C176: *Safety and Health in Mines Convention, 1995*. ILOLEX Database of International Labour Standards. Retrieved 4 Jan. 2010, Available at: http://www.ilo.org/ilolex/english/convdisp1.htm
ILO. n.d. Small-scale mining and the ILO. Retrieved 3 Jan. 2010. Available at: http://www.ilo.org/public/english/dialogue/sector/papers/mining/smallscalemining.pdf
Kimberley Process. 2009. *Annual Global Summary: 2008 Production, Imports, Exports and KPC Counts. Rough diamond statistics*. Retrieved 19 Dec. 2009, Available at: https://mmsd.mms.nrcan.gc.ca/kimberleystats/public_tables/Annual%20Summary%20Table%202008.pdf
Labonne, B. & Gilman, J. 1999. *Towards building sustainable livelihoods in the artisanal mining communities*. Tripartite Meeting on Social and Labour Issues in Small-scale mines, Geneva, 17–21 May 19. ILO.
Lang, C. 2007. *L'Or Camerounais mal exploité*. Le Messager. Retrieved 10 Nov. 2008, Available at: http://www.cameroon-info.net/cmi_show_news.php?id=19684
Leach, M., Mearns, R. & Scoones, I. 1999. Environmental entitlements: Dynamics and institutions in community-based natural resource management. *World Development* 27(2): 225–247.
Limbong, D., Kumampung, J., Rimper, J., Arai, T. & Miyazaki, N. 2003. Emissions and environmental implications of mercury from artisanal gold mining in north Sulawesi, Indonesia. *Science of the Total Environment* 302(1–3): 227–236.
Maconachie, R. 2009. Diamonds, governance and 'local' development in post-conflict Sierra Leone: Lessons for artisanal and small-scale mining in sub-Saharan Africa? *Resources Policy* 34(1–2): 71–79.
Newman, H.R. 2009. The Mineral Industries of Cameroon and Cape Verde. *USGS 2008 Minerals Yearbook*. Retrieved 19 Dec. 2009, Available at: http://minerals.usgs.gov/ minerals/pubs/country/africa.html#cm
Republic of Cameroon. 2001. *Loi No. 001-2001 du 16 Avril 2001 portant Code Minier de la République du Cameroun*. Sale, C. 2006. *Time for action. Editorial. Information Review for the Promotion of Mining Activities in Cameroon (Bimonthly)*. Yaoundé, CAPAM: 3.
Sandker, M., Campbell, B.M., Nzooh, T., Sunderland, V., Amougou, L.D. & Sayer, J. 2009. Exploring the effectiveness of integrated conservation and development interventions in a Central African forest landscape. *Biodiversity and Conservation* 18 (11): 2875–2892.
Shandro, J.A., Veiga, M.M. & Chouinard, R. 2009. Reducing mercury pollution from artisanal gold mining in Munhena, Mozambique. *Journal of Cleaner Production* 17(5): 525–532.
Siegel, S. & Veiga, M.M. 2009. Artisanal and small-scale mining as an extralegal economy: De Soto and the redefinition of "formalisation". *Resources Policy* 34(1–2): 51–56.
Sinding, K. 2005. The dynamics of artisanal and small-scale mining reform. *Natural Resources Forum* 29(3): 243–252.
Tarras-Wahlberg, N.H., Flachier, A., Fredriksson, G., Lane, S., Lundberg, N. & Sangfors, O. 2000. Environmental impact of small-scale and artisanal gold mining in southern Ecuador - Implications for the setting of environmental standards and for the management of small-scale mining operations. *Ambio* 29(8): 484–491.
Tieguhong, J.C. & Ndoye, O. 2007. *The impact of timber harvesting in forest concessions on the availability of Non-Wood Forest Products (NWFP) in the Congo Basin. Forest Harvesting Case Study 23*. FAO.

Tieguhong, J.C. & Zwolinski, J. 2008. *Unrevealed economic benefits from forests in Cameroon. IUFRO Conference (IUFRO Unit 4.05.00—Managerial economics and Accounting)*. Ljubljana, Slovenia.

Tieguhong, J.C., Ingram, V. & Schure, J. 2009. *Study on impacts of artisanal gold and diamond mining on livelihoods and the environment in the Sangha Tri-National Park (TNS) landscape*. Yaounde: CIFOR, IUCN.

Tschakert, P. 2009. Recognising and nurturing artisanal mining as a viable livelihood. *Resources Policy* 34(1–2): 24–31.

UNESC 2003. *Reports on selected themes in natural resources development in Africa: artisanal and small-scale mining and technology challenges in Africa. Third meeting of the committee on sustainable development. Addis Ababa, 7–10* October 2003. Economic Commission for Africa.

Vlassenroot, K. & Bockstael van, S. (eds) 2008. *Artisanal diamond mining: perspectives and challenges*. Gent: Academia Press.

Walle, M. & Jennings, N. 2001. *Safety and health in small-scale surface mines. Working Paper (WP.168)*. Geneva: ILO.

Wiersum, K.F. 2009. *Assessment of the influence of institutional factors on management decisions by small farmers in the Amazon: Results of ForLive Working Programme 2*. Wageningen: Forest and Nature Conservation Policy group.

World Rainforest Movement. 2002. *WRM Bulletin N° 54, January*. Retrieved 30 Jan. 2009, Available at: http://www.wrm.org.uy

WWF. 2008. *Mining concessions over forests and national parks in Southeast Cameroon*. GIS Unit Jengi Aug. 2008.

Geological Resources and Good Governance in Sub-Saharan Africa – Runge & Shikwati (eds)
© 2011 Taylor & Francis Group, London, ISBN 978-0-415-58267-4

Legal and fiscal regimes for artisanal diamond mining in Sub-Saharan Africa: Support for formalisation of artisanal diamond mining in Central African Republic

J. Hinton
Small Scale Mining Consultant, Entebbe, Uganda

E. Levin
Estelle Levin Ltd., Cambridge, UK

S. Snook
Tetra Tech ARD, Burlington, VT, USA

Contributions from J.P. Okedi, A. Surma & C. Villegas
Estelle Levin Ltd., Cambridge, UK

ABSTRACT: Artisanal Diamond Mining (ADM) in the Central African Republic (CAR) is an important livelihood for around 400,000 women and men and its formalisation holds significant development potential. The Property Rights and Artisanal Diamond Development (PRADD) Project in the CAR employs an effective formalisation process, yet licensing costs present an obstacle. This study compares fiscal as well as legal and institutional mechanisms to support legalisation via case studies from 10 countries. Results indicate that low license costs significantly increase formalisation and can actually yield greater government revenues than higher fees, royalties and taxes. Other fiscal success factors include regional tax harmonisation, simple reporting procedures, investment in ADM support services and tax collection at the point of export. Conducive fiscal provisions must nevertheless be coupled with legislation that sensibly recognises capacity constraints of miners and existing work arrangements on the ground, while the biggest challenge seems to lie in the institutional commitment needed to implement good policy.

1 INTRODUCTION

Artisanal diamond mining (ADM) is an important livelihood for around 400,000[1] women and men in the CAR and provides more than 60% of the country's export earnings, estimated at US$146.7 million in 2007.[2] While official statistics suggest that the mining sector in CAR contributes only 4–7% to national GDP, more than 95% of the country's alluvial diamond production is attributed to artisanal miners: the likelihood of substantial losses in government revenue is high due to production and sales outside of the formal chain of custody.[3]

Formalisation of artisanal diamond miners in CAR holds the potential to generate significant development benefits for the nation by contributing to the National Treasury, providing non-agricultural employment options, investing ADM revenues into agriculture, stimulating rural economies via small enterprise development, reducing rural-urban migration and enabling miners and their families to meet basic living requirements.

Despite this potential, ADM in many countries is typically characterised by rudimentary methods, serious occupational safety risks, environmental degradation, child labour, exploitative work arrangements, gender inequalities and illegality. Hundreds of thousands of women, men and children are nevertheless drawn into the ADM subsector by acute rural poverty, lack of viable alternatives and the hope of riches.

Licensing of artisanal miners provides an important vehicle to provide formal sources of financing and the advisory support needed to mitigate its negative consequences. This, in turn, can lead to increased diamond production and even greater contributions to the local and national economy.

However, in many artisanal and small-scale mining (ASM) countries, it has been widely demonstrated that costs of licensing, fees, rents and royalties are a common obstacle for informal miners and most will only obtain a license if they have the financial, technical and personal capacity to do so.[4]

The PRADD Project in the CAR has begun to show success in demonstrating an effective process to identify and formalise property rights in artisanal diamond mining zones. Despite this, artisanal miners still face barriers in purchasing an annual license (*patente*) for US$100 and, as in many countries, likely also face even more difficulties due to additional taxes, royalties, rents and bureaucracy.[5] In many jurisdictions, problems between different authorities and miners treated as "illegals" can increase the divide between miners and government and make future licensing even less likely.

Drawing from lessons learned in 10 ASM countries, this Comparative Study assesses how legalisation of artisanal diamond miners can be promoted through reduced costs of licensing, royalties, taxes and fees. It specifically examines what fiscal provisions are reasonable given the incomes and capacity of artisanal diamond miners in CAR and when and how the expected benefits of lower fees can increase government revenues through a greater number of licenses granted. A simple economic model has been developed to help the Government of CAR (GoCAR), donors and others evaluate expected changes in revenue streams and other benefits through increased licensing of artisanal miners. The study further highlights practical and achievable licensing requirements and the importance of government institutions in supporting formalisation.

1.1 *The importance of a supportive fiscal and legal regime*

Fiscal policy is crucial to ensuring that investment in economic development is stimulated while providing sufficient revenues to enable government to fulfil broader development objectives. The four main components of fiscal policy are (i) expenditure, budget reform, (ii) revenue (particularly tax revenue) mobilisation, (iii) deficit containment/financing, and (iv) determining fiscal transfers, particularly from higher to lower levels of government.[6] While this study examines revenue streams from licensed ADM, it also recognises the importance of fiscal transfers in supporting further licensing and improved development outcomes from ADM.

The business case for governments to support formalisation of ADM is clear[7]:

"Countries that create a legal and economic environment that supports integration of extralegal enterprises almost invariably prosper more quickly than those that do not. Furthermore, costs of imposing top-down authority over the extralegal economy are prohibitive, particularly when existing informal systems are viewed as legitimate at the grassroots level."

Important lessons can be learned from the experiences of countries such as Brazil, Colombia, Ghana, Indonesia, Laos PDR, Tanzania and many more that have, at one time or other, battled and failed to stop illegal ASM at great financial and reputational costs.[8] As demonstrated in case study countries, recent approaches *now* recognise the importance of incentive-based legislation and institutional support coupled with practical regulation and enforcement mechanisms.

1.2 *Overview of PRADD*

This chapter is adapted from a study funded through the PRADD pilot program launched in the CAR in 2007.[9] The PRADD project is supported by the US Department of State as one of its contributions to the Kimberley Process Certification Scheme (KPCS), and is managed by USAID. As designed by the USAID implementing partner, Tetra Tech ARD, the PRADD project aims to increase the amount of alluvial diamonds entering the formal chain of custody while improving the benefits accruing to mining communities through an approach of strengthening property rights. Being able to identify who owns the land on which a diamond is found, and making the right of artisanal miners to prospect and dig for diamonds more formal and secure, will create incentives for more miners to enter their production into the formal chain of custody, and will enable the government of CAR to track larger portions of diamonds from the point of extraction to market, and thereby meet the requirements of the KPCS. The PRADD project operates in cooperation with the Ministry of Mines, Energy and Hydraulics (MMEH) and is active in Lobaye and Sangha Mbaere Provinces.

According to Article 6 of the 2009 Mining Code all minerals in CAR are "the exclusive, unalienable and indefeasible property of the State." The Mining Code provides for compensation of customary land owners when their land is taken by the holder of a private mining title, but compensation is for the land, not for the resources. Customary rights do not apply to minerals. Corporations or individuals can obtain temporary rights to minerals. Rights to mine using semi-mechanised methods are granted through renewable three-year licenses. Artisanal miners must bear a license valid for a period of one year, but this license does not grant exclusive rights to minerals. The law also provides for an_ognizant_ion for artisanal exploitation, a renewable non-exclusive title which is valid for periods of two years.

The current body of laws and regulations governing mining rights in CAR contain a number of contradictions, and implementation is imperfect.

The result is that fewer than 5% of artisanal miners are licensed and the proportion of licensed miners has been sharply decreasing for the last 10 years.[10] This enables the Mining Police to treat nearly all artisanal miners as illegal miners. The implications for the proportion of diamonds from artisanal mines that enter the formal chain of custody are clear.

The Government of CAR is cognizant of these problems, and to assist in its deliberations about how best to rectify them, the PRADD project has provided tools, data and analysis. PRADD launched a survey on the issue of licensing over hundreds of miners, helped facilitate technical discussions through working groups of specialists from the key ministries of Mines, Trade, Finance and Small Enterprises, and helped facilitate a September 2010 fact-finding mission of Deputies of the National Assembly to the project's remote artisanal mining villages. One of the most critical tools of technical assistance to the government has been a comparative study of the taxation and royalty regimes in 10 artisanal mining countries, commissioned by PRADD in October 2010 and widely circulated among government members, international partners and representatives of the diamond trading houses, from which this book chapter is derived.

1.3 *Objectives*

The primary objective of this comparative study is to assess whether lowering the fee of the *patente* in CAR will encourage formalisation. The *patente* is the *carte d'exploitant artisan minier*, a permit which all miners must possess to mine within a designated Artisanal Mining Zone or as part of an artisanal cooperative.

In exploring this issue, the study further aims to identify factors contributing to success or failure in formalisation, and their potential relevance for CAR, by comparing legal regimes and fiscal frameworks in 10 ASM countries. It seeks to identify specific mechanisms that: facilitate the entry of artisanal miners into the formal system; generate significant financial benefits to the state; and apply these benefits to support development. Using a simple Excel model, the study also demonstrates a relationship between formalisation rates and the ratio between miners' incomes and licensing fees, rents, taxes and royalties.

1.4 *Methodology*

1.4.1 *Methodological approach*

The short time frame for the project (two weeks) confined research to desk-based methods including:

1. Documentary review and analysis of legal codes in 10 countries; government reports and statistics; consultancy reports and academic and non-governmental organisation (NGO) publications pertaining to ADM and ASM fiscal and legal systems and formalisation.
2. Statistical analysis of ADM populations (formal, semi-formal and informal), diamond production and trade and government revenues.
3. Consultation by email and telephone with experts and government officials in case study countries.

1.4.2 *Economic modelling*

A simple Excel model was developed to examine the relationship between government revenues and the number of licensed miners for future use by GoCAR. The model is an adapted version of that developed by Hinton (2009) to assess economic contributions of ASM in Uganda. Important variations include the use of basic correlations to assess potential influences of licensing cost on the number of legal miners. Correlations were developed using countries with comparatively good data on miners' incomes and production levels which were coupled with country licensing costs and percentage estimates of the formalised sector. While a number of assumptions and estimates were used to set up the model, refinement of statistics is easily undertaken by users.

The Economic Model is comprised of four spreadsheets:

1. *Main Statistics*: Where users enter data on the total number of licensed and unlicensed miners; annual total license costs; royalty rate and export tax rate and (optionally) the national gross domestic product (GDP) and foreign exchange earnings. Results are generated showing: non-tax revenue (NTR), tax and royalty revenues; value of officially reported production; % of GDP and Foreign Exchange Contributions; and other development benefits (e.g. local economic contributions from miners' incomes; direct, indirect and induced labour as well as indirect beneficiaries of ADM).
2. *License Costs versus % Licensed Miners*: This sheet prompts users to edit license costs, providing a prediction of licensing rates, associated government revenues and expected value of officially reported production.
3. *Graph—The Effects of License Costs*: Showing the relationship between license costs, percentage of licensed miners, non-tax and tax revenue to government and value of official diamonds produced are easily observed.
4. *Assumptions and Correlations*: Sources of estimates, as well as data informing the correlation

that can be refined as more information becomes available.

1.3.3 *Research limitations*

Research challenges included: limited data on miners incomes in different jurisdictions; lack of published literature on ASM fiscal regimes, in particular; variable availability of government data; and data limited to specific time frames, in some cases impeding clear linkages with legal reforms.

2 ADM AND THE DEVELOPMENT OF CENTRAL AFRICAN REPUBLIC

CAR is the world's 10th largest diamond producer based on value, and fourth biggest artisanal diamond producer, with annual diamond exports averaging 400,000 carats from 2000 to 2009.[11] In 2007 diamonds contributed 4–7% of CAR's GDP and approximately 40% of export earnings.[12] CAR's diamonds rank sixth in the world in terms of quality, with 75% being gems.[13] CAR's diamonds are alluvial and 95% of the sector is artisanal, with about 40–90,000 autonomous women and men artisanal miners forming the base for the mining activity in the country.[14] These miners, who manage and run the operations and earn about US$280 a month, employ up to 350,000 labourers (diggers and washers), who may earn up to US$50 per month.[15] Artisanal miners sell to about 160 certified collecting agents who, in turn, sell to two purchasing offices located in Bangui.[16]

Based on anonymous polls of a small sample of miners (around 230), formalisation rates vary from 5.6% to 12.1%.[17] Government documentation notes an average of just 1100 miners with the *patente* (carte *d'autorisation d'exploitant artisan minier*) in six out of seven diamond areas for 2009 and 2010.[18] On this basis, the Ministry demonstrates a formalisation rate of around 2% for 2009 and 2010.[19]

2.1 *Overview of the fiscal and legal framework in CAR*

CAR's Mineral Policy gives the GoCAR a mandate to promote responsible ADM through measures including, but not limited to: licensing of artisanal miners and ADM cooperatives; registration of diamond collectors and buying/export businesses; enforcing legal dealings through *la Brigade Minière*; maintaining records and a database to document mineral production and trading; and supporting ADM by enabling cooperatives to export directly, providing training to artisanal miners and providing other technical and material assistance.[20] Four decentralised offices are responsible for implementing the Mineral Policy at a regional level.

By all accounts, this policy mirrors that of many jurisdictions and is in line with international best practice. And yet the formalisation rates remain low.

2.1.1 *The mining code*[21]

CAR's artisanal mining sector is currently governed by the national Mining Code (*Code Minière*), law No. 9-005 of April 29, 2009, which has reshaped and restructured the legal and institutional framework supporting the mining industry as outlined in the 2004 Mining Code. The mining sector in CAR is directly under the Ministry of Mining and Energy (*Ministère des Mines, de l'Energie et de l'Hydraulique*). Other authorities overseeing the diamond sector are the General Director of Mines (*Directeur Général des Mines*), and the Minister of Mines (*Ministre d'Etat Chargé des Mines*).

The law distinguishes between artisanal mining, semi-mechanised artisanal mining, small-scale mining, and industrial mining (Art. 1). It defines artisanal mining as: "All activity by which a physical person of Central African origin, in an artisanal exploitation zone delimited in area and depth to a maximum of 30 meters, extracts and concentrates mineral substances using non-industrial, manual tools, methods and processes that are limited in mechanisation." (Art. 1)

Importantly, Article 15 states that: "The State favours the evolution of artisanal to small-scale mining, by the regulatory route." However, according to Article 64, potential mining sites are first considered based on their potential for industrial operations. Artisanal exploitation is permitted solely of small-scale mining deposits, where industrial exploitation is not viable due to technical or economic limitations. If a site is deemed suitable only for small-scale extraction, the Minister of Mines can designate an artisanal mining zone (AMZ), after consultation with the General Director of Mines. However, as of the end 2010 a decree on this last point had not yet been passed, so in effect there are no official AMZ's yet.[22] In practice the whole country is open to artisanal mining.

Most artisanal diamond miners in CAR are still clandestine. BECDOR (Control and Evaluation of Diamonds and Gold Office) data indicates that a mere 2% of miners are registered.[23] A principal concern is that illegality within the mining industry increases other illegal activity, and creates a wider economic environment of fear, fraud and lack of rule of law, which takes a toll on society, discourages investors and reduces the government's income.

An artisanal miner can become legal through three avenues:

1. Obtaining a miner identity card (*carte d'exploitant artisan minier*) and operating within a designated Artisanal Mining Zone. The card is valid for one year, and can be renewed without limitation (Art. 64). This is the *patente*.
2. Where miners wish to obtain a mining title over a specified area (outside of a designated AMZ), they must possess a *patente*, organise into a cooperative comprising at least ten licensed artisanal miners and obtain an artisanal mining exploitation license (*autorisation d'exploitation artisanale*), issued by the Minister of Mines (Art. 1, 66, 67). The license is valid for two years and can be renewed twice for the same period of time. It can cover an area of up to 62,500 square meters. The cooperative must respect health and safety, preserve the environment, commercialise their diamonds legally and not damage agricultural operations or water sources. (Art. 69, 70, 71)
3. Prospecting licenses (*autorisation de prospection*) are issued by the General Director of Mines for a period of one year and are renewable once for the same period of time (Art. 62).

It is not clear whether an artisanal miner who bears a one-year license can mine without a two-year authorisation.

2.1.2 *The fiscal regime*

The current 2009 CAR Mining Code was specifically aimed, as was the previous 2004 one, at facilitating foreign investment access into the industry and increasing revenues from mining for the country.[24]

The minimum cost per year for obtaining official documentation necessary to be considered a legal artisanal miner is 58,650 FCFA (US$132):

- *Patente* fee is 46,850 FCFA (US$105)
- Production notebook is 2000 FCFA (US$4.50)
- Minimum five mine worker's cards at 2000 FCFA (US$4.50) each is 10,000 FCFA (US$22.50).

Artisanal miners claim that this amount is too high for them because of the uncertainty of diamond production levels.[25] For miners organised into a cooperative and seeking an *autorisation d'explotiation artisanale*, they must pay for the *patente* for each miner, as well as a number of other fees associated with forming a cooperative, and the *autorisation* fee 100,000 FCFA (US$224) (Art. 16), in addition to a surface fee of 5000 FCFA (US$11) per hectare per annum (Art. 18). This is a total of 703,500 FCFA (US$1580). A prospecting license costs 100,000 FCFA (US$224) (Art. 62).

The 2009 Code clarifies the regulations for artisanal mining and offers some exemptions from taxes and fees for those miners. For example, under Article 132 any holder of an exploitation license is exempt for three years from the minimum tax (MFIs), contribution of the *patente* (CP), and the Contribution to Social Development (CDS). However, this benefit is limited to operations lasting longer than 10 years, and those making less than a 10-year commitment (that would include most artisanal miners) receive only a one-year tax break.

Other taxes and fees include:

- A 7% royalty is levied on the diamonds at the point of production and must be paid by the holder of the exploitation license, otherwise a penalty is applied (Art. 18).
- Precious stones *cut* and sold on the domestic market are subject to Value Added Tax (VAT) and an Artisanal Development Tax (TDA).
- Any *cut* gems for export are subject to the same export taxes as when sold through the Import-Export Purchasing Office—*Bureau d'Achat Import-Export*. They are also subject to an Artisanal Development Tax (TDA) (Art. 161).

2.2 *Key issues related to formality*

Key issues in CAR's ADM sector include:

- The majority of ADM—up to 98%—is unlicensed.[26]
- Up to 50% of diamond production is believed to exit the country illegally.[27]
- The sector is characterised by "a lack of respect for the law."[28]
- The *patente*, designed for registering diamond *miners*, is being taken out by diamond *buyers*, who hold the majority of the *patentes* in some Communes.[29]
- The majority of artisanal miners live "in the bush," far from urban centres, which makes controlling the sector practically impossible.[30]
- Other obstacles to formalisation are the fact that the fee is annual, from January to December; that the miners see the land as theirs and so do not see why they should register with the state; and that less buyers are financing miners owing to the recent collapse in diamond prices making it harder for miners to afford to mine.[31]

Consequently, the GoCAR is considering the following measures to incentivise miners to acquire the *patente*:[32]

1. Lowering the price of the *patente*.
2. Changing the structure of payment (introduce tranche payments and for a fixed twelve month period).
3. Introducing a system of "kelemba" and micro-credit.[33]
4. Massive sensitisation efforts.
5. Encouraging private investment to aid artisanal miners to organise into cooperatives.

2.3 *The development potential of ADM in CAR*

ADM provides an important source of labour intensive, non-agricultural rural work and almost 60% of foreign exchange earnings yet its informal and formal development contributions are likely to be far more extensive.

2.3.1 *Current contributions of ADM*

The current contributions of ADM are mostly invisible, yet, even in its largely informal state, it plays an important role in local and national development, particularly given that between 40,000 and 90,000 artisanal miners are estimated to be active in CAR, of which the majority mines diamonds.[34] The basic economic model provides interesting insights. When indirect labour, induced labour and CAR fertility rates are considered, *an estimated 2.8 million women, men and children, about two-thirds of the population, directly and indirectly rely on the ADM subsector.*[35]

Furthermore, with estimates placing average incomes at US$723 per year per miner, if only half of incomes are spent on local goods and services (an extremely conservative estimate), more than US$28.9 million may be injected into local economies annually, particularly in the regions of Berbérati, Upper Kotto and Sangha. Local demand by miners, who are often wealthier than others in rural economies, creates markets for locally grown or supplied products and increases the cash component of household incomes. Local formal and informal businesses, as a consequence of ADM-injected capital, may contribute an additional US$86.7 million to local economies.[36]

This is mainly because a substantial percentage of miners' incomes from ADM (if not all) is spent in communities where they live and work. In many countries, this investment may be directed towards the purchase of basic tools and supplies needed to mine (supporting small shops and enterprises), the basic amenities needed to sustain life (often providing some cash incomes to local farmers) while many miners are likely to invest some surplus into agricultural and other small businesses.[37] Incomes from ADM can also enable miners to send their children to school and meet household health costs, both of which are crucial to long-term, sustainable development. Indeed, *in some countries, an artisanal miner can contribute 15–20 times more to GDP than a person involved in farming and fishing.*[38]

Taking a conservative estimate of economic multiplier effects of ASM, *revenues from even informal artisanal mining combined with its spin-off economic enterprises may inject as much as US$144.7 million into the economy.* As one of the pillars of many national poverty reduction strategies, increasing the cash component of household income in ADM areas and regions can provide an essential foundation for national growth and development.

2.3.2 *Potential contributions of ADM*

Given the suspected informal contributions of ADM, it is astonishing to consider the development opportunities of a formalised ADM subsector. CAR's artisanal diamond production averages approximately 400,000 carats per year and is valued at US$60.4 million per annum.[39] Assuming only half of this passes through official channels[40] and based on a point of production levy of 7%, this would amount to losses in tax revenue on the order of US$2.1 million per annum.

Based on correlations between miners' incomes and license costs found in other countries, *a US$5 license fee could result in legalisation of more than 60,000 artisanal miners, comprising ~82% of the current estimated ADM workforce* (Fig. 1).

Statistics suggest a US$5 license fee this would yield *official* diamonds exports of US$82.4 million, comprising ~4.2% of the nation's GDP and bolstering foreign exchange earnings. *Even at low royalty (1.5%) and export tax rates (3%), as much as US$3.5 million could be generated in tax and non-tax revenue* (Fig. 2).

While the results of the economic model are promising, a key lesson learned from other jurisdictions is that low license costs, royalties and export taxes *can* stimulate licensing. However, immediate results are unlikely, except in areas where sufficient on-site intervention and support is provided. Consequently, fiscal provisions must be harmonised with sensible mining policies and the institutional mechanisms needed for implementation.

Such institutional support could yield even *greater* benefits in terms of development. For example, efforts to introduce appropriate, intermediate, "step up" technologies could boost diamond production even more while increasing miners'

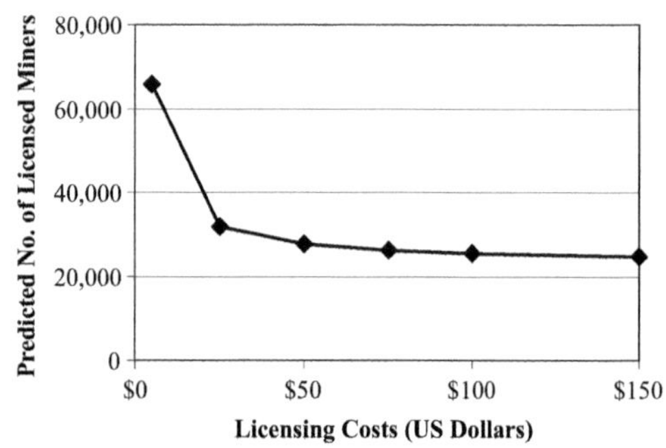

Figure 1. License costs (US$) vs. predicted number of licensed miners.

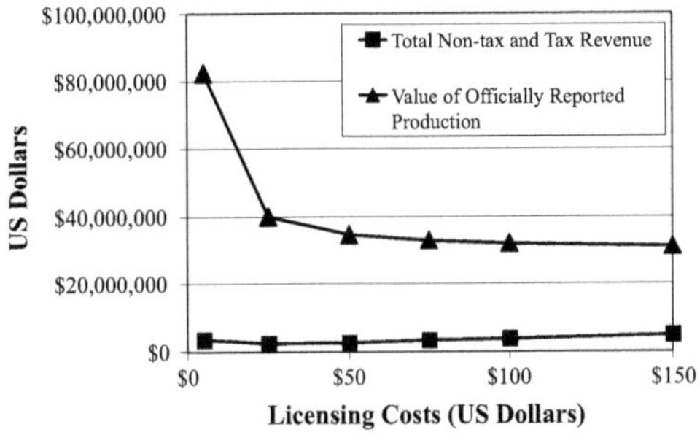

Figure 2. Licensing costs vs. total non-tax and tax revenue and value of officially reported production (US$).

capacity to mine in a safer and environmentally responsible manner.

Training in basic gem valuation and simple business skills can boost miners' incomes leading to multiple spin-offs in terms of micro- and small- enterprise development and growth of market gardens, fish farms and other agricultural ventures. Closer ties with mines authorities can further support linkages with other arms of government, such as agencies responsible for health, education and infrastructure. Reduced licensing fees and taxes would mark the first critical step towards achievement of the full development potential of ADM in CAR. Outcomes of these much needed reforms, however, shall ultimately be determined by incentive-focused mining legislation, institutional commitment, adequate financing for implementation and accountability for performance.

3 FISCAL AND LEGAL APPROACHES TO ARTISANAL MINING

This diagnostic assessment analyses fiscal and legal regimes from a number of case study countries that have experienced different degrees of success in legalising ASM and increasing its formal contribution to economic development. While recognising that every country and context is different, the assessment seeks to identify *key success factors and constraints in encouraging formalisation* in order to inform reforms to CAR's fiscal policy and legislation in the interest of best practice.[41]

Formalisation is founded on the principle that extralegal systems, social arrangements and organisational structures exist for legitimate reasons and the law itself should be an evolving *and enabling* instrument that reflects the changing ways people live (Siegel & Veiga 2009). Specifically, *formalisation* is the process of integrating rather than controlling extralegal enterprises by recognising local arrangements in legislation, reducing barriers to legalisation and creating clear benefits from participation in the formal system (Pelon 2005).

Legalisation refers to the status of becoming legal (e.g. obtaining a license) and complying with law. Legalisation of ASM is believed to largely depend on the process of formalisation.

The ADM fiscal and legal regimes for five other ADM countries are summarised below. Case studies were selected on the basis of the diversity of models and approaches, perceived relevance to the CAR situation and availability of useful data to build upon. Analysis is then also informed by approaches in five other jurisdictions in Africa, Asia and Latin America where useful best practice insights can be drawn.

3.1 *Artisanal diamond mining jurisdictions*

3.1.1 *Democratic Republic of Congo (DRC)*

Between 75 and 95% of DRC's diamonds are deemed to come from informal artisanal miners, principally in the Kasai Provinces.[42] In 2009, DRC produced around 22 million carats of diamonds, predominantly of industrial quality, with a total value of US$226 million.[43]

Relevant components of DRC law include:[44]

- "Artisanal Mining" includes simple non-industrial methods of exploitation with the use of basic tools and processes and can be done by any Congolese national who holds a valid Artisanal Exploration Card (Art. 5).
- Exploitation must take place within a designated AMZ, and cannot be deeper than 30 meters. Working outside of this zone is technically illegal.
- Groups of artisanal miners who wish to within an AMZ are required to form a cooperative and seek the consent of the Minister of Mines. Each cooperative member must hold an Artisanal Exploitation Cards, granted by the Head of the Provincial Division of Mines for one year (renewable). The organisation must have a non-profit character (*Reglement Minier*, Art. 234).
- The Card does not authorise the holder to sell or process minerals and it can be withdrawn if standards are violated (Art. 112). Officially all diggers should carry this card, but very few do.[45]

The formalisation rate of artisanal diamond miners in DRC is very low for two main reasons.[46] Firstly, the state institutions which govern the sector have huge capacity constraints, even to the point that they are not able to print enough cards. Secondly, there are many disincentives for legalisation. For example, although the card is supposed to cost US$25, miners often find that they are obliged to pay more. They also face logistical difficulties in reaching issuing centres, are not penalised for *not* holding a card, and many believe that there are no real benefits to having a card. Furthermore, the card only allows the digger to mine within a certain zone but miners are highly mobile, requiring purchase of multiple cards. One suggested approach is to validate the card for an entire province and/or drastically reduce this fee in order to make the cards accessible to the miners at a lower cost.[47]

The DRC example shows that low-cost licensing alone is not enough. The local situation of miners must be well understood to design effective systems.

3.1.2 *Guyana*

Guyana's diamond exports were worth US$14.6 million in 2009, accounting for just 1.9% of Guyana's exports owing to depression in world diamond markets and a move of ASM from diamonds to gold.[48]

With formalisation of the entire artisanal mining sector in mind, Guyanan authorities have taken a number of proactive, successful steps to formalise the industry at all stages of the production chain (from the miners to the exporters). Because of the simplicity of their tracking tool and low-cost fees, almost 100% of the production units (dredges) are legalised, although the *artisanal miners* who work on the dredges remain largely informal as many refuse to get their cards.[49] Key success factors include:

- Licensing fees (permits) of only US$5 for diamond diggers and US$50 for dredges (production units).
- A trader/exporters license costs only US$75. Royalties (3%) are paid based on a standard value of US$75 per carat, regardless of quality. An average Guyanan diamond is valued at US$100 per carat and exporters see this as a fair and simple royalty system.[50]
- A simple tracking system is used based on a system of mine site and buyer-side reporting.
- The license is issued for the main production unit, the dredge, which must be registered and licensed either at GGMC headquarters or at a regional office.

The vast majority of diamond mining in Guyana is done using mechanised jigs called dredges or *resumidors*.[51] Once a dredge is registered, an entry is made in the GGMC's master ledger, and the administrative clerk opens a dredge file where all production records are centrally kept in a numbered system. There were 3683 registered dredges in Guyana as of April 2006. Dredge owners are required to maintain weekly production sheets containing information on the dredge's location, hours, diesel use, and, most importantly, its production of diamonds. Coordinated and consistent field checks by Guyanan authorities contribute to the accuracy of the data being processed.[52]

Under this system, buyers can only buy from registered miners, diamonds have been tracked from source to export, government royalties are consistently paid and auditing is possible. While sourcing is restricted and subject to participation in the tracking system, exporters can sell to whomever they choose.[53] To address illicit financing, the GGMC has instituted reforms requiring traders to show, via bank or other money transfer records, a clear and legal source for the funds with which they purchase *Garimpeiros*' (small-scale miners') diamonds.

While measures used in Guyana are generally hailed as a success, Guyana has a number of attributes that may affect direct application of this model to other countries.[54] First, Guyana has a small population (approximately 800,000 people) and small and manageable surface area. Secondly, diamond quality is relatively consistent and does not require individual valuation. Thirdly, the government has a good level of capacity relative to other diamond-producing countries. Fourthly, Guyana has a track record of establishing competent and semi-autonomous state agencies for key economic sectors such as mining. The GGMC, for example, was established in 1979. Lastly, the system is believed to work because it targets and works through production units, i.e. dredges, which is believed to be far more manageable, traceable and accessible than targeting individual miners.

3.1.3 *Liberia*

In 2009, Liberia officially produced 28,368 carats of diamonds worth approximately US$11.25 million, with an average per carat value of around US$400.[55] Approximately 98% of Liberia's mineral exports come from ASM and it supports employment of at least 100,000 artisanal miners.[56]

Liberia faces a challenge with regard to formalisation of ASM. Only 8–12% of the country's approximately 100,000 artisanal miners work under a Class C License.[57] The low numbers of legal miners and largely illegal exports are due to a combination of lack of incentives, many disincentives, lack of enforcement, low awareness of the laws and procedures and, in many cases, limited capacity to fulfil them. Specifically:

- It is logistically difficult for ASM to formalise. Since licensing is centralised in Monrovia, miners must invest time and money to travel to the capital.[58] Intense rainy seasons restrict transport making it much easier for Monrovia-based business people to get licenses rather than miners.
- Mining agents and inspectors often have limited training, are paid low salaries, and are not

provided the funds, allowances or even transport required for the execution of their duties.[59]

- Many aspects of the Mining Code are not compatible with the reality of how mining takes place on the ground.[60] For example, the standard plot size allowed is 25 acres which "forces the artisanal miner to look for financial support to effectively mine such a plot, thereby increasing levels of indebtedness and poverty." Requirements concerning cooperative formation and profit sharing create additional challenges.[61]

In response to this, the Ministry of Lands, Mines and Energy (MLME) has undertaken a number of progressive measures to develop the ASM sector. These include adopting a new Mineral Policy (2010) that is consistent with international best practice; enacting legislation supporting implementation of the KPCS; establishing a transparent licensing system and clarifying access procedures; training mines inspectors; and establishing regional field offices for the Government Diamond Office (GDO).

MLME is now in the process of bringing into force new regulations specific to ASM (Class C Licenses). While both the Mining and Minerals Law (MML, 2000) and its Ch. 40 Amendment to the MML (2004), which is specific to diamonds and the KPCS, include many conducive provisions, some constraints to "best practice" nevertheless exist.[62] Different proposals for Class C Regulations are currently under review, one of which complies with the MML (2000, 2004) and provides interim measures for licensing artisanal miners. Key and unique features suggested will be shortly published in an upcoming USAID report but broadly include reducing the license fee, simplifying the application process, ascribing rights to miners to receive advisory support from government, and prohibiting the use of heavy, earth-moving equipment.

3.1.4 *Sierra Leone*

Since the turn of the decade, the Government of Sierra Leone (GoSL) has been restructuring its mineral sector with the intent of encouraging its formalisation and compliance with the KPCS. Sierra Leone was the first "conflict diamond" country to trial a certificate for its diamond exports in 2000, before enforcing the KPCS in 2003. Around 150–200,000 artisanal miners are estimated to be active, most of whom mine diamonds and many of which have transitioned into gold owing to the recent economic collapse.

The 2004 Mining Code was recently replaced by Parliament passing the "Mines and Minerals Act 2009" and the country is in the process of replacing its 2005 "*Policy Measures relating to Small Scale and Artisanal Mining and Marketing of Precious Minerals*" with its new Artisanal Mining Policy. The regulations governing artisanal mining are set out in the Mines and Minerals Act, but the marketing of diamonds, including export and associated taxes and fees, is presently still covered by the 2005 *Policy Measures*.[63]

Some key aspects of the new legislation include:

- Artisanal miners are awarded a mineral right of up to half a hectare, and small-scale licenses can be from 1 to 100 hectares.
- Only indigenous Sierra Leoneans may apply for and be issued artisanal mining licenses, whether they are applying as individuals or in an organisation.
- Artisanal mining licenses are issued by the Mines Department, and only if accompanied by a "certified copy of the agreement between the applicant and the Chiefdom Mining Allocation Committee or the rightful occupiers or owners of the land over which the artisanal mining license is granted" (Art. 88.2).
- Artisanal miners do not have to pay any royalty on their production (exporters pay a 3% royalty at point of export, which is consistent with the regional export taxes on diamonds in neighbouring countries).
- Whilst the new Act sets out a schedule of fees for artisanal miners, the old fees are still being applied.[64] Miners only pay a fee to government at the point of application for a license. Under the previous legislation, the total cost to license an acre for artisanal mining was around Le800,000 (about US$270 by 2005 exchange rates).[65]

Since 2004, artisanal production has been declining as a proportion of total exports, mainly because of large areas of land taken up by larger operators, the consolidation of artisanal operations into "small-scale" units (driven by decreasing accessibility of remaining reserves), and a move by many post-war miners to return to their traditional livelihoods. At the same time, gains in formalising the sector have been threatened as many artisanal diamond miners have returned to mining without licenses (though with permission from traditional authorities to whom they pay customary 'surface rent'[66]), disillusioned that formalisation does not bring the benefits, such as security of tenure, that were originally intended.[67]

Based on this experience—and further to the near collapse of the country's artisanal diamond sector due to the worldwide recession—in early 2009 the Ministry of Mines made the decision to lower the cost of licensing to Le500,000 plus surface fees (Le100,000) (US$156).[68] As the sector is presently recovering, the Ministry is now considering returning them to Le800,000.[69]

In 2004, Levin conducted research with artisanal diamond miners in Kono District to determine what motivated them to formalise their activities, or not. She found that the structure of a miner's livelihood and assets determines his/her decision to mine legally or illegally.[70] Understanding *who* your miners are, how they use the mining, and what it brings them is therefore key to designing the right legal structures and incentive structures for formalisation.

3.1.5 *Tanzania*

Tanzania's mining sector has grown dramatically in recent years, from 15.6% in 2006 and 10.7% in 2007 and comprising approximately 3.5% of GDP.[71] Diamonds officially represented approximately 1% of total mineral exports in 2008 and are worth approximately US$22.3 million, of which the most came from industrial-scale production.[72] Artisanal mining in Tanzania employs an estimated 500,000 artisanal miners producing coloured gemstones, diamonds, gold and other commodities; by comparison, formalised mining amounts to 8000 jobs.[73] Despite the sector's potential to fuel rapid economic growth, concerns that minerals have not contributed enough to improving the lives of the poor are widespread, particularly for those living in the vicinity of the mines themselves.

Consequently, a number of revisions have been adopted in the new Mining Policy (2009) and Mining Act (2010), interesting aspects of which include:

- The government will set up a new Mineral Development Fund, whose purpose will be, among other objectives, to support artisanal and small-scale miners.
- Only local Tanzanians will be able to mine gemstones and non-Tanzanians wanting to be involved in mining must form joint ventures with Tanzanian citizens.
- Mining royalties on rough diamond and coloured gemstones were increased from 5 to 7%; and cut diamond and coloured gemstones were reduced from 3% from 0%, providing an important incentive for in-country value addition.[74]
- Government will set aside specific areas that will function as buffer zones to reduce conflicts between corporate (large-scale) and artisanal miners.

Most artisanally mined diamonds are mined informally at and around the Mwadui mine,[75] a 146 hectare site located in northern Tanzania where ADM has taken place since the early 1900's. Supported by a large and well-developed financing and buying network, prior to 2008, an estimated 20,000 carats of the area's average 80,000–120,000 carats were thought to be illegally produced and sold.[76] Artisanal mining is the most significant income source for 75% of area families, however, because of its informality, little revenue reaches local government coffers and officially makes little impact on the area's development.[77]

Up until late 2008, the Williamson Diamond Mine at Mwadui was part of a joint venture between the Tanzanian government and De Beers, although an estimated 20,000 artisanal mining were and continue to be active. Much attention was focused on the Mwadui Community Diamond Partnership, a process which stalled when De Beers' stake in the Williamson Diamond Mine was taken over by Petra Diamonds.[78] Plans had included the creation of a digger cooperative to provide credit access for diggers, allocation of ADM areas as well as initiatives to introduce transparency in diamond prices and, in partnership with the Tanzania Ministry of Energy and Minerals, creation of a Diamond Valuation Centre.

3.2 *Other jurisdictions*

A number of other countries have taken action to support formalisation of ASM. Useful components that have led to some successes are described.

3.2.1 *Madagascar*[79]

It is estimated that 100,000 to 150,000 Malagasey gold miners produce one to two tons of gold annually.[80] Recent changes in Madagascar's Mining Code (2005) sought to encourage the formalisation of ASM through substantial reductions in license and permitting fees, simplifications in procedures and outreach to artisanal mining communities, including reinforcing the capabilities of local authorities and sensitisation of ASM on the new law. Important aspects of the Madagascar example include:

- A gold miner or group pays only 10,000 Ariary (US$5.50) for an annual permit (*Carte d'Orpailleur* or gold washer's card) granting the individual or mining association the right to pan gold in a washing channel (river beds or recent alluvial deposits).
- The card is available only to individuals over 18 years old of Malagasy nationality or local groupings of legally-established gold washers, payments go to their "Commune" which is a unit of local government that is authorised to grant permits and is responsible for ensuring that gold washers employ safe and environmentally protective practices (Art. 85).
- Collectors (local buyers) must obtain a *Carte de Collecteur d'Or from the Commune* at a price of 100,000 Ariary (US$105) per annum for each *Commune* where they are registered. The 2%

royalty on production, paid by the Collector based on their purchase price from the producer (requiring a jointly signed invoice on-site).

Individual production and incomes range from US$1–5 per day;[81] internationally, alluvial gold miners produce approximately 0.2 gram/day which, at local prices, would sell for approximately US$3, roughly giving a ratio of monthly income to annual licensing costs of 10.9.

Barriers to licensing still exist. These include high rates of illiteracy (discouraging documentation of sales), confusion of collection procedures for royalties returned to the *Commune* and respective governments, lack of *Commune* capacity to promote and advise miners in safe, environmentally responsible practices and delays in mapping the "gold panning corridors" in which artisanal panning would be permitted, included on other mining concessions. Still, in areas with outside assistance there is a high rate of registration. Of 1500 gold panners in the *Commune* of Antanimbary in 2006, 1383 had obtained their permits. Declines the subsequent year were largely attributed to heavy rains and the distance to the administrative centre while 40 of 55 registered collectors renewed the following year.[82]

The Madagascar experience shows that a concerted sensitisation campaign, combined with grassroots intervention and demonstrated benefits to the local community can encourage formalisation.

3.2.2 *Uganda*[83]

While Uganda is not a diamond producing country, almost 200,000 women (45%) and men (55%) are engaged in artisanal mining of gold, tin, coltan, wolfram and a range of industrial minerals (limestone, stone aggregate, salt, clay etc). In an effort to support formalisation, the Government of Uganda, with support from the World Bank, undertook massive training and sensitisation campaigns targeting almost 200 trained trainers and over 1000 artisanal miners between 2007 and 2009. In order to ensure both women and men realized benefits from the program, the committee included a gender expert and gender was mainstreamed throughout the Campaign.

In the short time since implementation, results have included:

- Granting of over 80 prospecting licenses, requests for 50 location licenses (and granting of 10) to trainees.
- Formation of more than 20 local small scale mining associations, three regional associations and a National Artisanal and Small Scale Miners Association (NASMA).
- Demonstrated improved practices in savings, increased selling prices, value addition activities and miner-initiated measures to improve environmental management and health and safety.
- Distribution to miners of contact information of licensed mineral dealers and outreach to dealers has also resulted in almost tripling of royalties over the past two years and doubling of non-tax revenue within a four-year period.

A National Strategy for the Advancement of Artisanal and Small Scale Mining in Uganda marked the culmination of this project, included in which are detailed work plans, budgets and performance monitoring and evaluation frameworks.[84] As identified in the National ASM Strategy, major constraints to formalisation relate to bureaucratic and costly centralised licensing (~US$350 exclusive of transport and fees for assistance),[85] lack of miners' awareness of and ability to comply with licensing requirements and, despite Mineral Policy commitments, the absence of institutional roles and mandates in the Mining Act (2003) and Regulations (2004). Recommendations for specific legal, policy and institutional reforms are accordingly being reviewed in the Strategy. Successful outcomes from training and outreach activities thus far attest to the benefits of "walking the talk" of good policy while doubts in long-term financing of institutional support threaten future progress.

The Uganda experience shows that success lies in extensive services, training and outreach to artisanal miners, including the importance of gender mainstreaming.

3.2.3 *Ghana*

In 1989, when a Small Scale Mining Implementation Committee was formed to oversee "The Regularisation of Small Scale Gold and Diamond Mining Project," formalisation efforts commenced with the demarcation of eight small-scale mining districts.[86] Between 1989 and 2002, milestone efforts included:

- Provision of extension services supplemented by the hiring of district officers and mines wardens, who were charged with carrying out the Small Scale Gold Mining Law (PNDC Law 218).
- Training of district officers in the Mining Law, health and safety issues in ASM, and geology before deployment. Subsequent "Training of Trainers" courses were provided to the officers, including environmental management, health and safety, basic bookkeeping and project planning and management.[87]
- Subsequent government partnerships with development organisations in order to extend support through "Rent-A-Pump" and "Hire-Purchase" (or rent-to-own) Schemes, technical assistance in introducing Chinese Hammer mills

(which are now in widespread use), pilot testing of hard rock and alluvial mining equipment, a program to make geological information available to small-scale miners (resulting in a number of suitable demarcated areas) and reclamation of three degraded sites.[88]

- Subsequent projects saw adaptation of the Mercury Law to enable miners to legally purchase small quantities of mercury for gold mining and training in the safe use of mercury.

Earlier efforts to support ASM produced variable results but demonstrated outstanding commitment on the part of government and some notable progress was achieved. Between 1998 and 2002, officially reported production rose from 2%–7% for gold and 40%–80% for diamonds.[89] And yet, only 620 licenses were granted in this period, to which miners attribute to a complex licensing process that can take up to six months.[90]

Perhaps more than any other country, Ghana exemplifies that (i) there is no "best practice" without practice; (ii) a simple licensing process is paramount to encouraging miners to register; and (iii) formalisation and legalisation of ASM is a long process requiring steadfast government commitment to creating incentives through grassroots interventions.

3.2.4 *Peru*

Lack of proper consultation with women and men miners and other ASM stakeholders (i.e. sole reliance on top down approaches) can easily lead to failures in formalisation. Peru successfully avoided this pitfall in its Law 27651 "Formalisation and Promotion of Small Scale and Artisanal Mining," a remarkable example of constructive consultation and engagement where formal proposals for reforms of both laws and regulations were received from ASM associations and were largely enshrined in legislation.

Peru's mining sector rapidly expanded since the early 1990s. In 1993, mining accounted for almost 3% of GDP and increased to 11% by 2000.[91] In 2000, it contributed almost half of foreign exchange earnings with export revenues of US US$24 billion.[92] By 2005, mining accounted for more than 60% of the country's total export revenues. From a near totally informal sector in 2000, by 2008 50% of artisanal miners were operating formally, thanks to facilitative legal frameworks.[93]

Prior to 2002, the mining code in Peru was initially developed to facilitate large-scale, multinational investment in the country. Between 2001 and 2002, artisanal miners organised themselves to persuade their government to revise the mining code in order to also make it appropriate to the way that they mined; the system at the time simply precluded licensing of artisanal miners. In 2002 the code was changed to recognise ASM on the basis of title area and production capacity, and made provisions in line with artisanal mining realities.[94]

The success of the Peru experience lies in a number of factors[95]:

- Strong political will to support formalisation.
- A receptive environment and the desire of artisanal miners to be formal, as well as the establishment of a Swiss-funded development intervention, Proyecto GAMA, which was designed to formalise the sector.[96]
- Efforts to identify similar interests among otherwise polarised parties.
- Organising talent and initiative of artisanal mining leaders; and a close connection between traditional leaders and their base, who were also willing to include women miners as leaders.
- A constructive multi-stakeholder environment and the presence of impartial external stakeholders who could act as mediators and facilitators, and provide funding and other resources for advocacy activities.

3.2.5 *Philippines*

Republic Act No. 7942, known as the "Philippine Mining Act of 1995," regulates mineral resources development in Philippines. Section 42 of the Act states that small-scale mining is to be governed by Republic Act No. 7076 (Philippines People's Small Scale Mining Law, 1991) and other pertinent laws.

Almost 100% of the Philippines' industrial minerals and up to 80% of its gold are produced mainly through ASM.[97] More than 300,000 artisanal and small-scale miners are active in the Philippines, two-thirds of which are engaged in gold mining. In recognition of its economic and social significance, the government has instituted a number of laws pertaining to gold panning and sluicing (PD 1150), mining of small deposits (PD 1899), identification and segregation of ASM zones (RA 7076) and ASM mine safety rules (AO No. 97-30).

Interesting features of these laws include:

- Miners can obtain a range of renewable permits, which can be granted for one to three year periods contingent on limited production, non-mechanisation, explosives bans and exclusion of child labour.[98] Permits are commodity specific.
- PD 1899 requires that all gold be sold directly to the Central Bank or its official buying stations in gold areas.
- When a small-scale mining area is designated within an existing mining right, PD 1899 also exempts small scale miners from annual work obligations, payments and payment of fees, rents and property taxes, and they are further afforded a reduced royalty rate of 1.5%.

Decentralised Small-Scale Mining Offices undertake most ASM-related regulation, monitoring, and technical assistance functions. Much of this work is funded by the 15% share in government revenues from mining via the People's Small-scale Mining Protection Fund. Funds are mainly allocated towards information dissemination and training of small-scale miners on safety, health and environmental protection, and the establishment of mine rescue and recovery teams, including the procurement of rescue equipment necessary in cases of emergencies such as landslides, tunnel collapse, etc. The fund is also accessible to the small-scale miners in case of accidents and/or serious unforeseen events.

Bugnosen (2004) contends that a number of ASM-related legislative measures have failed, while others have succeeded. Failures include: attempts to designate ASM areas; permitting activities within existing concessions; overly restrictive provisions for obtaining permits; and the need for multiple permits depending on the stage of operations (mining, permitting and marketing). Successes have been observed in terms of gold rush control measures that have enabled tax collection, environmental protection in these areas, and efforts to inhibit damaging sand and gravel operations. The emergence of "contract mining" wherein formal companies purchase minerals from ASM producers, has been deemed promising, although local indigenous communities have expressed concern over their capacity to stimulate uncontrolled activities.[99]

The Philippines experience demonstrates how the decentralisation of ASM governance, coupled with extension services to help artisanal miners manage their occupational health and safety and environmental responsibilities, can encourage formalisation.

4 CRITICAL COMPONENTS OF A CONDUCIVE FISCAL AND LEGAL REGIME

This section outlines the main fiscal, legal and institutional lessons learned from other countries to inform CAR's efforts to develop a regime that supports formalisation—and increased development contributions—of its ADM subsector.

A key finding is that reduced licensing costs, royalties and export taxes *can* stimulate licensing, however immediate results are unlikely, except in areas where sufficient on-site intervention and support is provided. Consequently, fiscal provisions must be harmonised with sensible mining policies and the institutional mechanisms needed for implementation.

4.1 *Fiscal provisions*

Fiscal provisions that support legalisation of ASM *can* generate revenues to the state and national economy overall. For example, part of Ugandan formalisation efforts has been to provide even unlicensed miners with contact information of licensed mineral dealers combined with outreach to dealers. This has been cited as one of the main factors that led to almost tripling of royalties since 2008 and doubling non-tax revenue within a four-year period.

While tax and non-tax revenues from ASM can provide some motivation to government, it is important to recognise that *the primary intent of formalisation is to unleash its overall development potential to expand revenue potential from wider economic sources*. In the past, many countries (e.g. Brazil, Lao PDR, Ghana), via heavy-handed regulation and enforcement in the interest of state revenues, have led only to further marginalisation of ASM, driving miners "deeper into the jungle." In effect, overzealous attention to state revenue generation alone often results in continued losses in terms of taxes, fees and royalties and limited improvements to the ASM subsector, thereby undermining its development contributions.[100]

Many countries now seem to recognise the fiscal *and* broader development potential of ASM and have adopted a number of related measures in order to support legalisation. Supportive fiscal provisions include: (i) low-cost licensing of producers as well as buyers/traders; (ii) low tax rates and regional harmonisation of taxes; (iii) payment or royalties and taxes by mineral dealers rather than miners; (iv) other fiscal incentives to legalise; and (v) returning a portion of revenues to ASM affected communities.

(i) Low cost licensing of producers and buyers
A number of case study countries have instituted low-cost licensing of artisanal miners, which when combined with other legal and institutional mechanisms, have resulted in comparatively high rates of legality. In Guyana, annual license fees of only US$5 for diamond diggers, US$50 for dredges and US$75 for traders and exporters combined with low royalties (3%) on a set standard diamond value (US$75) have been major factors that have led to legalisation of almost 100% of the sector.[101] Madagascar's low-cost, locally administered permit model enabled gold panners to legalise activities at a price of only US$5.50. In one region of Madagascar where panners were intensively sensitised, 1383 of 1500 gold panners obtained permits for their activities.[102]

Another remarkable example can be drawn from Sri Lanka's gemstone sector. When heavy machinery is not used, gemstone mining licenses can be obtained from the National Gem and Jewellery Authority (NGJA) for a fee of only US$10 for licensing plus US$10 for site reclamation, which is refundable pending acceptable restoration of land.[103] The one-year licenses are issued within two to three weeks of application and require either two-thirds share in the land or a leasing agreement with the owner. The NGJA further encourages ASM by holding auctions for gem-bearing sites identified by government geologists on crown land or in water bodies. These legal conditions resulted in issuance of 3702 gem mines licenses in one year alone while the country boasts a high proportion of legal miners (~80%).[104]

In some cases, low cost permits have *not* yielded as favourable results, particularly if coupled with other inappropriate legal or institutional provisions. For example, DRC's well-meaning, low-cost registration card (*carte d'exploitant artisanal* at US$25) fails to recognise the broad areas covered by migratory artisanal miners, thereby requiring them to register in multiple locations at great financial and bureaucratic costs and with no notable benefit.

The case for low-cost licensing is nevertheless strong and is further supported by the lack of success in countries with high license costs. In Uganda, only ~10% of miners are working under a *location license*, which costs around US$350 and is granted for mining operations whose capital investment is below US$5000. In Sierra Leone, where a license cost around US$270 in 2005, at that time an estimated 48% of miners were operating legally in the diamond heartland of Kono where, supported by international programs, government efforts to formalise the sector were intensively focused.[105]

(ii) Low tax rate and regional tax harmonisation

As shown in Mano River countries, harmonising export taxes can help discourage smuggling, however it does not necessarily discourage illegal mining. This is because informal miners are able to sell to formal miners and dealers, so laundering their diamonds into the official chain.[106] In comparison, a low tax rate (3%) *combined with* a simple tracking system in Guyana, however, is believed to be a key factor in increased reporting and reduced smuggling as well as reduced informal mining.

(iii) Payment of royalties and taxes by mineral buyers/dealers

Some countries (e.g. Liberia, Ghana, Uganda, CAR) still, under law, require payment of royalties on production by artisanal miners themselves, adding an additional level of bureaucracy that few miners manage to fulfil. However, in Philippines and Uganda, the law obligates a dealer (exporter) to pay royalties on production from informal miners before export. While this scenario is far less complicated for artisanal miners, costs are often passed to the miner during price negotiations who, in the case of the Philippines, are also obligated to pay a 1.5% royalty to the claim owners or owners of private lands. In this way informal minerals enter the formal chain and still get taxed.

It is interesting that in Madagascar, local gold buyers (*collecteurs*) pay royalties based on the purchase price from the producer (using a jointly signed invoice) while export taxes are paid by exporters based on a more realistic valuation. Although a practical mechanism, low literacy levels of both buyers and sellers often prevent declaration of sales.[107]

(iv) Other fiscal incentives for formalisation

A number of countries have identified other fiscal measures to promote legalisation as well as progressive improvements to the performance of ASM. In Ghana, and based on 2002 data, all licensed small-scale miners are exempted from payment of taxes and royalties for the first three years of operation.[108] In Uganda, a zero import duty is charged for importation of mining equipment; a policy which helps advance the sector.[109] A number of jurisdictions have legislation that provide for discretionary powers to allow for temporary or one-time waiving of certain types of taxes, typically royalties. This is usually only applied for projects experiencing short-term financial downfalls, particularly if it is due to uncontrollable circum-stances (e.g. commodity price drops).

A number of governments have instituted financing schemes for ASM with varying results.[110] The Government of Zimbabwe has likely implemented the greatest number of financial programs for ASM including for loan-based support.[111] Unfortunately, many loans were used to set-up "ghost mines" or fronts for buyers of gold and gemstones from illegal artisans. In Mozambique, the government administered Mineral Development Fund (*Fundo do Fomento Mineiro*, FFM) provides financing to small-scale miners yet funding criteria is out of reach for most artisanal miners, while support is nevertheless achievable for more advanced small-scale operations. The Fund is challenged with common misuse of funds and a lack of monitoring of implementation and use of funds.[112] By contrast, the Namibian Mineral Development Fund has largely been a success. It has provided US$92 million in loans and US$9 million in grants for large and small-scale projects. With low interest rates, an ample repayment period (five-years plus a two-year grace period), sufficient management resources and minimal bureaucratic requirements, over 90% of loans have been repaid. However, again, this program mostly aimed at medium- and large-scale producers.[113]

Mining Ministries are *not* commercial banks or microfinance institutions (MFIs) and so are often challenged by inadequate experience and skills to administrate these programs effectively. Two recent programs rely on *existing* financial institutions while raising their awareness of the needs of the ASM subsector. In Papua New Guinea, support from the Japanese Government has allowed the Social Development Fund to focus on building business and microfinance skills in ASM communities. In the target region of Wau, over 25% of trainees have opened accounts at a local MFI.[114] In Nigeria, Loan Guarantees are funded by the Nigerian Government, but builds on available financing from commercial banks. The Ministry of Mines and Steel Development (MMSD) is responsible for providing the technical support needed to verify feasibility, ensure proper use and therefore guarantee the loans. Intensive sensitisation of banks, including branches in ASM areas, has resulted in the establishment of "Mining Desks" in some banks.[115]

It is important to note that most of the earlier financing schemes have targeted small- to medium-scale rather than *artisanal* miners. Although coverage of local savings and credit cooperative associations (SACCOS), MFIs and banks is sparse in many remote areas, building on existing financing programs with approaches adapted for the rural poor with *supportive loans guarantees and reasonable interest and payment period requirements* are a start. Grassroots support for numeracy, literacy, organisation formation, business skills development and group savings can help artisanal miners take crucial first steps out of poverty.[116]

(v) Distribution and use of financial revenues

Revenue sharing describes arrangements whereby mineral taxes and other revenues are collected by the central government, with a certain portion of the revenues returned back to the areas in which mining occurs.[117] Measures to re-distribute wealth generated from mining *and* increase the benefits to and capacity of miners and their communities have become an important component of fiscal policy and are recognised as an additional formalisation incentive.

(vi) Royalty sharing for local development

An increasingly common mechanism to redistribute benefits is via royalty sharing. Typically, communities around mining areas bear the brunt of environmental impacts and social disturbances (e.g. HIV/AIDS, increased price of goods, etc.) associated with mining. Different countries distribute these revenues differently. For example, under Uganda's Mining Act (2003), sharing of royalty is 80% for government, 17% for local governments (intended to equalise provision of services and infrastructure across a district) and 3% for owners or lawful occupiers of land subject to mineral rights (to compensate for negative side effects associated with mining). A major shortfall is lack of requirements for use and accountability for returned revenues.

In other cases, such as Ghana and Sierra Leone, community development funds are targeted towards development projects in mining-impacted communities. Under Ghana's Mineral Development Fund (MDF), 9% of royalties paid by a mining project are to be divided between the district assembly (the local political administrative unit) and the local traditional authorities.[118] Main complaints concerning the MDF relate to the amount of funds and delays in their release; lack of information about amounts transferred (between institutions and with communities); inadequate reporting procedures by traditional institutions concerning use of funds; absence of auditing mechanisms and omission of the MDF from legislation.[119]

While challenges of local government in collecting royalties or MDF monies from the national treasury are often cited, the most common challenge relates to local government capacity to plan and account for their use, which could theoretically include responsiveness to ASM (thereby generating greater revenues). In some cases, these funds are disbursed without clear specifications and it is suggested that—much like many Health Ministries track and provide guidelines and support to local government for use of health related transfer payments—the same capacity development and procedures are needed for transfers from the minerals sector.

In Sierra Leone, 25% of the 3% export royalty (0.75% of total export value) is returned twice yearly to Chiefdom Development Committees, who disburse 15–40% to district councils and 5% to town councils through the Diamond Area Community Development Fund. This is done to incentivise local authorities to encourage miners to not just pay surface rents to traditional authorities, but to legalise their activities with central government too. Nearly US$1 million was returned to chiefdoms in 2007. Chiefdoms receive amounts in proportion to their share of artisanal mining license numbers out of the national total, though the National Advocacy Coalition on Extractives has recommended that this be tied to local production levels instead. In practice, strengthening of local governance and administration as well as improving accountability mechanisms is typically needed to ensure real benefits are felt on the ground.[120]

(vii) Funding ASM formalisation and development

A portion of proceeds is sometimes applied directly to respond to the needs of artisanal miners. Support delivery mechanisms in the form of Small-Scale Mining Units or Departments have been established in Mozambique, Papua New Guinea, DRC (under SAESSCAM), and Philippines. In all cases, as with the Community Development Funds in Ghana and Sierra Leone, implementation of what appears to be good policy has been hindered by a lack of resources to these units and limited accountability for their performance.

The most obvious mechanism to finance ASM subsector support activities is via a levy on fees, royalties and export taxes. Although it has not yet been institutionalised, in Uganda's *National Strategy for the Advancement of Artisanal and Small Scale Mining*, detailed gender-responsive work plans and budgets were developed to determine that only 5% of mineral royalties was needed to finance a cross-section of training, outreach, regulation and data collection activities needed to responsibly manage and develop the ASM subsector.[121] The Strategy recognises that these efforts would support further legalisation, including by women miners who constitute 45% of the ASM workforce. These activities would, in turn, generate progressively increasing royalty returns. This direct link between institutional performance and financing of activities was supported by a clear institutional monitoring and evaluation framework that promotes accountability and outcomes by individuals as well as units and departments responsible for undertaking the work.

Besides royalty sharing, a number of other mechanisms exist to finance government support for ADM licensing and communities affected by ADM.[122] For example, government institutions can institute "trading accounts" wherein products or services (e.g. maps, reports) are provided at a fee to the minerals sector. This, as well as creation of a quasi-independent, self-financed Mines Authority or Minerals Commission, typically relies on large inputs from a highly active exploration and large scale mining sector, as found in Ghana. The GGMC in Guyana is an autonomous body and therefore self-financing with any surpluses rendered to the National Treasury while its operations also rely heavily on large-scale mining. National Facilities can also be created to subsidise costs by linking complimentary and overlapping functions of other government agencies (e.g. laboratory services, documentation and statistics requirements, enforcement). However, unless coordination and collabouration agreements are extremely solid and capacity to understand and commit to the minerals sector exists in other agencies (e.g. police, statistics bureaus), which is rarely the case, these arrangements tend to function poorly.

The Philippines *People's Small-scale Mining Protection Fund*, derived from 15% of royalties on mineral production, is used mainly for information dissemination and training of small-scale miners on safety, health and environmental protection, and emergency response measures.[123] Philippines has faced a number of challenges, such as designation of ASM areas or permitting activities within existing concessions and overly restrictive requirements. However, it has been successful in terms of gold rush control measures that have enabled tax collection and environmental protection in highly degraded areas.[124] A similar financing approach could also be used to fund support for organisation and registration of miners, training in diamond valuation and pricing, introduction of intermediate, safe methods and other key needs.

These examples demonstrate that many countries *have* recognised the importance of redistribution of benefits from taxes, royalties, rents and fees from ASM. While many countries have overzealously given attention to generating revenue from ASM via heavy handed regulation and enforcement alone, it has been demonstrated throughout the world that *top-down enforcement and policing without clear benefits does not work and there must be clear local-level incentives to encourage compliance.* The next sections, therefore, address the legal and institutional mechanisms needed to complement fiscal provisions to develop a functioning ASM subsector.

4.2 *Legal and regulatory requirements*

ADM policies, laws and regulations are only useful if they are realistic to the way people mine artisanally, if artisanal miners have the capacity to obtain licenses and benefit from them and if they are enforced. Key lessons drawn from other countries relate to the capacity of both the ADM subsector and those institutions and agencies responsible for executing policy and legislation. In terms of legal and regulatory frameworks, key success factors primarily relate to: (i) recognition of the diversity and different categories of miners; (ii) legislating mandates of mining institutions; (iii) reconciling international, national and local priorities; and (iv) ensuring ASM laws are realistic to the existing structures of production and trade.

(i) The diversity and different categories of miners
ASM varies in terms of organisation of work (independent, small teams or production units, associations etc); scale of activities; permanent, migratory or seasonal nature of work (determined by their overall livelihood strategy and motivation for mining); and degree and capacity for mechanisation, among others. In some countries, this requires more than one category of ASM license to account for the in-country diversity *and* provide

a step-up opportunity as activities become more formal. Increased capacity brings increased legal obligations, which in turn should bring increased benefits.

For example, Levin's work in Sierra Leone in 2004 demonstrates how different types of miners have different reasons for choosing (or being obliged) to mine legally or illegally. In many countries, women constitute large proportions of the workforce and can, compared to their male counterparts, face even greater barriers in terms of literacy, mobility (e.g. to travel to regional offices) and autonomy, among others factors.[125] "*Despite claiming to be 'gender neutral,' many policies, even mineral policies, can affect women and men differently and can actually serve to worsen gender inequalities.*"[126] Understanding *who* your miners are, how they use the mining, and what it brings them is therefore key to designing the right incentive structures for formalisation.

Some countries develop ASM legislation with the intent of promoting mechanisation, thereby integrating "artisanal" with "small-scale" miners in a single category. This practice can serve to exclude and further marginalise the artisanal majority (especially people who mine on an occasional basis and subsistence miners, who often include a large proportion of women, where obstacles to licensing can be more pronounced) in favour of the few relative "elites" who are positioned to take advantage of such efforts. For example, a small stationary jig used by many diamond miners in South America costs on the order to US$30,000–75,000,[127] well out of reach of average CAR miners who earn ~US$723 per year.[128] Promotion of mechanisation must also consider that measures are usually needed to minimise the effects of their introduction. For example, siltation of more than 300 km of the Tapajos River in Brazil, a tributary of the Amazon River, is largely attributed to widespread use of hydraulic monitors and intensive dredging activities.[129] In Burkina Faso, introduction of one mill for ore grinding effectively put 300 women out of work, while a combined crusher-grinder unit decreased labour requirements from 425 to 14 people.[130]

(ii) Legislate institutional mandates

"*Informality begets informality. Unless ASM support is formally enshrined in (mining authorities) work programs and budgets, ASM is unlikely to make much progress towards formalisation.*"[131] Many countries outline ambitious and well-crafted policy objectives related to ASM yet do not provide a legal mandate for mining institutions to execute them.

Tanzania's Mineral Policy outlines ASM-specific roles, objectives and strategies for each of: provision of extension services, licensing and related transparency measures; financial services; health and safety; women's issues; child labour issues; institutional framework (including information flows; roles of regional offices; and establishment of training centres). Although an implementation plan for the policy is currently being developed, the Ministry of Energy and Mines has not enacted law to institute these measures.[132]

Recognising the importance of institutional support, under Article 90 of the Nigerian Minerals and Mining Act (2007), the government through the Ministry commits to provide services to registered mining cooperatives in: prospecting and exploration; mineral testing and analysis; assistance in mine planning and design; teaching adequate technical skills; making proper links and guarantees related to plant-for-hire and equipment leasing-to-own; introducing appropriate mineral processing methods; providing EIA reports and detailed guidelines for waste and water management; introducing health and safety procedures; and holding regular workshops on legal, marketing and business skills.

Defining institutional roles, functions and obligations of mining departments, units and offices in mining law, and even more specifically in regulations, not only provides a clear, legal mandate, but provides a basis to lobby for and receive funds from the central government to fulfil this mandate. Furthermore, this can serve to reduce discretionary powers and can hold management, departments and their officers accountable to fulfil legally defined mandates while providing a legal basis for monitoring and evaluation of performance.[133]

(iii) Reconciling international, national and local priorities

Institution of policy measures without sufficient attention to context and consequence can have negative repercussions, potentially exacerbating artisanal miners' poverty and driving the minerals trader deeper underground. For example, the Government of Ghana halted diamond exports between 2006 and 2007 until charges were cleared concerning the country's role in harbouring stones from Côte d'Ivoire. Effects on communities in the main diamondiferous area were severe.[134] Similarly, recent moves in Tanzania to prohibit export of Tanzanite without in-country value addition appear to serve the interests of national development. However, the infrastructure needed to cut and polish the gems is, at present, extremely insufficient to meet the demands given current Tanzanite production levels.[135]

In both cases, the policy decisions of government may have received much international acclaim and are, by all appearances, sound. However, the impacts at the village level can prompt gemstone miners, who largely subsist hand-to-mouth, to strengthen informal channels of diamond dealing, thereby worsening the illicit minerals trade and encouraging smuggling.

(iv) Ensuring ASM laws are realistic to the existing structures of production and trade
Successes and constraints in Liberia,[136] Sierra Leone,[137] DRC[138] and Peru[139] all point to the importance of formalising existing arrangements that work on the ground, rather than attempting to force artisanal miners to adopt a legal model that may be ideal from a government perspective, but is unachievable for artisanal miners. Extensive consultation with artisanal miners to understand: a) how and why they presently mine, b) what operating in legal structures can bring them and their communities, and c) what obligations would be reasonable on their part is a prerequisite for developing a legal and fiscal regime that is appropriate. Providing a sense of ownership over the process (as in Peru) brings much greater legitimacy to legislation and increases likelihood that miners will abide by it.

4.3 *Institutional requirements*

The success of any fiscal measures ultimately hinge on a conducive legal framework and political and institutional commitment to their implementation. *The main challenges in each case study country largely relate to inadequate disbursement of funds to fulfil mandates, weak governance (particularly at local levels), and a lack of transparency and accountability for performance.* As such, understanding the key institutional arrangements, administrative structures and mechanisms as well as needs in terms of performance monitoring and evaluation provide insight into how fiscal measures can effectively be implemented.

(i) Institutional arrangements
Many countries face the challenge of spreading mandates across a wide range of institutions, making accountability for implementation of mining policies more difficult, particularly if coordination is poor. Very few countries seem to have succeeded in effectively sharing the information needed to track minerals (for example between revenue authorities collecting export data and mining departments compiling production statistics).

Quasi-independent "authorities" or "commissions" that partially or wholly generate operating funds through their activities are, in principle, more likely to fulfil their functions. However, these typically require well-established revenue generation systems to "get off the ground" (as is the case for Ghana's Mineral Commission whose large scale mining sector has been well established for several years) and effective means to ensure proper use of funds.

(ii) Regional versus central management
Regional management makes sense in large countries or where there are high concentrations of artisanal miners. However, good management can be hindered by a lack of institutional capacity and resources, especially if minerals sector revenues are not effectively disbursed from central to local government. While it may be administratively sensible to place responsibility for an area under a single individual, group or organisation (local government or otherwise), using such a system to register, monitor and regulate a large number of miners dispersed across an area also requires consistent application of laws, and indeed *understanding* of the law. This can be a challenge if the law is to be administered by local officials who typically have little experience in the technical and organisational needs of ASM.[140] Mechanisms that ensure capacity and good governance at this level are therefore important.

This includes the need for verification procedures to be observed by and between central and regional mines offices to prevent abuse of low cost ASM licensing systems by buyers seeking to maintain control of a site. In countries such as Uganda, CAR and Guyana, for example, a significant proportion of small-scale mining licenses are actually held by mineral dealers rather than producers themselves. In the latter case, many of the claim blocks in diamond areas have been bought by Georgetown traders and by established Guyanese mining firms, rather than by actual miners.

(iii) Administration, regulation and enforcement
Administrative mechanisms to collect taxes and royalties differ from country to country. A unique model is found in Ghana, which produces an average of 500,000 carats of diamonds from alluvial deposits, about 10% of which are gem quality.[141] Official exports in 2009 were valued at only US$7.32 million yet the government supports local buying through the Precious Minerals Marketing Company Ltd (PMMC), a state-owned enterprise overseen by the Ministry of Lands, Forests and Mines. The PMMC has 70 staff that serve offices located in all gold and diamond regions of the country. Companies who receive a license for buying and exporting diamonds (at a cost of US$30,000) are housed in the PMMC Diamond House Building in Accra, where they undertake buying. PMMC handles export documentation, sealing of parcels and transport of diamond parcels to the airport.[142] Export taxes are set 2.3% of value, well below that of other countries in the region.

Other diamond and gemstone countries, like Madagascar, DRC and Sierra Leone, have also set up dedicated offices to manage exports and collect royalties. In Sierra Leone, this job falls to the Gold and Diamond Department, based in Freetown where most exporters have their offices. Under the previous administration this came under the authority of the National Revenue Agency, but was returned to the Ministry of Mineral Resources by the new Mines and Minerals Act (2009).[143] The GDD levies a 3% export tax, which is based on the highest of three valuations done by the exporter, the GDD's valuator and an independent valuator. Like the PMMC, it manages all procedures relative to compliance with the KPCS.

(iv) Chain-of-custody tracking

Where incentives to formalise the export stage have worked reasonably well, a major weakness with the KPCS has been the inability of nations to formalise the mining and trading stages and the possibility of illegally mined diamonds entering the legal chain.[144] Effective, simple mechanisms to report production and track it through to export are needed to not only ensure KPCS compliance, but also to promote official reporting, payment of royalties and taxes and evaluate the effectiveness of changes to fiscal and legal provisions.[145]

Inspiration has come from Guyana's successful experience with diamonds, where the simplicity of the approach seems to set the standard:[146]

- Each week, the operator of each dredge fills out a "production sheet." This includes the location where they are working, working hours, diesel consumption and weekly diamond production.
- A copy of the production sheet accompanies the dredge's weekly production as they are sold to buyers in the field and then exporters in the capital city.
- Each production sheet in an exporter's parcel is checked by a representative from the Guyana Geology and Mines Commission (GGMC) and entered into a database.

This basic system has many benefits. It is simple enough to follow and, because GGMC keeps a record of all dredges (including the pump size, number of workers and their weekly production data), production as well as export statistics have effectively been tracked since 2003. While organised, mechanised dredging is much different than practices and arrangements in CAR, this model suggests that focusing on the mining units (groups, teams, sites) provides a practical means of administration. In the case of a large number of relatively dispersed miners, Madagascar's *Commune* system of registration may be a more workable example if local technical and management capacity is sufficiently developed.

Although KPCS procedures in Liberia are well-established and known to many registered buyers, they are far more complex which, together with high trading license costs, is cited as the main reason why only ~10% of the 750 initially licensed brokers and traders have renewed their licenses. Procedures under the KPCS require site managers or license holders to report production to regional offices once a diamond is found, receive a voucher for sales, return to the mine site and use the voucher to sell the diamond to a local buyer, who also returns to the regional office to submit a copy prior to continuation to the Government Diamond Office in Monrovia (requiring further valuation and paperwork). Given incomes of miners (~US$66 per month), transport costs and access to regional offices and literacy constraints of miners, it is, perhaps not surprising that only ~12% of artisanal diamond miners work on Class C licenses[147].

Different outcomes from the Guyana and Liberia examples suggest the advantage of simplified procedures for registering production (e.g. making it a weekly or regular simple activity rather than an "as and when" one and having the report accompany the sales for official corroboration at export, rather than reporting to government at point of production). Issues such as basic literacy and numeracy must nevertheless be considered in many contexts.

5 CONCLUSION

ADM is an important livelihood for around 80,000 women and men in the CAR and provides more than 60% of the country's export earnings.[148] Currently, it is estimated that more than 2.8 million women, men and children directly and indirectly rely on ADM while economic spin-off effects may inject as much as US$144.7 million into local economies, thereby stimulating local businesses while providing a non-agricultural source of rural employment.

Legalisation of artisanal diamond miners in CAR holds even greater promise in terms of national development. While official statistics suggest that the mining sector in CAR contributes only 4% to national GDP and more than 25% of the country's alluvial diamond production is attributed to artisanal miners, the likelihood of substantial losses in government revenue due to smuggling and illegal exports is high.[149] Legal operation should bring benefits such as access to financing, training and support that can increase diamond production and stimulate multiple spin-offs in terms of micro- and small-enterprise development and growth of market gardens, fish farms and other agricultural ventures.

In numerous ASM countries, it has been widely demonstrated that the costs of licensing, fees, rents, taxes and royalties are a common obstacle for informal miners. The primary objective of this Comparative Study was to assess whether lowering the fee of the *patente* can encourage legalisation of artisanal miners in CAR. Comparisons between artisanal diamond mining fiscal regimes from Liberia, Sierra Leone, Guyana, Tanzania and DRC, and valuable experience from several other ASM countries provide interesting insights into mechanisms that can support entry of artisanal miners into the formal system.

5.1 *Fiscal provisions: Key success factors*

Certain elements of fiscal regimes have emerged as key factors in formalisation:

- *Low-cost Licensing*: Guyana has reportedly formalised almost 100% of its sector (in terms of production units), largely due to low-cost licenses for both miners and buyers. Almost 80% of gemstone miners in Sri Lanka operate legally at a fee of only US$10 per license, which is typically issued within only two to three weeks of application.
- *Incentives for Payment of Royalties and Taxes*: Again, Guyana's low royalty (3%) paid on a standard diamond value of US$75 per carat (combined with an easy to administer reporting and tracking system) sets the standard. Payment of royalties by buyers (rather than miners) prior to export, as in Philippines, Sierra Leone and Uganda, also reduce the bureaucracy incurred by miners, who often face literacy, numeracy and logistical challenges. Ghana's PMMC provides assistance via provision of a dealers office in a central buying building and even helps prepare export documentation, seal parcels and transport diamond parcels to the airport.
- Additional provisions, such as zero import duties on mining equipment in many countries, can help mitigate costs of mechanisation, while Ghana also offers a three-year deferral on royalty payment for new small-scale mining license holders.
- *Re-investing in ASM and ASM Communities*: Measures to re-distribute wealth generated from mining and increase *benefits to* and the *capacity of* miners and their communities have become an important component of fiscal policy. Countries such as Mozambique, Papua New Guinea, DRC, and Philippines have established Small-Scale Mining Units or Departments in order to provide outreach to artisanal miners. As with Community Development Funds in Ghana and Sierra Leone, which intend to bring benefits through grassroots development projects, implementation of what appears to be good fiscal policy has been hindered by a lack of resources to these units and limited accountability for their performance.
- Most countries cannot rely substantially on the direct tax and non-tax revenue from ADM to finance extension services and regulatory mechanisms, particularly in the early phases of formalisation. Rather, revenues from the entire minerals sector should be used to support the advancement of ASM and thereby its contribution to broader economic growth.
- While a semi-autonomous, self-financing authority or minerals commission can provide a useful implementation mechanism for such support, these tend to function only in cases where a significant large-scale mining sector exists and can provide considerable input. Particularly when a largely informal subsector is the starting point, the most obvious mechanism to finance ADM subsector support activities is *subsidising* implementation costs via a levy on *all* mineral royalties and export taxes with awareness that formalisation takes time and dedicated commitment.

In terms of next steps for CAR, a simple economic model derived from case study data suggests that lowering the *patente* cost to US$5 license fee could encourage legalisation of more than 65,000 artisanal miners and yield *official* diamond exports of US$82.4 million, thereby providing an additional 4.2% of the nation's GDP and bolstering foreign exchange earnings. Even with a reduction of royalty (1.5%) and export tax (3%) rates, this could generate as much as US$3.5 million in tax and non-tax revenue. Remarkably, by comparison with the US$5 license fee, a *patente* cost of US$50 would expect to attract only 28,000 miners to the legal system, while officially reported diamond production could *decrease* by over US$47 million and generate ~US$0.8 million *less* in tax and non-tax revenue.

It is important to recognise that fiscal measures alone are insufficient to achieve such impressive outcomes. Thus, it is not surprising that, with respect to legalisation success, "high performing" countries, such as Guyana and Sri Lanka (where a license costs only 0.1–0.3 months of a miners income) have spent years putting in place extensive support mechanisms to complement their fiscal and legal frameworks. "Mid-range" performers such as Madagascar, whose license requires only 0.1 months of a miner's income, have fiscal and legal measures in place but have done only marginal outreach to ASM areas, thereby stifling uptake. "Lower performing" countries, such as Liberia and Uganda (who charge 3.9–5.3 times the monthly income of a miner for a license) have only very recently begun to revisit their legislation and take steps towards establishing regulation, enforcement and extension service mechanisms that are appropriate given the nature of their respective ASM subsectors.

While sensible fiscal measures *do* seem to increase the legality of ASM and generate significant benefits for the state and economy, success of these measures largely depends on the capacity of miners and traders to comply with legal requirements and whether it provides any real advantage. Progress is often slow, except in areas where sufficient on-site intervention and support is provided. Consequently, fiscal regimes must be harmonised with a conducive legal framework and the institutional measures to support implementation.

5.2 *Supportive legal and institutional frameworks: Key success factors*

For any artisanal miner, the benefits of legal operation *must* be greater than costs of licenses, rents, taxes and royalties. However, not only must artisanal miners see advantages from operating legally, they must have the *financial, personal and technical capacity* to fulfil bureaucratic and legal obligations.

Effective legal and institutional models for the minerals sector consider the following:

- *The Diversity and Different Categories of Miners*: Different types of miners have different reasons for choosing (or being obliged) to mine legally or illegally. Understanding *who* miners are, *how* they use the mining, and *what* it brings them is central to designing the right incentive structures for formalisation. Legal frameworks may require more than one category of ASM license to account for the diversity *and* provide a step-up opportunity as activities become more formal. Increased capacity brings increased legal obligations, which in turn should bring increased benefits.
- *Legislate Institutional Mandates*: Many countries outline ambitious and well-crafted policy objectives related to ASM yet do not provide a legal mandate for mining institutions to execute them. ASM support must be formally enshrined in mining authorities' work programs and budgets. Defining institutional roles, functions and obligations of mining departments, units and offices in mining law, and even more specifically in regulations, not only provides a clear, legal mandate, but provides a basis to lobby for and receive funds from Central Government to fulfil this mandate. It can also reduce discretionary powers and aid accountability.
- *Reconciling international, national and local priorities*: Institution of policy measures without sufficient attention to context and consequence can have negative repercussions, potentially exacerbating artisanal miners' poverty and driving the minerals trade deeper underground. Certain policy decisions of government may receive much international acclaim and, by all appearances, be sound. Consideration to the impacts at the village level are essential, however, what is logical for government, may not be logical for miners and may actually impede development.
- *Making artisanal mining laws realistic to the existing structures of production and trade*: Various examples point to the importance of formalising existing structures, rather than attempting to reconfigure artisanal mining into a legal model which may be ideal from a government perspective, but unachievable for artisanal miners. Extensive consultation with artisanal miners is a prerequisite for developing a legal and fiscal structure that is appropriate and ensuring that the miners have a sense of ownership over the process, thereby increasing the likelihood it will be abided by.
- *The Right Institutional Arrangements*: Quasi-independent "authorities" or "commissions" that partially or wholly generate operating funds through their activities are seemingly more likely to fulfil their functions. However, these typically require well-established revenue generation systems to "get off the ground" and effective means to ensure proper use of funds.
- *Regional versus Central Management*: Regional management makes sense in large countries or where there are high numbers of artisanal miners. However, good management can be hindered by a lack of institutional capacity and resources at the local level. Mechanisms to increase government capacity and promote good governance at this level, such as outcome-based performance monitoring of government activities, are therefore important.
- *Administration, Regulation and Enforcement*: Various diamond and gemstone countries, including CAR, have set up dedicated offices to manage exports and collect royalties. This has helped to formalise the export phase but does not necessarily induce the legalisation of mining activities. Achievable license criteria and formalisation incentives coupled with simple mechanisms to report production and track it through to export can help promote official reporting, payment of royalties and taxes and evaluate the effectiveness of changes to fiscal and legal provisions. The Guyanese system is instructive. However, organised, mechanised dredging is much different than practices and arrangements in CAR. In the case of a large number of relatively dispersed miners, Madagascar's *Commune*

system of registration may be a more workable example if local technical and management capacity is sufficiently developed.

5.3 *The final factor*

Based on experiences in other jurisdictions, reduced licensing costs, royalties and export taxes *can* stimulate licensing, particularly if supported by well-conceived mining legislation. However, *any* business must generate more profits than the costs of operation in order to succeed. This principle is no different for artisanal diamond miners. The benefits of legal operation must be greater than costs of licenses, rents, taxes and royalties. However, not only must artisanal miners see advantages from operating legally, they must have the *financial, personal and technical capacity* to fulfil associated bureaucratic and legal obligations.[150]

For these reasons, the harmonisation of practical, achievable fiscal and mining legislation underpins an effective ADM framework but must, in any event, be coupled with institutional support to build miners' capacity to comply and create related incentives. As shown in multiple countries, even given the most well-conceived and perfectly crafted legislation, the biggest challenge to formalisation seems to lie in implementation.

Reduced licensing fees and taxes would mark the first critical step towards achievement of the full development potential of ADM in CAR. Outcomes of these much needed reforms, however, shall ultimately be determined by incentive-focused fiscal and mining legislation, institutional commitment, adequate financing for implementation and, most importantly, accountability for performance.

6 ADDENDUM

The comparative study was widely circulated in CAR government circles. Key personnel in the Ministries of Finance, Trade, and Small Enterprises, the Economic Advisors of the Office of Prime Minister, and key deputies of the National Assembly adopted the view that a lowered license fee would bring in artisanal miners in numbers sufficient to increase the State's revenues and contribute to economic growth. On December 27-28, 2010 a group of diamond collectors met the Minister of Mines who decided that the reduction of two license fees—the miners' as well as the collectors'—would increase diamond production. The Minister took the issue to the President, who agreed. Upon request of the President and the Minister of Mines, the Minister of Finance officially decreased the two licenses in a ministerial order. Effective the first month of 2011 the fee for artisanal mining licenses was reduced by 36% and for collectors licenses by 32%.

ENDNOTES

1 Sebastien Pennes, email to Levin, 6th October 2010. He notes that there are approximately 40,000–90,000 miners; total of 400,000 including artisanal labourers (diggers, washers).

2 Mbendi Information Services 2010a; Pangea Diamond Fields Plc 2010. The *Etats Généraux* workshop proceedings document repeatedly notes that the contribution of diamonds to GDP is about 4%.

3 Mbendi Information Services 2010b; Diamond Development Initiative 2010; Economist Intelligence Unit 2006. The EIU reports that diamonds composed 42% of exports from CAR in 2004.

4 Hinton et al. 2003, pp. 99–115; Hinton 2006.

5 This bureaucracy can include paperwork (and paying for assistance to complete it including preparation of map-sheets, environmental briefs and other requirements). Other costs include transport to government offices and accommodation and food while waiting for applications to be processed.

6 ASARC, 2007.

7 Siegel & Veiga 2007.

8 Hinton 2009. Siegal 2009, Pelon 2005.

9 The study is available at http://usaidlandtenure.net/library

10 The Ministry of Finances, Direction of Taxes, which is responsible for delivering licenses, recorded 1244 one-year licenses issued in seven of the eight mining regions during the period January to September 2010. The *Etats Generaux du Secteur Minier* organized by the Ministry of Mines in 2003, estimated there are 70,000 artisanal miners in CAR.

11 KPCS 2009.

12 ARD 2007, and Data from Ministère des Mines (2010).

13 Chupzei et al. 2009; KPCS 2009.

14 Diamond Development Initiative 2010 and Sebastien Pennes, email to Levin, 6th October 2010.

15 Wardell Armstrong 2008 and Sebastien Pennes, email to Levin, 6th October 2010.

16 Chupzei et al. 2009.

17 Sebastien Pennes, email to Levin, 6th October 2010.

18 Data from Ministère des Mines (2010).

19 *Les Papiers Officiels pour devenir Artisan Minier Legal.*

20 ARD 2007.

21 All translations from the French are the author's own.

22 Sebastien Pennes, comments to authors, 13th October 2010.

23 *Les Papiers Officiels pour devenir Artisan Minier Legal.*

24 IMF 2010.

25 *Les Papiers Officiels pour devenir Artisan Minier Legal.*

26 *Les Papiers Officiels pour devenir Artisan Minier Legal.*

27 Chupzei et al. 2009, after Matip 2003. See page 30.

28 ARD 2010.

29 ARD 2010.

30 ARD 2010.

31 ARD 2010.

32 ARD 2010.

[33] "Kelemba" is a type of "group savings and lending mechanism" which has "proven to be effective in increasing income and security, particularly among women managing small enterprises." ARD 2008, p. 14.

[34] Sebastien Pennes, email to Levin, 6th October 2010. He notes approximately 40,000–90,000 miners and up to 400,000 artisanal labourers (diggers, washers).

[35] Due to lack of reliable data, commonly used multipliers of 2.5 and 2 for indirect and induced labor (*after* Hinton 2009), respectively, were reduced to 1.5 and 1.0. Average number of household dependents in ASM areas was conservatively estimated at 4.6 dependents per miner based on average fertility rate.

[36] Source of Incomes from Chupzei 2009. Some remittances may be sent outside of mining areas, however it is probable that the bulk of expenditures are local to sustain day-to-day needs. Thus, the extent of local economic contributions are more likely to be underestimated than overestimated. Multipliers of economic contribution in ASM areas in other countries are calculated at 2.5. Due to lack of data, a multiplier of 1.5 was used (Priester 2010; Hinton 2009).

[37] Hinton 2006.

[38] Hinton 2009.

[39] Source: Data from Ministère des Mines (2010) based on averages from 2001–2009.

[40] Chupzei et al. 2009, after Matip 2003. See page 30.

[41] *Legalization* refers to the status of becoming legal (e.g. obtaining a license) and complying with law. Legalization of ASM is believed to largely depend on the process of formalization. *Formalization* is founded on the principle that extralegal systems, social arrangements and organizational structures exist for legitimate reasons and the law itself should be an evolving *and enabling* instrument that reflects the changing ways people live (Seigel, 2009). Specifically, formalization is the process of integrating rather than controlling extralegal enterprises by recognizing local arrangements in legislation, reducing barriers to legalization and creating clear benefits from participation in the formal system (Hinton 2009).

[42] Wardell Armstrong 2008.

[43] KPCS 2009.

[44] DRC's artisanal mining sector is governed by the national Mining Code (*Code Miniere*), law No. 007/2002 of July 11, 2002, and the protocol for its application contained in Decree 038/2003 of 26 March 2003. (DRC 2003; Pole Institute 2010).

[45] Partnership Africa Canada and CENADEP 2007.

[46] PACT 2010.

[47] World Bank 2008.

[48] KPCS 2009; Bank of Guyana 2009.

[49] Shawn Blore, email to Jennifer Hinton, 4th October 2010.

[50] Blore 2006.

[51] Shawn Blore, email to Jennifer Hinton, 4th October 2010.

[52] Elbow 2010.

[53] Blore 2006.

[54] Elbow 2010.

[55] KPCS 2009. Note that an average per carat value of US$400 is very high for Liberia (and for alluvial producers generally). Average value was just US$122.47 in 2007 and US$210.43 in 2008. Neighbouring Sierra Leone, which is traditionally understood to have a higher quality of gem than Liberia, had an average value of US$234.50 in 2007 (making it the highest amongst alluvial producers in Africa), and US$266.05 in 2008. KPCS 2007 and 2008.

[56] Garrett & Lintzer 2010; Garrett & Carstens 2008.

[57] Garrett & Carstens 2008. In 2008, the MLME issued just 877 Class C licenses. (Temple 2010).

[58] Temple 2010.

[59] Temple 2010.

[60] Temple 2010.

[61] Temple 2010.

[62] Hinton 2010a.

[63] Samuel Koroma, Government Gold and Diamond Office, telephone call with Levin, 4th October 2010.

[64] Andrew Keili, CEMMATS, email to Levin, 5th October 2010. Under the new Act, total cost is 550,000 Leones per half acre (US$143), which is nearly US$300 per acre.

[65] GoSL 2005. Licensing fees for an artisanal or cooperative mining lease per acre, per year: Le200,000. Other fees: rehabilitation fees (Le200,000), Mine's Manager's certificate (Le100,000), monitoring fees (Le100,000); and payments to traditional authorities, including surface rent of around Le100,000.

[66] Andrew Keili, CEMMATS, email to Levin, 5th October 2010. To get their license, artisanal miners must also pay "surface rent" to local chiefdom authorities, the price of which varies by chiefdom and according to the social position of the miner. In cases where miners choose not to get a mining license from central government, they generally pay surface rents to "formalize" their activities at the local level, although their activities remain technically illegal.

[67] Estelle Levin, interviews with GOSL officials and artisanal miners 2007, 2009.

[68] Based on June exchange rate of 3850 Leones = US$1. (Andrew Keili, CEMMATS, email to Levin, 5th October 2010).

[69] Andrew Keili, CEMMATS, email to Levin, 5th October 2010.

[70] Levin 2005, p. 79.

[71] U.S. Geological Survey 2008a.

[72] KPCS 2008; MaFCFArlane, M. 2008; Levin, Mitchell & MaFCFArlane 2008, Annex 12.

[73] Tanzania Ministry of Finance and Economic Affairs 2008.

[74] Mining Journal 2008.

[75] KPCS 2008; MaFCFArlane, M. 2008; Levin, Mitchell & MaFCFArlane 2008, Annex 12.

[76] MFCFArlane 2008.

[77] MFCFArlane 2008 citing Progress report Feb 2007.

[78] MFCFArlane, 2008.

[79] Levin 2007 and based on further research in Madagascar in 2008; ONG Green 2006.

[80] U.S. Geological Survey Minerals Yearbook 2005.

[81] Levin 2007.

[82] ONG Green 2006.

[83] Hinton 2009.

[84] Hinton 2009.

[85] Based on the monthly incomes of an average Ugandan gold miner, the ratio of license cost to miner's monthly incomes in Uganda is approximately 2.3.

[86] Yakubu 2003.
[87] Yakubu 2003.
[88] Yakubu 2003.
[89] Aryee 2003.
[90] Azameti 2003.
[91] *Trade and Environment* 1994.
[92] U.S. Geological Survey 2008.
[93] CASM 2008.
[94] Medina 2003; Hruschka 2003.
[95] Hruschka 2003, p. 24.
[96] Hruschka 2003.
[97] Bugnosen 2004.
[98] Ibid.
[99] Bugnosen 2004; Caballero 2004.
[100] Caballero 2004.
[101] Blore 2008; Shawn Blore, email to Jennifer Hinton, 4th October 2010.
[102] ONG Green 2006.
[103] Dharmaratne 2004.
[104] Dharmaratne 2004.
[105] In 2005, nearly 2400 artisanal mining licenses were granted in Sierra Leone (Levin & Gberie 2006). At the time, total ADM population was estimated at 200,000, with the vast majority (perhaps 150,000) being in Kono District, the ADM heartland. (Levin 2005) ADM licences cover up to 50 laborers; a typical number would be closer to 30, giving a rate of 48% legality.
[106] Based on interviews with artisanal miners and dealers in Sierra Leone by Estelle Levin 2004, 2006, 2007.
[107] Levin 2007.
[108] Hilson 2002.
[109] Hinton 2009.
[110] Hinton 2009.
[111] Dreschler 2001.
[112] Dreschler 2001.
[113] Malango 2004.
[114] Hayes & Van Wauwe 2009.
[115] Hayes & Van Wauwe 2009.
[116] Hinton 2009.
[117] Okedi 2010.
[118] International Study Group 2010.
[119] International Study Group 2010.
[120] National Advocacy Coalition on Extractives 2009, p. 34.
[121] National Advocacy Coalition on Extractives 2009.
[122] National Advocacy Coalition on Extractives 2009.
[123] Bugnosen 2004.
[124] Bugnosen 2004.
[125] Hinton et al. 2003.
[126] Hinton 2010b.
[127] Blore 2008.
[128] Chupzei et al. 2009, p. 24.
[129] Hinton 2003.
[130] Jacques et al. 2002. For further information on the relationship between mechanisation and formalization, see Priester et al. 2010.
[131] Hinton 2009.
[132] Personal Communication, Government Officer (anon), Ministry of Energy and Minerals of Tanzania.
[133] Hinton 2009.
[134] Hilson & Clifford 2010.
[135] Personal Communication, Government Officer (anon), Ministry of Energy and Minerals of Tanzania.
[136] Temple 2010.
[137] Mitchell, Garrett & Levin 2008.
[138] Mitchell, Garrett & Levin 2008.
[139] Hruschka 2003.
[140] Levin, interviews with local authorities in Madagascar, 2007, 2008.
[141] Precious Minerals Marketing Company Ltd. 2010.
[142] Precious Minerals Marketing Company Ltd. 2010.
[143] Samuel Koroma, Government Gold and Diamond Office, telephone call with Levin, 4th October 2010.
[144] Global Witness 2008; Mitchell, Garrett & Levin 2008.
[145] The same issue of supply chain traceability applies to tin, tantalum, and tungsten ore and gold exports from DRC, and has inspired a number of regional and international initiatives to develop chain of custody systems from mine to smelter with a view to curbing illegal mining. A Regional Certification Mechanism to track minerals from source and aid the collection of revenues by national governments is being developed by the Intergovernmental Conference for the Great Lakes Region. See Levin, 2010 and Blore and Smillie, 2010.
[146] After Blore 2008.
[147] Hinton, J. 2010a.
[148] Mbendi Information Services 2010a.
[149] ARD 2007.
[150] Hinton et al. 2003.

REFERENCES

ARD 2007. *The Mining Sector in Central African Republic with a Focus on Information Collection and Management Systems and Procedures*, Unpublished Report.

ARD 2010. Note de Travail, Discussions Préliminaires relatives à la Question de la Patente d'Artisan Minier. *Salle de Réunion du Ministère des Mines, vendredi 10 septembre 2010, 10h. Internal Document.*

Aryee 2003. Overview of artisanal mining and its Regularisation in Ghana. *Presentation at the Second CASM Annual General Meeting, Elmina, Ghana, Sept 8–10, 2003.* Retrieved from www.casmsite.org/programmes_learning_Elmina.htm

Azameti, E. 2003. Regularisation—a view from the bottom. *Presentation at the Second CASM Annual General Meeting, Elmina, Ghana, Sept 8–10, 2003.* Retrieved from www.casmsite.org/programmes_learning_Elmina.htm

Bank of Guyana 2009. *Annual Report*. Georgetown, Guyana: Bank of Guyana: p. 115 Retrieved from http://www.bankofguyana.org.gy/bog/images/Reports/annrep2009.pdf

Blore, S. 2006. Triple Jeopardy: Triplicate Forms and Triple Borders: Controlling Diamond Exports from Guyana. *PAC occ. Paper*, 14.

Blore, S. 2008. The Misery and the Mark-up: Miners' wages and diamond value chains in Africa and South America. In: Koen Vlassenroot and Steven Van Bockstael (Eds).: *Artisanal diamond mining. Perspectives and challenges.* Egmont. Gent: Academia Press: 66–92.

Blore, S. & Smillie, I. 2010. *An ICGLR-Based Tracking and Certification System for Minerals from the Great Lakes Region of Central Africa*. Partnership Africa Canada. Ottawa, Canada March 2010.

Bugnosen, E. 2004. Small Scale Mining Legislation in Philippines: A General Review. *Presentation at the Third CASM Annual General Meeting, Colombo, Sri Lanka, Oct 12–14, 2004*. Retrieved from www.casmsite.org/programmes_learning_Colombo.htm

Caballero, E. 2004. Philippines: Traditional Artisanal Gold Mining In Indigenous Communities And Their Current Concerns. *Presentation at the Third CASM Annual General Meeting, Colombo, Sri Lanka, Oct 12–14, 2004*. Retrieved from www.casmsite.org/programmes_learning_Colombo.htm

CASM 2008. Small Stories: 12 Stories About Small-Scale Mining. *Communities and Small-Scale*. Washington, DC.

CASM 2008. Short Stories. World Bank: 32 p. Washington D.C. Retrieved from: http://www.artisanalmining.org/userfiles/file/CASMshortstoriesBooklet_FINAL_low.pdf

Dharmaratne, P. 2004. Legal, Fiscal, Institutional And Infrastructural Development Of The Gem Industry In Sri Lanka. *Presentation at the Third CASM Annual General Meeting, Colombo, Sri Lanka, Oct 12–14, 2004*, Retrieved from www.casmsite.org/ programmes_learning_Colombo.htm

Diamond Development Initiative 2010. *Property Rights and Artisanal Diamond Development (PRADD) Pilot Project*. Retrieved from http://www.ddiglobal.org/login/Upload/mod5PRADD.pdf

Dreschler, B. 2001. *Small-scale Mining and Sustainable Development within the SADC Region*, Commissioned by MMSD: 165p. Retrieved from http://www.iied.org/mmsd/

Economist Intelligence Unit 2006. *Central African Republic: Mining*. London: EIU Ltd.

Elbow, Kent 2010. *Memo on the Diamond Development Initiative: Workshop on the Guyana Model*. p. 7.

Etats Généraux workshop proceedings document.

Garrett, N. & Carstens, J. 2008. Implementing Transparency in the Artisanal and Small Scale Mining Sector. *Report to BGR*: p. 122.

Garrett, N. & Lintzer, M. 2010. Analysis of Past and Current Interventions in the artisanal diamond mining sector in Liberia. In: K. Vlassenroet & S. Van Bockstael: *Forthcoming book*. Egmont.

Global Witness 2008. *Loupe Holes: Illicit Diamonds in the Kimberley Process, 28th October 2008*. London: Global Witness. Retrieved from http://www.globalwitness.org/media_library_detail.php/674/en/loupe_holes_illicit_ diamonds_in_the_kimberley_proc

Hayes, K. & Van Wauwe, V. 2009. Microfinance in Artisanal and Small Scale Mining. *Background Papers: 9th Annual CASM Conference, Communities and Small Scale Mining*: 1–8.

Hentschel, T. et al. 2001. *Global Report on Artisanal and Small Scale Mining, Report commissioned by the Mining, Minerals and Sustainable Development Project*. IIED and WBCSD publication.

Hentschel, T., Hruschka, F. & Priester, M. 2002. *Global Report on Artisanal and Small Scale Mining*. Report commissioned by the Mining, Minerals and Sustainable Development Project, IIED and WBCSD publ.: p. 67.

Hilson, G. 2002. A Contextual Review of the Ghanaian Small Scale Mining Sector. *MMSD/IIED Report*, 76: p. 29.

Hilson, G. 2008. Mining and Rural Development: The trajectory of diamond production in Ghana. In: K. Vlassenroet & S. Van Bockstael (2008) *Artisanal diamond mining: perspectives and challenges*. Egmont Institute: Belgium.

Hilson, G. & Clifford, M.J. 2010. A 'Kimberley protest': Diamond mining, export sanctions, and poverty in Akwatia, Ghana,. *African Affairs*, 109, 436: 431–450.

Hinton, J. 2003. Organisation and Formalisation of Artisanal Mining in the Tapajos Region, Brazil. *Report to CETEM/IDRC*: p. 55.

Hinton, J. 2009. National Strategy for the Advancement of Artisanal and Small Scale Mining in Uganda. *Report to Ministry of Energy and Mineral Development.*

Hinton, J. 2010a. Recommendations for Class C Mining Regulations and their Implementation and Future Amendments to the Mining and Minerals Act in Liberia. *Report to USAID Governance and Economic Management Project*: p. 90.

Hinton, J. 2010b. National Guidance Strategy for the Promotion of Gender Equity in Mining. *Report to Ministry of Energy and Mineral Development*: p. 70.

Hinton, J.J. 2006. *Communities and Small Scale Mining: An Integrated Review for Development Planning.* In Press, CASM (World Bank) publ: p. 214.

Hinton, J.J., Veiga, M.M. & Veiga, T. 2003. Clean Artisanal Mining Technologies, A Utopian Approach? *Journal of Cleaner Production*, 11, 2: 99–115.

Hinton, J.J., Veiga, M.M. & Beinhoff, C. 2003. Women and Artisanal Mining: Gender Roles and the Road Ahead, Ch. 11. *Socio—economic Impacts of Artisanal and Small—scale Mining in Developing Countries*. Balkema, ed. G. Hilson, Rotterdam.

Hruschka, F. 2003. *Review of the Organisation Process of the Artisanal Miners in Peru (2000–2002)*. Unpublished.

International Monetary Fund 2010. *Central African Republic Country Report, No.10/21*. Washington, DC: IMF.: p. 75 Retrieved from http://www.imf.org/external/pubs/ft/scr/2010/cr1021.pdf

International Study Group 2010. *Africa's Mining Regimes: Framework Report.* AU and UNECA, unpublished report: p. 217.

Jacques, E., Zida, B., Billa, M., Greffi é, C. et al. 2002. Artisanal And Small-Scale Gold Mines In Burkina Faso: Today And Tomorrow. *Unpubl. BRGM report*: p. 18.

Jha, Raghbendra 2007. Fiscal Policy in Developing Countries: A Synoptic View. *ASARC Working Paper 2007/01*. Australian National University, Australia and South Asia Research Centre: p. 39 Retrieved from http://ideas.repec.org/p/pas/asarcc/2007-01.html

KPCS 2007. *Annual Global Summary: 2007 Production, Imports, Exports and KPC Counts*. Retrieved from https://kimberleyprocessstatistics.org/static/pdfs/AnnualTables/2007GlobalSummary.pdf

KPCS 2008. *Annual Global Summary: 2008 Production, Imports, Exports and KPC Counts*. Retrieved from https://kimberleyprocessstatistics.org/static/pdfs/AnnualTables/2008GlobalSummary.pdf

KPCS 2009. *Annual Global Summary: 2009 Production, Imports, Exports and KPC Counts*. Retrieved from https://kimberleyprocessstatistics.org/static/pdfs/AnnualTables/2009GlobalSummary.pdf

Levin, E. 2005. *From Poverty and War to Prosperity and Peace? Sustainable Livelihoods and Innovation in Governance of Artisanal Diamond Mining in Kono District, Sierra Leone*. University of British Columbia, Vancouver.

Levin, E. 2006. Reflections on the Political Economy of Artisanal Diamond Mining in Sierra Leone. In Gavin Hilson (ed.): *Small Scale Mining, Rural Subsistence, and Poverty in West Africa. Experiences from the Small-scale Mining Sector.* Intermediate Technology Development Group Publishing.

Levin, E. 2007. *Ethical Assessment of Artisanally-Mined Gold. Antanimbary, Madagascar,* May 2007. Urth Solution. Unpublished report.

Levin, E. 2008. *Scoping Study for Fairtrade Artisanal Gold: Tanzania*. Unpublished report. Medellin: Association for Responsible Mining.

Levin, E. 2010. *Mineral Certification Schemes in the Great Lakes Region: A Comparative Analysis*. Bujumbura: GTZ.

Levin, E. & Gberie, L. 2006. *Dealing for Development? A Study of Diamond Marketing and Pricing in Sierra Leone. FULL REPORT.* March 2006. Ottawa: PAC and DDI. At http://www.diamondfacts.org/pdfs/media/perspectives/case_studies/DDI_A_Study_of_Diamond_Marketing_and_Pricing.pdf

Levin, E., Mitchell, H. & MaFCFArlane M. 2008. *Feasibility Study for the Development of a Fair Trade Diamond Standard and Certification System*. Unpublished report. Transfair USA.

MaFCFArlane, M. 2008. Annex 12. Mwadui Community Diamond Partnership. Case study report for Transfair USA. In: Levin, E., Mitchell, H. & MaFCFArlane M. 2008: *Feasibility Study for the Development of a Fair Trade Diamond Standard and Certification System.* Unpublished report. Transfair USA.

Malango, V. 2004. *Country Gemstone Experiences: Namibia*, Presentation at the Third CASM Annual General Meeting, Colombo, Sri Lanka, Oct 12–14, 2004. Retrieved from www.casmsite.org/ programmes_learning_Colombo.htm

Mbendi Information Services 2010a. *Mining in Central African Republic- Overview.* Retrieved 25 September 2010 from http://www.mbendi.com/indy/ming/af/cr/p0005.htm

Mbendi Information Services 2010b. *Diamond Mining in Central African Republic–Overview.* Retrieved from http://www.mbendi.com/indy/ming/dmnd/af/cr/p0005.htm

Medina, G. 2002. *Peru's New Artisanal Mining Law*, Presentation to the First CASM Annual General Meeting, Ica, Peru, Sept. 23–28, 2002, www.casmsite.org/ programmes_learning_Ica.htm

Mining Journal 2008. Political debate in Tanzania—A supplement to Mining Journal. *Mining Journal*, August: 5.

Mitchell, H. Garrett, N. & Levin, E. 2008. *Regulating Reality: Reconfiguring Approaches to the Regulation of Artisanally Mined Diamonds*. Egmont Institute: Belgium. At http://www.resourceglobal.co.uk/index.php?option=com_docman&task=doc_details&gid=46&Itemid=41

National Advocacy Coalition on Extractives (NACE) 2009. *Sierra Leone at the crossroads: Seizing the chance to benefit from mining*. Retrieved from http://www.nacesl.org/NACE_Sierra%20 Leone.pdf

Okedi, J.P.O. 2010. *Mineral Investment for Sustainable Development in Uganda, An analysis of issues and challenges in the Fiscal and Regulatory Regimes.* Research Paper, University of Sydney, Australia. Unpubl. p. 22.

ONG Green 2006. *L'experience d'Antanimbary dans le Cadre du Projet de Gouvernance des Ressources Minerales (PGRM): Appui à la Gestion Décentralisée des Ressources Minerales de la Commune Rurale D'Antanimbary'Maevatanana.* Presentation given at the 6th Annual CASM Conference, Antananarivo, 11th November, 2006.

Otchere, F., Veiga, M.M., Hinton, J.J., Farias, R. & Hamaguchi, R. 2004. Transforming Mining Open Pits Into Fish Farms, *Natural Resource Forum*, 28: 216–228.

PACT 2010. *PROMINES Study: Artisanal Mining in the Democratic Republic of Congo.* Retrieved 4th October 2010 http://pactworld.org/galleries/resource-centre/PROMINES%20Report%20English.pdf

Pangea Diamond Fields Plc. *Central African Republic: Political and Economic Climate & Mineral Industry, Policy and Regulations*. Retrieved from http://www.pangeadiamondfields.com/car_profile.htm

Partnership Africa Canada and CENADEP 2007. *Diamond Industry Annual Review: Democratic Republic of the Congo.*

Pole Institute 2010. *Blood Minerals: The Criminalization of the Mining Industry in Eastern DRC.* Goma: Pole Institute.

Precious Minerals Marketing Company Ltd. 2010. *Diamond Production and Marketing Services*. Retrieved from http://www.pmmcghana.com/diamondproductionandmarketing.html

Priester, M., Levin, E., Carstens, J., Trappeniers, G. & Mitchell, H. 2010. *Mechanisation of Artisanal Alluvial Diamond Mining*. Diamond Development Initiative.

Seab Gems Ltd. *Mining Opportunities in Tanzania: Primary Prospecting License.* Retrieved from http://seabgems.com/Mining%20Opportunities%20Tanzania/Primary%20Prospecting%20 Licence.html

Siegel, S. 2007. *The Needs of Miners: Political Ethics, Mercury Abatement, and Intervention in Artisanal Gold Mining Communities.* PhD Dissertation, University of British Columbia. p. 226.

Siegel, S. & Veiga, M. 2009. Artisanal and small-scale mining as an extralegal economy: De Soto and the redefinition of formalisation. *Resources Policy*, 34, 1–2: 51–56.

Tanzanian Ministry of Finance and Economic Affairs 2008. *The economic survey 2007.* Dar Es Salaam, Tanzania: Ministry of Finance and Economic Affairs, p. 208.

Temple, P. 2010. *The Legal and Non-legal Framework*. In: K. Vlassenroet & S. Van Bockstael, Forthcoming book. Egmont Institute.

Tieguhong, J.C., Ingram, V. & J. Schure 2009. *Study on impacts of artisanal gold and diamond mining on livelihoods and the environment in the Sangha Tri-National Park (TNS) landscape, Congo Basin.* CIFOR and IUCN. Yaoundé, Cameroon.

Trade and Environment 1994. *Case Study: Peruvian Mining*, Case 357. Retrieved from http://www1.american.edu/TED/perumine.htm

U.S. Geological Survey 2005. The Mineral Industry of Madagascar. *U.S. Geological Survey Minerals*

Yearbook–2005. Retrieved on 11 April 2007 from http://minerals.usgs.gov/minerals/pubs/country/2005/mamyb05.pdf
U.S. Geological Survey 2008a. *Minerals Year Book Tanzania*. Retrieved from http://minerals.usgs.gov/minerals/pubs/country/2008/myb3–2008-tz.pdf
U.S. Geological Survey 2008b. *Minerals Yearbook: Peru*. US Geological Survey publ., p. 15.
UNECA 2002. *Compendium on Best Practices in Small-scale Mining in Africa*. UNECA publ., Addis Ababa, Ethiopia: p. 112.
USAID 2008. *Property Rights and Artisanal Diamond Development Pilot Program—Central African Republic. Progress Report (May–July 2008)*. September 2008. Unpublished report.
Veiga, M.M. & Hinton, J.J. 2002. Abandoned Artisanal Gold Mines in Latin America: A Legacy of Mercury Pollution, *Natural Resources Forum*, 26: 13–24.
Wardell Armstrong 2008. In: K. Vlassenroet and S. Van Bockstael (2008) *Artisanal diamond mining: perspectives and challenges*. Egmont Institute: Belgium.
World Bank 2008. *Democratic Republic of Congo: Growth with Governance in the MiningSector.* Retrieved 4 October 2010 from http://siteresources.worldbank.org/INTOGMC/Resources/336099-1156955107170/drcgrowthgovernanceenglish.pdf
Yakubu, B.R. 2003. *Regularisation—technical approach and its shortcomings.* Presentation at the Second CASM Annual General Meeting, Elmina, Ghana, Sept 8–10, 2003, www.casmsite.org/programmes_learning_Elmina.htm

COUNTRY LEGISLATION

Central African Republic

Mining Code 2009 (RCA Code Minier 2009)—Law n°9-005 from April 29, 2009.
Ordonnance n°83.024 du 15 mars 1983 fixant les conditions de possession et de détention et réglementant l'exploitation et le commerce de l'or et des diamants bruts; Papiers officiels [TBC].

Democratic Republic of the Congo

DRC Mining Code (RDC Code Minier 2002)—LAW No. 007/2002 of JULY 11, 2002.
DRC Mining Regulations 2003 (Reglement Minier 2003)—DECREE NO 038 / 2003_OF 26 March 2003.
Family Code (Code de la famille)—Law No. 073/84 of 17 October 1984.
Handbook of Environmental Legislation, 2003a.
Guide for the Mining Investor, 2003.

Ghana

Small Scale Gold Mining Law 218,1989,PNDC Law 218.
The Mercury Law, 1989, PNDC L 217.

Liberia

Mining and Minerals Law 2000.
Minerals and Mining Law 2000, Chapter 40 amendment passed in 2004.
Mining and Minerals Act 2006, Act 703 of 31 March, 2006.

Madagascar

Mining Code 1999, Law No. 99 022 of August 30, 1999.
Mining Regulations 2000, Law No 2000-170, May 15, 2000.
Revised Mining Code 2005, Law No. 2005-021 of October 17, 2005.

Mozambique

Mining Development Fund Statute (Decree nr. 17/2005) dated 06/24/2005.

Namibia

Minerals Development Fund of Namibia Act of 1996.

Nigeria

Minerals and Mining Act 2007.

Peru

Law 27651, 2002, "Formalisation and Promotion of Small Scale and Artisanal Mining;" General Mining Law of Peru, Supreme Decree No. 014-92-EM, Official Gazette 4 June 1992.

Philippines

People's Small-scale Mining Act of 1991—Republic Act No. 7076.
Mining Act of 1995 - Republic Act No. 7942.
PD 1150 - Gold panning and sluicing regulation.
PD 1899 - Mining of small deposits.
RA 7076 - Identification and segregation of ASM zones.
AO No. 97-30 - ASM mine safety rules.

Sierra Leone

Regulations to the Mines and Minerals Act 2009.
The Mines and Minerals Act 2009 in Supplement to the Sierra Leone Gazette Vol. CXLI, No. 3, January 7, 2010.
Schedule of Fees, First Schedule: Fees, Expenditures and Penalties, 2009.
Details of Policy Measures relating to Small Scale and Artisanal Mining and Marketing of Precious Minerals" 2005.

Tanzania

Mining Act 2010, Act No. 14/10 from April 24, 2010.
The Mineral Policy of Tanzania 2009, The Ministry of Energy and Minerals, July 2009.
The Mineral Policy of Tanzania 1997, The Ministry of Energy and Minerals, October 1997.

Uganda

Mining Act 2003.
Mining Regulations 2004, Statutory Instruments Supplement No. 38 to The Uganda Gazette No. 57 Volume XCVII dated 5th November, 2004 Printed by UPPC, Entebbe.
Mining Regulations 2004, First Schedule.

Geological Resources and Good Governance in Sub-Saharan Africa – Runge & Shikwati (eds)
© 2011 Taylor & Francis Group, London, ISBN 978-0-415-58267-4

Reflections on capacity building for women in small-scale mining within the CEMAC zone

I. Boukinda
CEMAC/Department of Human Rights, Good Governance and Human and Social Development, Bangui, Central African Republic

J. Runge
Department of Physical Geography, Goethe-University Frankfurt & CIRA, Centre for Interdisciplinary Research on Africa, Frankfurt am Main, Germany

ABSTRACT: Within the CEMAC zone and equally in other Sub-Saharan regions, artisanal and nowadays increasingly commercial mineral exploitation has a critical influence over human development. That's why mining cooperatives and international companies play an important role in areas affecting gender equality and the rights of women. There is evidence that the effects on rural communities affected by mining often represent a direct denial of their basic human rights, especially their rights to prior, free and informed consent, self determination, land use and livelihoods. Women are particularly hit by mining projects in the informal, artisanal small-scale, as well as in the semi-professional to professional mining sector. Facing the third MDG goal some reflections on CEMAC's responsibility and ownership in this important forward-looking economic field are highlighted and propositions are made how women could be best supported by a project in small-scale mining.

1 INTRODUCTION

The third Millennium Development Goal (MDG) is to 'promote gender equality and empower women'. This can be seen not only as a goal to be achieved, but also as a strategy and an effective way of alleviating poverty and promoting sustainable development. But despite the declarations and commitments of governments worldwide in favour of gender equality, progress has been hesitant, even slow.

Women have always played a predominant role in the social organisation of mining in African countries in general and in the CEMAG zone in particular, and the proportion of women is significantly high in small-scale mining. These women generally work in very difficult conditions and are consequently exposed to the risk of contamination.

Mining villages usually sprout up spontaneously in remote areas close to extraction sites and are often of a makeshift nature. Isolated sites in the bush do not have health centres or medical staff. They often face problems of malnutrition, alcohol, drugs, a rapid decline in morals at some sites (prostitution, drug-taking, criminality, sexually transmitted diseases, etc.). Added to this, the food situation is insecure, and family authority and structures are weakened and disrupted.

2 THE ROLE AND WORK OF WOMEN AT MINING SITES

In Africa, women play an essential role in society, particularly in rural areas where the majority of small-scale mining takes place. They manage the production of food, the procurement of household fuel and trading in basic commodities.

However, the key role played by women has not been recognised in the field of small-scale mining. Traditionally in Africa, it is the men who often assume the role of head of the family, while the women assume that of mother and wife. This distribution of roles leads to a gender-based division of labour. Furthermore, in their role as mothers, women are responsible for providing food, water and fuel and for running the household. They are also responsible for looking after children and elderly members of the family. All these tasks limit their ability to venture beyond the confines of the family and make it very difficult for them to carry out regular mining work.

Some of those who are paid in gold have diversified their activities to include trading in gold and have become increasingly involved in mining activities, working as dealers in mining regions. But it is not easy for women to relocate to mining sites because this can have very severe repercussions on the family structure, affecting family and community life and creating problems of isolation.

The women who own small-scale mines usually have no mining experience. They have often come into contact with the sector through relatives or because they live near to mining areas. Their lack of expertise and knowledge of the mining industry often represent an obstacle to the effective and constant management of their mining activities.

3 WOMEN'S LEVEL OF INVOLVEMENT IN SMALL-SCALE MINING

The extent to which women are involved varies from country to country, and depends on the ore being mined, the type of processing taking place and also certain socio-economic factors. Although they are under-represented in the large-scale mining industry, women play an important role in the small-scale industry. Statistics on small-scale mining rarely provide reliable data on the direct involvement of women in extracting and processing the ore or their indirect involvement as suppliers of goods and services.

Estimates on the level of female participation in small-scale mining are vague, but sufficiently consistent to indicate that women play an essential role that varies depending on the region and the work carried out. In some countries, women hold permits or are entrepreneurs. In most countries, a large proportion belong to less structured groups, and the vast majority are employed to do menial tasks that require no skills, in particular processing ore or providing goods and services to mines and mining communities.

The increase in the number of women involved in small-scale mining is a result of several factors. These include the impact of structural adjustment programmes, the closure of state-owned mines, low basic commodity prices, a shortage of jobs in the public and private sector, trade and commerce, agriculture and inflation, which have resulted in many people, particularly women who previously relied on subsistence farming, looking for other paid work or an extra job to improve their quality of life or, more frequently, just to survive. In rural areas, working in small-scale mining is often the only way of earning money without having to move to the nearest town. Most of the people who work in mining do so through necessity.

Although in small-scale mining some women perform 'high-level' functions as entrepreneurs, owners, employers, equipment hirers or dealers in gold or precious stones, in large-scale mining they are in a minority.

Women usually transport the ore, load it, crush it, sieve it, wash it and provide the mine and the mining community with goods and services.

4 PROBLEMS EXPERIENCED BY WOMEN IN MINING

Small-scale mining can provide long-term work for women, but only if the conditions are right. Despite efforts made to improve the sector, the issue of inequality between men and women still exists. Consequently, women experience additional problems when attempting to increase the extent of their participation. Regardless of the region or type of mining, women who work or want to work in small-scale mining doing anything other than menial tasks encounter the same problems. These include:

- A lack of access to credit and funding and less technical expertise than their male colleagues. Since they are not regarded as heads of household, women own neither goods nor revenue in their own right and have great problems obtaining credit. Even if they succeed, they usually need permission from the head of the family (a man). Without credit, women are generally confined to mining activities that just about enable them to survive.
- A lack of representation and support. Legal and financial obstacles hinder women from participating in small-scale mining or restrict them to poorly paid activities often carried out in an unstructured or even illegal context.
- A lack of management and administrative skills. Long-winded, complex administrative procedures, a lack of technical knowledge and, in many cases, a lack of basic education prevent women from obtaining credit. In addition, financial institutions, which are not accustomed to giving women loans, err on the side of caution, and women living in rural areas are usually not even aware of the existence of development agencies that provide credit guarantees.
- The fact that the majority of women in rural areas are illiterate is a handicap to performing many activities, particularly those involving the completion of mining-related administrative formalities.
- Time constraints are another unfavourable factor for women, who usually have three demands on their time: household chores, childcare and helping to support their families.
- The division of power that is traditional in Africa often relegates women to household chores and a position of inferiority. This reduces their mobility and independence. In some regions, they need their husband's permission to perform activities outside the family home.

5 SOME SOLUTIONS TO THE PROBLEMS IDENTIFIED

In view of the significant number of women working in small-scale mining and the problems they face, there is all the more need to be aware of the contribution that women make or could make. This objective can only be achieved if there is more and better-quality information, which should be gathered with the help of all those concerned. Measures required include the implementation of programmes developed jointly with women miners, public bodies, NGOs and lending institutions to address the following areas:

- Training in management skills. Their lack of education means that women are unable to benefit from the growing number of technical assistance and financial aid programmes aimed at this sector. Such assistance could perhaps be better focused to more closely meet the needs of women.
- Creating infrastructure to improve the health of women and children. In view of the essential role women play at a family level, it is vital that sanitation be improved to take account of the environment and the situation at the workplace and in the community. Setting up schools would also have a very powerful, direct and immediate impact on improving living conditions for women. In addition, improving hygiene and safety in mines should be encouraged, and steps taken jointly with the public authorities, the private sector and NGOs to slow the spread of sexually transmitted diseases, particularly HIV.
- Setting up cooperatives is not necessarily the best way of increasing permanent activity; people used to working independently may have trouble adapting to the discipline of cooperatives. Solutions to this issue should be thought through carefully.
- Raising the awareness of women and the community about small-scale mining and improving conditions for women in the home.

6 THOUGHTS CONCERNING A PROJECT ON WOMEN IN SMALL-SCALE MINING IN THE CEMAG ZONE

In the mining sector, there are very few targeted surveys or socio-specific data to enable a full, systematic analysis to be made of the needs of men and women. Statistics on small-scale mining rarely reveal the extent of women's participation in extracting and processing ore or as suppliers of goods and services. We are therefore unable to provide a reliable picture of the situation of women in mining, since their needs vary from country to country. However, it is evident that women have always played an important role in small-scale mining.

In this context, an CEMAG project aimed at this sector should include funding for data collection in order to conduct a country-by-country analysis of statistics on women in mining. This would help us to identify the areas and needs of women involved in this activity. A study is therefore needed if we are to have a minimum of information on women working in the mining sector and also on NGOs working to improve the conditions of women in mining.

The results of this study could lead to strategies being implemented and support being provided for technical training or simply literacy training for women to strengthen their capacity in this field.

Such a project could focus on three aspects: resources, training, technical advice.

6.1 *Mission*

The purpose of the project is to help women working in mining become better organised to access the financial, technical and marketing services they need to carry out mining activities that are economically and commercially viable and environmentally sustainable and to improve the living standards of women performing associated activities, which will also be developed, such as small-scale trading.

The project will play an important proactive role in regulating and improving small-scale mining, the living conditions of the population and the quality of the rural environment. It will achieve this by improving small-scale mining activities in a climate of gender equality and favourable technical, environmental, legal and social conditions.

6.2 *Objectives*

The project aims to introduce support systems for small-scale miners. This support would take the form of a regular information service for small-scale mines, providing advice on extraction and processing methods, training, the environment, health and safety issues and market access.

6.3 *Specific objectives*

The project aims to promote the development of female miners, not only by setting up women's cooperatives, but also through education, professional training and relevant management practices.

Steps have been taken to establish a regional association aiming to protect these women's interests. It would be affiliated to national associations and should help to raise awareness and promote measures throughout the region to improve the role of women in small-scale mining.

It will exert pressure with a view to mobilising support and gaining recognition for women in mining at a national, regional and international level.

It will identify the problems women miners face with regard to training, technology and literacy and mobilise the necessary resources to meet these needs.

It will organise training workshops for small-scale craftswomen and set up a regional branch to organise campaigns to raise awareness and provide focus groups with knowledge and advice.

It will identify income-generating activities at mining sites and infrastructure projects involving, for example, wells, dykes, schools and clinics.

It will promote networking with public and private companies and with development partners.

6.4 *Expected results*

- Assistance has been provided to small-scale mining and industry to encourage more women to participate in the mining sector. Support measures have been implemented to develop income-generating activities for women.
- State organisational and institutional capacity has been strengthened based on a vision to boost women's participation in mining and so reduce poverty.
- Resource people are trained to promote gender integration in the national mining sector.

The anticipated results of this project include increased support for gender awareness by teams and partners and the development of partners concerned with women working in the mining sector at both the country and sub-regional level.

7 CONCLUSION

Any project focusing on women in small-scale mining must factor in the priorities needed to achieve greater involvement by women in these activities in order to ensure sustainability. It should therefore encourage more women to participate in the sector as entrepreneurs. This would involve taking account of all the following factors: resources, training, technical support and the environment in which these women work, nationally and sub-regionally. This will allow women to earn more money to finance other long-term activities, contribute to a sustainable economy and help reduce poverty.

REFERENCES

Ardayfio-Schandorf, E. & Wrigley, C. 2001. *Women in natural resources management in Ghana*. UNU/INRA.

Dreschler, B. 2001. Small scale mining and sustainable development within die SADC region. *Report to MMSD*. http://pubs.iied.org/pdfs/G00735.pdf

Hinton, J. 2006. Women in Artisanal and Small Scale Mining in Africa. In: Lahiri-Dutt, K. & Macintyre, M. (eds.): *Women Miners in Developing Countries: Pit women and Others*.

Labonne, B. 1996. Artisanal mining: an economic stepping stone for women. *Natural Resources Forum*. (22)2: 117–122.

Ranchot, S. 2001. Research Topic 3: Mining and Society Gender and Mining: Workplace. *Report to MMSD*. http://pubs.iied.org/pdfs/G00605.pdf

Geological Resources and Good Governance in Sub-Saharan Africa – Runge & Shikwati (eds)
 ISBN 978-0-415-58267-4

Best practices working in partnership in response to HIV and AIDS in mine site Lake Zone in Tanzania

L. Ndeki, P. Sekule, K. Kema & F. Temu
African Medical and Research Foundation, Mwanza, Tanzania

1 INTRODUCTION

HIV and AIDS epidemic forced low-and middle-income countries to struggle to cope with the overwhelming task of providing prevention, care, treatment and support services in resource-challenged settings. This challenge presents an opportunity for the private sector to become more involved in HIV and AIDS response, collaborating with governments that are already providing financing and services to millions in need.

Private sector companies are an increasingly important component in addressing the HIV and AIDS epidemic globally. Based on the experience of companies operating in AIDS-affected countries in Sub-Saharan Africa, it has become very clear that not only is addressing HIV and AIDS at a corporate level the "right thing to do" in humanitarian terms, but also in economic ones. If the epidemic is not addressed could have devastating results for both the company and country in general including; increasing health care costs, reducing workforce productivity and market consumption, even destabilising the entire political climate of particularly hard-hit countries.

In response to HIV and AIDS epidemic in Tanzania, Public-private partnership has proved to be successful in improving access and utilisation of mobile and migrant workforce and the surrounding community to HIV prevention care, treatment and support services. The National Multisectoral Strategic Framework (2008–2012) has emphasised on importance of scaling up of public-private partnership to complement to the national HIV and AIDS response.

Workforce and the community for example in mining areas in Tanzania have an inseparable HIV risk pattern due to the mutual co-existence. Majority of the workforce living in the community and the community depend on them for generating income through variety of economic activities. Prevailing interaction signify this inseparable risk pattern and necessity initiation of public-private partnership in response to the epidemic.

AMREF in Tanzania in implementing its strategy has been emphasising on strengthening of partnerships through building the capacity of both community and health systems and ensuring that health policy and practice create space for the community to pro actively participate and engage in strengthening the health system and promoting their own health.

AMREF-Mine Health Initiative in collaboration with Tarime District Council and Barrick North Mara has been spearheading implementation of Public-Private Partnership model in response to HIV and AIDS epidemic in Tarime District in Mara region. The region has a high prevalence rate of 7.7% compared to the National rate of 5.7% (THIMS 2007/08). Baseline survey done in 2006 in North Mara, Tarime District found that there is significantly high HIV prevalence among women working in recreational facilities of 14.7%. Prevalence among mine workers was 10.1% and for women and men community members were 7.1% and 2.6% respectively. Condom use was lowest in community whereby 48.2% of community respondents and 49.8% of mine workers reported to had never used condom in the past 4 weeks prior to the interview, indicating a high risk of contracting HIV among mine workers and surrounding community members.

This partnership is also addressing other health challenges including STIs, TB and malaria challenges in the area. Similar partnership is also being implemented in Geita District, Mwanza region with District council and Anglo Ashanti Geita Gold Mine.

2 STRATEGIES

This comprehensive model has proved to be successful in improving access and utilisation of mine workforce and surrounding communities to integrated Sexual and Reproductive Health (SRH) services. Both mine workforce and surrounding communities at present have an access to:

- HIV Health education and advocacy for behavioral change.
- Established Peer educator's Scheme (PHEs).

- Integrated SRH service including Counseling and Testing service, Sexual Transmitted Infections (STIs) management, family planning service and opportunistic infection management.
- Outreach SRH service particularly to most vulnerable groups; Artisan miners and people living in remote and hard to reach areas.

3 RESULTS

The progress survey indicated decline in HIV prevalence compared with baseline results which current stood at 2.3% (men), 8.3% (women) and 3.5% (mine workers) respectively in North Mara by 2008. (North Mara External evaluation 2008). Also more than 70% of Barrick North Mara mine workers has accessed HIV counseling and testing service. Moreover After the intervention the proportions not using condom had dropped to 28.4% among community and 26.9% among mine workers.

Established HIV information center in North Mara has been experiencing an increasing number of people demanding for various SRH services. About 10,948 accessed various SRH service at the centre, among them 6785 received counseling and testing service and 5204 were screened for STIs in the three years period (2006–2008).

The partnership has also improved access of People living with HIV to care treatment and support services by facilitating referral network and supporting both provision of Anti Retroviral Therapy at the District hospital and home based care using established Post Test clubs. Monitoring and evaluation process has revealed that the programme has been successful in sensitising the target population on SRH issues including HIV and AIDS whereby on average each individual was reached more than 10 times during three years period 2006–2008).

"*At the beginning, the attendance was rather poor with about 700 clients in the whole of first year. After intense sensitization through the PHEs scheme, many women and men are now coming for HIV test. Similarly, we are experiencing an increase in the number of clients requesting for condoms*" (AMREF project Officer).

"*Before the project started, some of us used to hear about HIV in the mass media like the radio and news papers and therefore we thought of it as a strange condition, far from our communities. PHEs have opened the minds of many and people have now realized that HIV is within their communities*" (community Member).

"*We used to talk to mine workers at their work sites and even outside the mines. They showed deep interest in the talks and asked very pertinent and practical questions and also asked for reading materials like brochures to further broaden their understanding of the issues*" (mine PHE).

"*Training by AMREF has helped some of us as well as our customers to change our sexual behaviours. Personally, I can abstain from sexual intercourse for even three months, something which was almost impossible in the past*" (women working in recreational facility).

"*The program has reached everywhere in peoples' lives so if anyone is indulging in risky sexual behavior it is not because of lack of knowledge or skills but it is either due to irresponsible personal decisions, alcohol intoxication or poverty. The community is quite aware of malaria prevention but due to poverty; people cannot afford to buy mosquito nets*" (KI, ward executive officer).

4 CONCLUSION

This indicated that Public Private Partnership has successful compliment the national effort in response to HIV and AIDS and other communicable diseases in Lake Zone mining industry. This best practice model could be replicated and implemented in other areas with similar situation in Tanzania and worldwide in response to HIV and AIDS epidemic. Basing on the experience and existing HIV and AIDS structure built by the partnership, there is a window of opportunity of addressing other health related challenge in the area including Maternal newborn and child health issues. The area still has low coverage of mosquitoes treated bed nets use, low utilisation of family planning service and prevailing gender inequalities including gender based violence which to a large extent hindering women access to SRH services.

5 LESSONS LEARNT

- Partnership saves to cover all areas at the same time and it avoids duplication of activities.
- Partnership makes strong decision on implementation activities.
- Partnership draws on entrepreneurship skills.
- Partnership enables Government and Private sector maximise efforts.
- Partnership builds sense of owned community. resources supported by project that brings results of impact of project to be sustainable.

Geological Resources and Good Governance in Sub-Saharan Africa – Runge & Shikwati (eds)
© 2011 Taylor & Francis Group, London, ISBN 978-0-415-58267-4

HIV/AIDS in the informal mining sector evidenced by rapid antibodies tests—the Bossoui village case study (Lobaye, Central African Republic)

J. Runge
Department of Physical Geography, Goethe-University Frankfurt & CIRA, Centre for Interdisciplinary Research on Africa, Frankfurt am Main, Germany

C. Ngakola
Renforcement de la gouvernance dans le secteur des matières premières en Afrique Centrale, REMAP-CEMAC, Bangui, Central African Republic

ABSTRACT: The Central African Republic (CAR) ranks among the countries most affected by the HIV/AIDS pandemic in the Economic and Monetary Community of Central Africa (CEMAC). In comparison with the rest of the country where data are mostly compiled on an average basis by prefecture, it was hypothesised that prevalence of HIV/AIDS in artisanal mining communities will be significantly higher than for the rest of the population. More than 3000 inhabitants of the diamond village Bossoui in the SW of CAR (Prefecture Lobaye) were informed and sensitised during several field campaigns (HIV workplace policy) by the German Technical Cooperation (GTZ). Subsequently, between 2008–2009 almost two thirds of the village's population of around 5000 persons had been voluntarily, anonymously and free of charge HIV tested using HEXAGON HIV rapid tests (Human GmbH). Seroprevalence in Bossoui is high and varies between 8% and 23%. Men and women working in the mining sector are both affected (18%) by the virus, whereas people working in other business like agriculture and trade reach lower figures of 14% in seroprevalence. The age classes of 41–60 years show the highest prevalence (23%), but also minors up to 15 years already reach a seroprevalence of 12%. It is recommended to set up more HIV/AIDS workplace programmes in rural mining areas and to support the public health sector in CAR.

1 INTRODUCTION

Around 33 million people worldwide are infected by HIV/AIDS, and more than two-thirds live in Africa (CEMAC 2010). The Central African Republic (CAR) ranks among the countries most affected by the HIV/AIDS pandemic in the Economic and Monetary Community of Central Africa (CEMAC) (UNDP 2005). CAR owns vast mining resources like diamonds, gold, uranium, iron and others. However, since its independence in 1960 the mining sector always remained informal and more or less non-transparent. Up to now (2010) no bigger, international mining companies have reached to establish professional, industrialised open pit mining in the country. Therefore the artisanal (informal) diamond mining constitutes the key mineral and represented 35% of CAR's export earnings in 2007. The sector directly employs about 80.000–100.000 mineworkers which means that 400.000 to 500.000 people (9 to 12% of the population) directly depend on the resources generated by mining (ADF 2008).

Even there is a high potential in geological resources, CAR is one of the world's least developed countries, ranking 171 among 177 countries on the human development index (HDI). Its socio-economic indicators remain below the averages of CEMAC and Sub-Saharan Africa, and the attainment of the millennium development goals (MDGs) by 2015 is unlikely. The incidence of poverty estimated at 67.2% in 2003, has hardly improved over the past years. The life expectancy was estimated at 44.7 years in 2007, approximately ten years below the African average. The HIV/AIDS prevalence rate, which stood at 6.2% in 2006 for the 15 to 49 years age bracket, is the highest in the CEMAC zone (ADF 2008).

2 RECENT SITUATION AND DATA

Present day statistics on an average prefectural basis show the following: the prevalence of the HIV/AIDS epidemic in CAR is generalised with a prevalence rate of 6.2%. According to a country wide study carried out by ADF (2008) it presents the following overall characteristics:

- it is higher among women (7.8%) than men (4.3%).
- it is more pronounced within the age group of 15–49 years (which is economically spoken the most active).
- it is more pronounced in urban areas where it reaches 8.3% against 4.7% in rural areas.
- people living in wealthy households are HIV infected with a prevalence rate of 8.3%; and 6% among the middle class. The prevalence among the poor reaches only 4.5%.

According to the ADF (2008) study it is quite surprising that people with a higher education in CAR are the most HIV infected reaching up to 13% of women with a secondary school level and 10.4% of the men with a higher level. Reliable data availability on other population groups mainly in rural areas such as farmers, pygmies, truckers, soldiers, hunters, displaced people and refugees as well as artisanal miners is still weak.

The most recent data on HIV prevalence covering CAR on a prefectural basis was compiled by UNDP (2005) including information from the Pasteur Institute Survey (i.e. data from 2003). These country wide results are presented in Figure 1. Prevalence by prefecture shows very pronounced geographical disparities ranging from a minimum of 2.6% in Nana-Gribizi to a maximum of 13.6% in Haut-Mbomou. Bangui, the capital of CAR, comes up to 7.8%. These data mainly were based on blood samples taken routinely from pregnant women who come for antenatal consultations, and who are regarded as the group which best reflects the prevalence of infection among the general sexually active population (ADF 2008).

3 BACKGROUND AND OBJECTIVES

In the framework of the sub-regional project REMAP *Renforcement de la gouvernance dans le secteur des matières premières en Afrique Centrale* (see contributions of J. Runge in this volume) on transparency and good governance in the Central African mining sector, an attempt was made by CEMAC and the German Technical Cooperation (GTZ) to implement HIV mainstreaming activities/workplace policies to artisanal miners in CAR. The project was financed and supported by the Federal Ministry for Economic Cooperation and Development (BMZ).

One important question at the beginning of the study on stake holders in the informal mining sector was to get more information on the overall HIV prevalence in selected CAR villages dominated by artisanal miners. Certainly there are many reasons for assuming increased prevalence in mining areas. Miners, the majority of whom are men between 18 and 49 years old, often have to live apart from their families in order to work on the mine site. Prostitution therefore flourishes in mining areas (CEMAC 2010).

Other factors likely to contribute to the spread of the disease in the mining sites are limited knowledge of the methods for preventing HIV/AIDS transmission; the persistence of erroneous ideas on

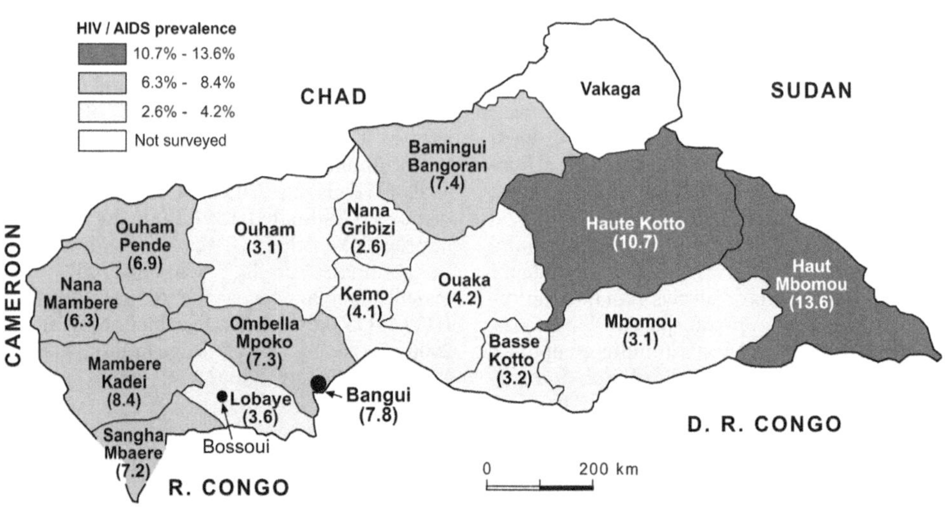

Figure 1. HIV/AIDS prevalence among men and women aged 15–49 by prefecture in CAR (Source: cited and redrawn from ADF 2008, p. 29).

the method of transmission of the disease (belief in witches), and risky sexual behavior such as the low use (or the total refusal) of condoms during casual sexual relations, multiple-partner sexual relations, and early sexual relations.

Risky sexual behavior is the key factor of transmission of the disease in the mining sites. Only half of those surveyed by ADF (2008) knew the three methods to prevent transmission, i.e. limitation of sexual relations to a single uninfected partner, the use of condoms during casual sex and abstinence. Moreover, the mining sites are generally strong poles of attraction for young and mobile labor in search of work. When occasionally mineworkers have earned some money, they irrationally spend their cash income on leisure, notably alcohol consumption and prostitution.

On the basis of these reflections it was hypothesised within the REMAP project that prevalence on HIV/AIDS in informal diamond mining communities must be significantly higher than in the rest of CAR, and a study series with HIV rapid tests was prepared.

4 STUDY SITE AND METHODS

The village of Bossoui with around 5000 inhabitants located in the Lobaye prefecture (Fig. 1), 38 km south of Boda, and 226 km west from the capital Bangui was selected as study side for applying HIV rapid tests. Collaboration with the USAID/ARD project on property rights in CAR's mining sector (see contribution by J. Hinton et al. in this volume) helped to establish contacts and trust to locals and to inform/develop sensitivity on the topic to set up the anonymous HIV/AIDS test series. The local dispensary in Bossoui with three nursing auxiliaries assisted in the HIV/AIDS testing.

For the study HEXAGON HIV 3rd generation immunochromatographic rapid tests for the detection of antibodies to human immune defiency viruses 1 and 2 were applied. 'HEXAGON HIV is intended for the rapid, qualitative detection of immunoglobulin (IgG, IgA, IgM) antibodies to Human Defiency virus HIV-1 and HIV-2 in human serum, plasma or whole blood as an aid in early diagnosis of AIDS. The test employs recombinant antigens representing immunodominant regions of the of the envelope proteins of HIV-1 and HIV-2. As the fingertip whole blood (20 μl) sample migrates through the absorbent pad of the test, anti-HIV-1 or anti-HIV-2 antibodies bind to the antigen-dye conjugates to form immune complexes. These bind to the respective capture antigens in the test lines "1" and "2" respectively and produce red-violet lines. Excess conjugate reacts in the control line "C" (Fig. 2) to denonstrate the correct function of the reagents' (HUMAN GmbH).

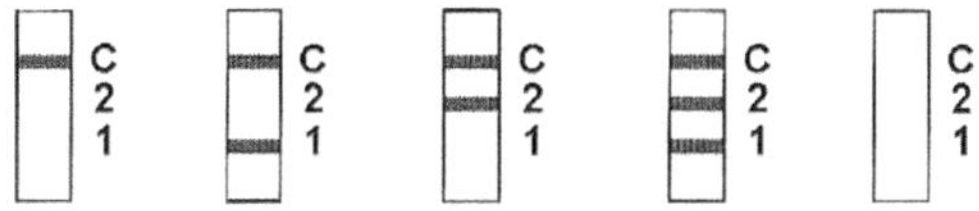

Figure 2. Interpretation scheme of the HEXAGON HIV rapid antibodies test (Source: HUMAN GmbH, Wiesbaden, Germany).

HEXAGON HIV tests cost less than 1 EUR each and give reliable results between 5–20 minutes. Strong reactive samples showed a HIV-1 or a HIV-2 test line within a few minutes. Weakly reactive samples required a longer incubation period up to a maximum of 20 minutes. Test must not be interpreted later than 20 minutes (HUMAN GmbH).

A test is "negative" (i.e. no HIV/AIDS infection) when the control line "C" appears in the upper part of the rectangular result window showing that the test has been carried out correctly and the reagents worked (Fig. 2, left). If no control line appears, even if a test line appears, the test has to be repeated with a fresh test (Fig. 2, right). One or two additional color bands next to the "C" line indicate a "positive" (i.e. HIV/AIDS infection); results for HIV-1 (band at "1") and/or HIV-2 (band at "2") respectively (Fig. 2).

Tested subjets in Bossoui had been individually informed about the tests and received a personal explanation on their results; it was confidential and remained unknown to others. Translation into Sango, CAR's national language, was offered when nescessary. In the case of a positive HIV result, it was recommended to re-check the rapid test results by applying an ELISA test at Boda hospital (cost FCFA 5000 or EUR 7.63).

5 RESULTS

In total 3021 villagers were successfully tested. Seroprevalence for HIV/AIDS in Bossoui is significantly higher compared to the prefecture level (see Fig. 1) in Lobaye that come up to 3.6%. Regarding the age distribution of HIV/AIDS infections (Fig. 3) it can be stated that infection rates are increasing from young person, 0–15 years old, with a seroprevalence of 12%, to 18% in the age class of 16–25 years. The majority of tested subjects lies in the age classes of 26–40 and 41–60 years with a seroprevalence of 16% and 23% (!), i.e. many of the adults in Bossoui life with the virus. The age class of elder people (over 60 years) is less prominent due to the reduced live expectancy and mortality and shows a seroprevalence of 8%.

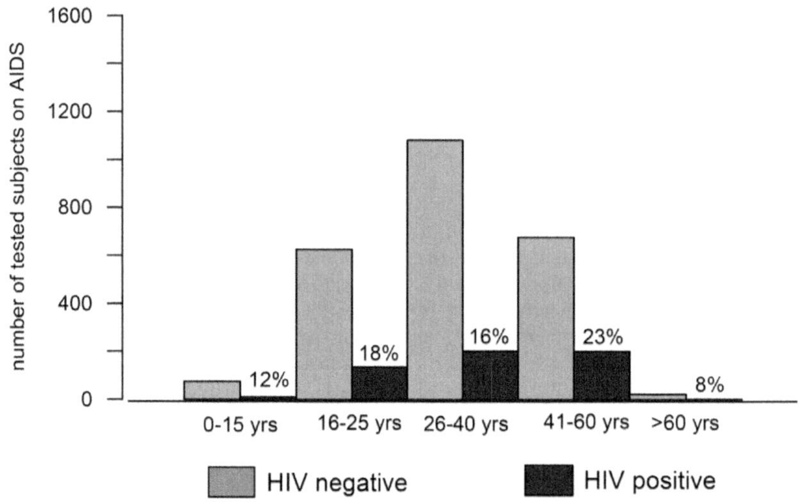

Figure 3. HIV/AIDS prevalence in Bossoui according to age classes (n = 3021, own inquiry).

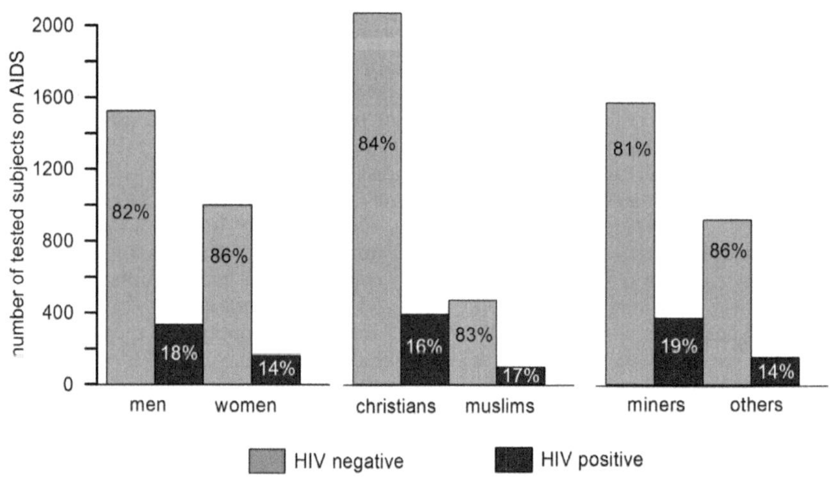

Figure 4. HIV/AIDS prevalence in Bossoui in relation to sex, religion and professional activity (n = 3021, own inquiry).

In Bossoui (Fig. 4) men generally do have higher infection rates (18%) than women (14%). Muslims who are often dealing with diamonds (i.e. as managers of buying houses) have a seroprevalence of 17% whereas Christians show a slightly lower figure of 16%. Comparing artisanal miners with people in other business (e.g. agriculture and trade), the seroprevalence of diamond diggers is significantly higher (19%) than those working outside the sector (14%).

More differences can be observed when comparing differences in gender relation: men and women working outside the mining sector have a lower seroprevalnec with 16% and 12% respectively. The small group of Muslims who are not involved in the mining sector show the lowest seroprevalence with 2% only, whereas Christians still come up to 16% in "positive" cases.

Men and women working in the mining sector both show high infection rates of 18%. Artisanal mining was often regarded as a mainly male dominated sector. But the results from Bossoui also evidenced that 1520 person in artisanal mining were men, and also a relative high number of 948 female miners were active in the diamond sector.

6 CONCLUSIONS AND RECOMMENDATIONS

Even if the mining sector in CAR stayed up to now informal (and perhaps will remain) for years, it contributes to the national income and export winnings of the country. If one intents to develop the sector in a sustainable way, contributions to improve public health of artisanal miners are important and the following points have to be taken into consideration.

First, it is essential to strengthen the overall organisation and transparency in the mining sector (by implementation of EITI standards and the KCS). An implementation of the operational plans of the HIV/AIDS control strategic framework in prefectures, especially those where mining zones are located; and a strengthening of the basic health infrastructure in the mining sites to promote voluntary anonymous screening and safe sexual behavior (condom availability, accessibility and distribution) is necessary.

Second, a strengthening of the education system in the mining zones and implementation of enrolment incentives to enable children to enroll in schools is indispensable. According to ADF (2008) the finalisation and implementation of the existing HIV/AIDS control action plan of the Department of Mines of the Ministry of Mines, Energy and Water, with the support of the World Bank and other partners is important.

To guarantee sustainability in investments from artisanal miners the setting up of a savings system to promote investment in the villages of origin of mine workers is necessary (miner's bank and micro credit facilities). These measures could contribute on the long run to improve living and health conditions in artisanal mining areas.

REFERENCES

ADF, African Development Fund 2008. *Central African Republic (CAR): Study on HIV/AIDS in the mining sector—assessment of factors of vulnerability*. Bangui, p. 41.

CEMAC 2010. Geological resources and good governance in Central Africa. *Report of the CEMAC conference, 24–25 September 2009, Yaoundé, Cameroon*, p. 35.

HUMAN GmbH (n.d.). Gesellschaft für Biochemica und Diagnostica mbH, Wiesbaden, Germany. www.human.de/data/gb/vr/1i-hivp.pdf

UNDP 2005. *Impact of HIV/AIDS on development in Central African Republic (CAR)*. Bangui.

Outlook

Geological Resources and Good Governance in Sub-Saharan Africa – Runge & Shikwati (eds)
© 2011 Taylor & Francis Group, London, ISBN 978-0-415-58267-4

Conclusion: The urgent need to include African people—where do we go from here?

J. Shikwati
IREN, Inter Region Economic Network, Nairobi, Kenya

J. Runge
Department of Physical Geography, Goethe-University Frankfurt & CIRA, Centre for Interdisciplinary Research on Africa, Frankfurt am Main, Germany

1 OBSERVATIONS AND REMARKS

The papers brought together in this book contribute to different aspects of the informal and professional mining sector in Sub-Saharan Africa. They try to clarify good governance, transparency and sustainability in an always highly sensitive political field. They also propose visions for the attainment of Millennium Development Goals (MDGs) by supporting sustainable development in the extractive industries sector and for the environment where it takes place.

Sustainable development and management in the mining economy are mainly based on the principles of maximum integration and openness to cooperation in conjunction with—in its best case—unconditional concurrent alignment of all core processes with the paradigm of sustainability in all its dimensions. By stakeholder integration and participation into the process, the informal and artisanal mining activities could be integrated into this system; in the same way as it is promoted today for professional multinational companies setting up large-scale, open pit mining sites.

Progress in good governance, transparency and economic as well as social and environmental sustainability in this sector depends on a positive mindset and willingness to change on the part of governments with their respective political leaders, and the companies active in the extractive industries sector. As the Kimberley Certification Scheme (KCS) has demonstrated since its setting up in 2002, it is possible to overcome the problem of established structures that are not fit for the purpose. The first essential is harmonisation of resource policy at various levels. For example, rigorous implementation of the Extractive Industries Transparency Initiative (EITI) enables patronage, clientelism and corruption in the resource sector to be curbed by a policy of transparency. However, this approach can only work if it is enshrined in legislation (and if it is not carried out in the sense of technocracy), so that effective sanctions can be applied when the rules are broken. A useful complement to the standardisation process is the decentralisation of political structures and decision-making processes, because decentralised systems can often be more easily adapted to conditions in the region.

However, before all participants can derive benefit from cooperative and participatory processes, they must first be accepted as legitimate stakeholders. Governments are expected to adopt a more open information policy towards their citizens—for example with regard to broader access to EITI reports and other relevant information. Other important elements of a sustainable resource policy are certification systems and implementation of measures to improve basic education and to train African resource experts. Businesses need to be encouraged to adopt international standards (such as the OECD principles for multinational enterprises) and develop their social and ecological profile. Cooperation with government—perhaps in the form of public private partnerships—can be a stabilising factor in such efforts.

Governments ought to consider NGOs and other societal groups' (e.g. the Publish What You Pay, PWYP philosophy) input as legitimate stakeholders to provide them with information and get involved in decision-making processes. They must likewise be involved from the outset in negotiations on concessions to defuse the potential for conflict and create a better climate for investment. By involving local stakeholder groups, commitments can be entered into for measures with development potential, such as cooperation between local businesses and foreign investors.

To promote the development potential of small-scale miners and address complex issues such as ecological and environmental standards or the prevalence of HIV/AIDS, it is helpful to set up interest groups. The small-scale mining sector will only develop for the benefit of African economies when the state creates legal conditions in which it can operate.

2 WHERE DO WE GO FROM HERE?

One of the most recurring mistakes in Africa has been the tendency to overemphasise one aspect of the economy to the other. Over-reliance on donor communities and development agencies' prescriptions has made African countries to focus on tourism and agriculture as key drivers of development at the expense of other sectors. A casual look at Africa's airport bookshelves and university bookshops will reveal more publications on tourism, agriculture, governance, human rights abuse and conflict among others. Contributing authors in this book attempt to stimulate positive debate to illuminate the fact that exploitation of geological resources is an additional driver to social and economic development in Sub-Saharan Africa.

Due to the mining sector's high set-up cost, long term focus and contribution to major armed conflicts on the continent; it has been treated as a taboo subject. Intrigues surrounding the exploitation of geological resources in Sub-Saharan Africa expose the fact that African governance institutions are still too weak to manage revenue and police wealthy mining corporations. This explains in part why international institutions such as the Extractive Industries Transparency Initiative (EITI) try to remedy the situation by urging voluntary transparency among mining players. Such transparency may however not bring good governance on the continent automatically unless individual citizens take steps to increase awareness about mineral exploitation, responsibilities of the state and organise themselves to build institutions of leadership that focus on rule of law and not rule of man.

3 ENVIRONMENTAL AWARENESS

Awareness of negative effects of environmental degradation in Sub-Saharan Africa has increased since 2002 when South Africa hosted the World Summit for Sustainable Development (WSSD). Public awareness is strategic to ensure that mining activities do not harm the environment and people. Mining causes irreversible impact on the environment through removal of a non renewable natural resource from one given spot. Negative environmental impacts occur in all phases of geological resource exploitation from exploration, disposal of waste rock and overburden, ore processing and plant operation, tailings (processing wastes) management, infrastructure (access and energy) to construction of camps and towns (Boocock 2002). Boocock (2002) discusses major impacts of abandoned mine sites that include acid mine drainage, loss of productive land, visual effects, surface and groundwater pollution, soil contamination, siltation, contamination of aquatic sediments and fauna, air pollution from dust, risks posed by abandoned shafts and pits, and landslides due to collapse of waste and tailings dumps.

The quest to generate revenue from geological resources calls for additional investment in environmental stewardship. It will not make sense for Sub-Saharan Africa countries to mint millions of dollars from natural resources but destroy the environment that sustains human and other biotic life. Increased environmental awareness issue has made developed countries to push for strict operational codes governing industry operations. African countries ought to take advantage of initiatives such as the financial sector's 'The Equator Principles' for managing environmental risks. The Equator Principles offer 10 key focus areas that the financing sector has to consider when giving support to a venture. They include review and category projects for financing on the basis of social and environmental due diligence; Social and Environmental Assessment; and Applicable Social and Environmental Standards among others (The Equator Principles 2006).

4 EMERGING ECONOMIES

Presentations in this book focus mostly on Africa's traditional economic allies—notably the Western countries—but little on the impact of emerging economies. The current surge in demand for geological resources is largely driven by emerging economies such as Brazil, Russia, India and China (BRIC). BRIC countries' appetite for geological resources is informed by the fact that they have been able to substantially reduce the number of people living below the poverty line. China, for example, reduced the number of people living on less than US$1.25 a day from 84% of the population in 1981 to 16% in 2005 (reducing number of poor people from 835 million to 208 million). India reduced the proportion of people living on less than US$1.25 from 60% in 1981 to 42% in 2005; Brazil from 17% to 7% (IFPRI 2010). Reduction in poverty levels increases the population of the middle class. This class in turn presents new demands on the economy as energy and product range requirements increase.

It will be wrong to assume that Africa's 1 billion people will remain stuck in poverty and not develop an economic interest for geological resources. Sub-Saharan Africa countries ought to plan for the reality and possibility of high demand for energy and range of products within its borders. The enthusiasm to export should be tampered with a serious quest to build domestic markets and industries. African governments must therefore focus less on asking for dollars in cash but demand for investment that can enable them exploit resources for the benefit of their people.

From emerging economies, Africa ought to learn that a government must offer services to its people; develop a policy of inclusiveness (India's development trend shot upwards as they abandoned class system and opened up the country to all castes); feed its people (Brazil is one of the leading agricultural innovators) as well as learn the international market system and how to play in it (China, a communist country has become a major capitalist player globally).

Most discussions on exploitation of natural and geological resources in Africa tend to fall in the trap of ignoring the African people. This book has certainly not escaped this trap. Most presentations outline Sub-Saharan Africa's geological resource exploitation and how it fits in the global economic systems but little on African people. It is our hope that our readers will find inspiration from this book to delve deeper into how African people can take charge of ensuring that geological resources drive their economies.

5 POLITICS AND RESOURCE EXPLOITATION

Politics in Sub-Saharan Africa ought to move from competition between individual personalities for power to competition in policies and strategies on how best to include the citizenry in resource management. The politics of exclusion breeds conflict on the continent. According to George Ayittey; civil wars cost the continent US$15 billion annually in lost output, wreckage of infrastructure and refugee crisis (Ayittey 2005). It does not make sense to invest in a governance mechanism whose result is to enrich a few individuals while destroying generations of people and the entire country. Africa ought to invest in analysing the social and economic burden of existing governance systems on the continent and review them accordingly.

Sub-Saharan Africa's quest to develop holistically evokes many theories and emotions. Emotionally, the Western countries' characterisation of the region as poor and incapable of mapping its own destiny, has spurred policies of aid handouts and lectures on good governance. The region's 50 year history since independence has witnessed one prescription after the other from the World Bank, the International Monetary Fund and individual western country capitals on political and development policy.

Awareness of the existence of untapped mineral wealth, under-exploited farmlands and underutilised aspects of creative economy ought to push Africans to bust the myth of aid and externally engineered policy prescriptions. While Africa should learn from her international partners, the final decision making ought to remain with the African people. The only help Africa needs is growing a class of African investors through whom the much sought after sound public policy and good governance will gain support and nourishment. Africa ought to partner with developed economies on the basis of long term country interests as opposed to short term individual (politician) interests. Developed nations are not under any obligation to create 'good visions' for Africans.

Geological resource rich countries must prioritise growing domestic and regional markets as a strategy to spur "cottage industries" in the sector. (Intra CEMAC recorded trade volumes are low at 2% of GDP and 1.5% of total trade of CEMAC countries.) A casual view of conflict prone countries in Africa indicates that they tend to trade less with their neighbors but more with outside markets (Economic Development in Africa Report 2009). Kenya's 2008 post election violence revealed how important it is for an African country to rely on another as a market—Kenya's neighbors could not allow the country to burn to ashes. Trade promotes interdependence and peace.

Africa's political elite ought to recognise the existence of the African people in their plans. Most often, when African governments negotiate with foreign investors, they view their own citizenry as an obstacle to development. As discussed in this book; a government will either cheat its own citizenry by offering low standard compensation for land acquired for mining purposes or simply evict them by force. This alienation gives rise to informal artisanal miners, civil society movements against resource exploitation by foreigners and at worst conflict. Efficient governance systems will save the continent losses incurred due to conflict.

6 INFORMATION IS POWER

Sub-Saharan African countries spend less than 1% of their GDP on research and development but spend heavily on arms. African countries spend an estimated US$12 billion annually in purchase of arms. Mineral exploration is therefore left in the hands of foreign companies that at times hide the truth from governments. Lack of knowledge on available geological wealth in a given country renders and otherwise resource rich countries only wealthy in potential. According to the late agriculture guru Norman Bourlag 'one cannot eat potential'. Research information can assist investors to transform potential into value and consumables.

Majority of contributors to this book are not from Africa. The dearth of serious research and academic work on the continent confines its wealth status to 'potential wealth'. In spite of our effort to contact over 200 geologists across the continent, it was difficult to get them to write a paper on applied geology and good governance. African research has relied heavily on donor community. The donors often have funded aspects that meet their (donors') interest to the detriment of the continent's quest to develop. Such funding has converted trained geologists to human rights, gender and political activists. Sub-Saharan Africa countries must urgently invest in training constituencies of experts to enable them manage and exploit resources efficiently. As long as Africans remain in the 'silent mode,' the continent will scarcely attain holistic development. African Governments must therefore allocate funds to research in their annual budgets as an additional tool to development.

7 FINANCING AND SECURITY FOR INVESTORS

The long term nature of geological resource exploitation calls for secure legal and fiscal regimes. The term "investor" in African parlance carries exclusionary undertones that portray people in Africa as 'non investors' while their foreign counterparts are the investors. African governments consequently proceed to implement policies that favor external investors but do little to promote the growth of homegrown investors. The so-called foreign investors find themselves mired in local political competitions as the excluded Africans attempt to join the investment club. To guarantee the security of any investor, it is important that governments nurture both indigenous and foreign investors. Inclusion of indigenous investors will motivate the local people and add a sustainability element in investment in terms of a secure policy and business environment for all. A policy environment that only pleases 'outsider' investors breeds challenges especially with the changing international system that has brought new players from the East who have a totally different approach from the West. While the Western investors rely heavily on their government to push for policies favorable to them; the Eastern investors rely on 'non interference in internal affairs of a sovereign state' to secure their interests.

The West and East approach to geological resource exploitation in Sub-Saharan Africa is also captured in the financing landscape. Historically, Western mining companies raise capital through the stock market, investment funds and equity markets. The Eastern model (especially China) rely on state owned corporations utilising state funds to finance their operations. The absence of African indigenous initiatives to finance geological resource exploitation is not healthy in the long run. Financiers may impose policies that may be unfavorable to the people and cause a potential threat to security of investors.

The geological resource exploitation financing landscape is dominated by the World Bank through the International Finance Corporation (IFC, mostly in Africa), the European Investment Bank, and the European Bank for Reconstruction and Development (mostly in Russia and Uzbekistan). Other private mining investors include the HSBC Investment Services Africa, Merrill Lynch, Barclays Bank through Barclays Capital, JP Morgan, Citibank and the Standard Bank. The domination of foreigners in financing the geological resource exploitation partly explains why Africans are largely in the artisanal sector because they are yet to be picked by the radar of international financing giants.

The global market system ought to be reformed to allow expanded financing options to drive value addition within countries endowed with geological resources by opening up the financing sector to indigenous entrepreneurs. For example, micro—financing aspects in sectors such as tourism and agriculture ought to be evaluated to determine how similar approach can be used in the mining sector. The continent has to be cautious to put in place policies that help spread wealth and not indebtedness.

8 SUSTAINABILITY AND HUMAN RIGHTS

The quest to exploit geological resources creates conflict in cases where human populations have to be moved to facilitate access to mineral wealth and the environment is polluted to their detriment. The respect for value of African human life should be demonstrated through sober policies that guide investments in global resource exploitation. Too much focus on natural resources by developed and emerging economies may not resolve this predicament unless Africans work towards claiming a value-added role in the global economic system.

The best way to advance the cause of human rights is to accord them freedom to contribute value to the global system. Such contribution has to be tampered with the awareness of the short and long term effect of human action to the environment.

Soil degradation, destruction of forests and environmental pollution all work against the initial goal of human beings to exploit nature to improve their quality and standards of living.

9 INTRA-COUNTRY APPROACH

To reap maximum benefit from geological resources, African countries ought to consider a joint investment approach. Such an approach will entail intra-country infrastructure development to increase connectivity across cities in the continent. The intra-country infrastructural development can take advantage of the existing Regional Economic Blocs (RECs) such as the East Africa Community (EAC), the Southern African Development Community (SADC), the Central African Economic and Monetary Community (CEMAC) and the Economic Community of West African States ECOWAS among others to offer a framework that will spur domestic markets and development. A joint approach can facilitate harmonies in mining policies to address aspects of good governance.

Intra country transportation corridors, such as roads, railways, and airports must be opened. Africa must invest in complimentarity as opposed to rivalry. CEMAC countries-Equatorial Guinea, Cameroon, Gabon, Central Africa Republic, Chad, and Republic of Congo; ought to take advantage of relative cultural homogeneity (ethnic communities that cut across country borders) to push for sound structures that will ensure geological resources bring prosperity to the people.

10 INSTITUTION BUILDING AND PEOPLE PARTICIPATION

People participation in governance issues (constitution making, taxes to government, defining leadership) is strategic to good governance. Governments in Sub-Saharan Africa ought to open up space to allow for participation of Non Governmental Organisations (NGOs). Initiatives such as Publish What You Pay (PWYP) are an important monitoring function and help maintain the prominence of critical issues in the minds of policy makers and international public. Civil society groups act as advocates for populations whose political and economic life is severely constrained by limited education and resources.

It is important for governments to equip its people and stakeholders in general with relevant information regarding geological resource exploitation and involve them in decision making processes. It's through such partnerships that citizens will be able to drive the solution process to challenges that confront them both as individuals and as a country.

11 USE OF INCENTIVES TO ATTRACT INVESTMENTS

Measurements and keeping of accurate records provide an incentive to both indigenous and foreign investors to prioritise accordingly. Measurements are not only confined to research data gathered on availability on geological resources, it has to include standards that govern operations of exploitation activities. Measurements such as the Kimberley Certification Scheme (KCS) and the Extractive Industries Transparency Initiative (EITI) have demonstrated ability to curb patronage, clientilism and corruption in the resource sector for countries and corporations that have adopted them. Countries should consider incorporating such standards in their legislation to ensure enforceability. Incentives also include putting in place simpler procedures for one to start business and obtain permits to operate in a transparent manner.

Institutions that embrace an environment for safe exploitation of resources and protection of knowledge and property of innovators ought to be built as part of an incentive drive to investors.

Natural resources can assist in a country's development if the population creates efficient means of profitable exploitation and responsive organisational structures that ensure delivery of public goods. The focus ought to be on a country's population; an education system that allows talent and knowledge expansion. A government that is accountable to its people and an open global market system (not the current system that relegates Africa to raw material exports).

12 HIV/AIDS IN MINING COMMUNITIES

Mining enterprises not only disrupt subsistence lifestyle of a given community but also bring along new challenges in societal norms. It is important while assessing the possible benefits of a mining enterprise to inbuilt safeguards for communities in terms of when the mines 'dry up of resources' and also against prevalent of HIV/AIDS. Since most mines are rarely situated in areas where workers come from, the staff leaves their families to go and work in the mines. Miners tend to have a low perception of risk because their daily jobs are risky, hence the tendency to engage in unprotected sex. Both government and mining companies have to be proactive in shielding communities from such dangers through education and awareness campaigns.

REFERENCES

Ayittey, N.B.G. 2005. *Africa unchained: The blueprint for Africa's future*, New York, Palgrave Macmillan.

Boocock, N.C. 2002. *Environmental impacts of foreign direct investment in the mining sector in Sub-Saharan Africa*, OECD http://www.oecd.org/dataoecd/ 44/40/1819582.pdf (accessed March 15, 2011).

Communauté Economique et Monetaire de l'Afrique Centrale (CEMAC) 2010. «*Resources géologiques et bonne gouvernance en Afrique Centrale*, 24–25 September 2009, Yaounde, Cameroon, p. 37, www.yaounde2009.net

IFPRI 2010. *The role of emerging countries in global food security.* http://www.ifpri.org/sites/default/files/publications/bp015.pdf (accessed March 15, 2011).

The Equator Principles 2006. *A financial industry benchmark for determining, assessing and managing social & environmental risk in project financing.* http://www.equator-principles.com/documents/Equator_Principles.pdf (accessed March 15, 2011).

Geological Resources and Good Governance in Sub-Saharan Africa – Runge & Shikwati (eds)
© 2011 Taylor & Francis Group, London, ISBN 978-0-415-58267-4

Author index

Colour plates

PLATE 1

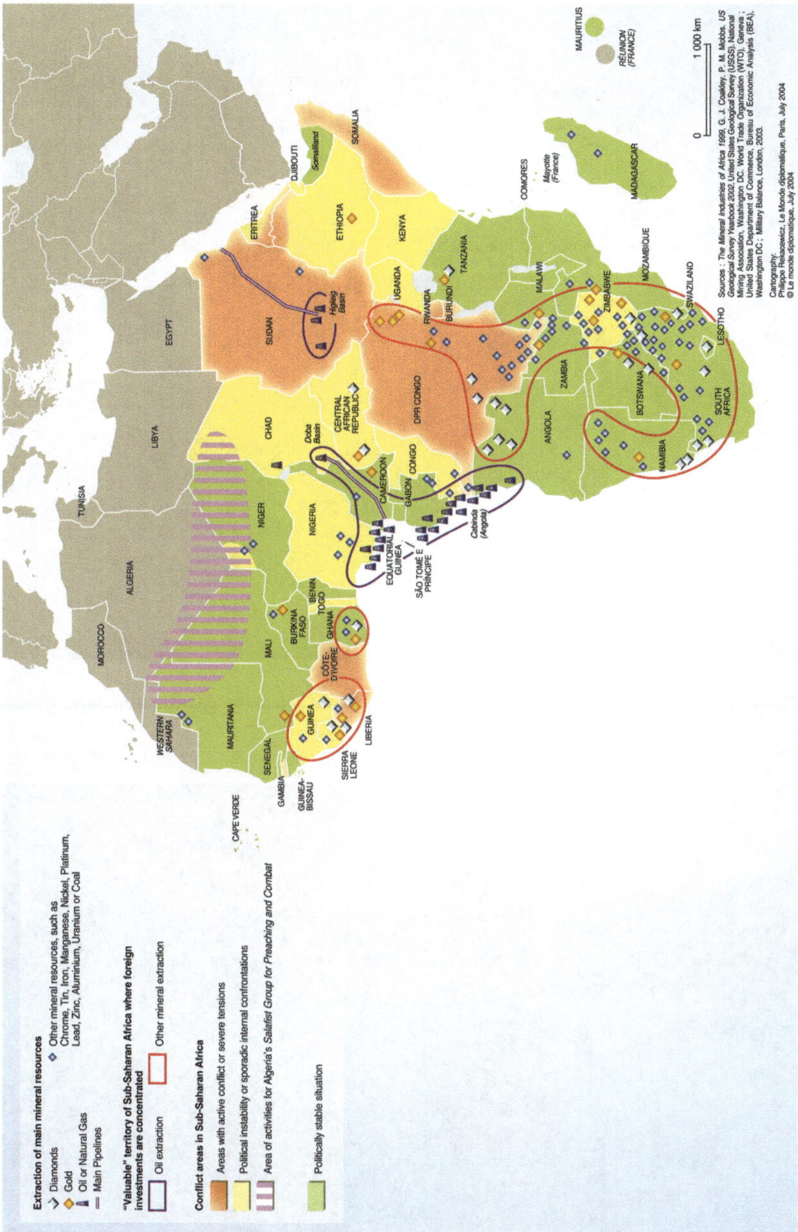

Figure 1. Shikwati, J., How geological resources can aid Africa's development (for text see contribution on pp. 11–12).

PLATE 2

Photo 1. Artisanal miners digging within river bed sediments (alluvia) in Central African Republic for finding diamonds (T. Bittiger, October 2010).

Photo 2. Artisanal miners blocking a river tributary close to Pessandro (Central African Republic) by an artificial dam (in the middle) allowing prospection and digging in the dry river bed (on the right). Also the deviation of natural river courses causes environmental damage and provokes land use conflicts with farmers and cattle breeders (J. Runge, July 2010).

PLATE 3

Photo 3. One of the last diamond cutters in Bangui, Central African Republic, working on raw diamonds (J. Runge, March 2009).

Photo 4. Sorting and valuation of raw diamonds at BECDOR (*Bureau d'Evaluation et de Coordination de Diamants et d'OR* / Office for Valuation and Coordination of Diamonds and Gold) which was set up in Bangui in 1995. Since 2003/2004 each diamond that is officially registered and taxed by BECDOR receives a certificate according to the Kimberley Certification Scheme (J. Runge, June 2008).

PLATE 4

Photo 5. Exploration program in a gold mining area carried out by mobile, lorry based drilling platforms. Bore holes reach depths between 100–120 metres (J. Runge, October 2008).

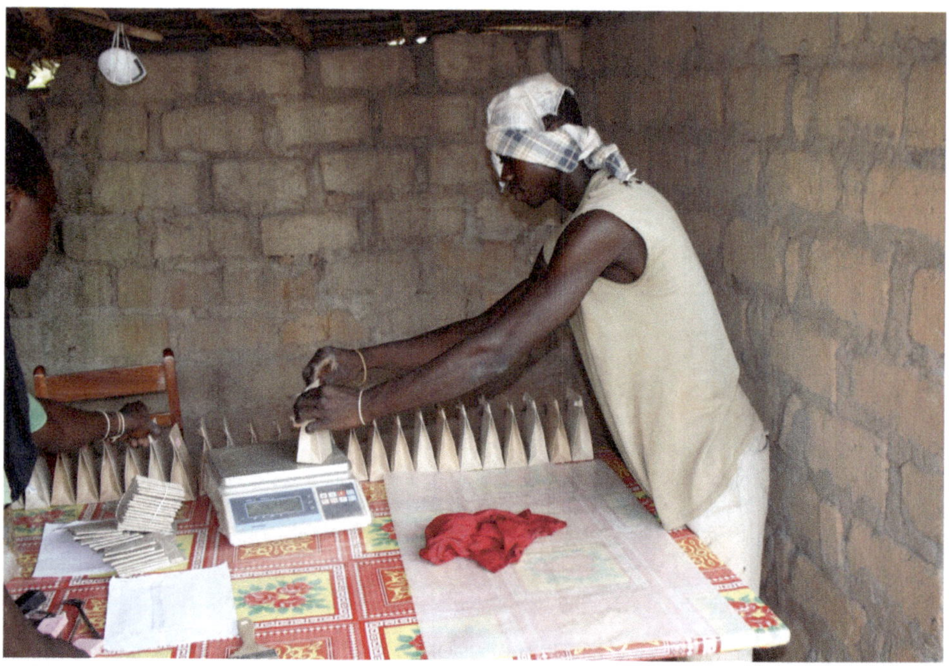

Photo 6. Preparing pulverized rock samples for geochemical analysis in a gold mining exploration area of South African Pan African Resources Ltd. at Bogoin, Central African Republic (J. Runge, June 2008).

PLATE 5

Photo 7. Highly mechanized exploitation of geological resources: Caterpillar heavy quarry truck in an open pit mine (J. Runge, June 2007).

Photo 8. Chrushing of rocks by heavy machines in the Tenke Fungurume Copper Mine in Katanga, D.R. Congo (T. Bittiger, October 2010).

PLATE 6

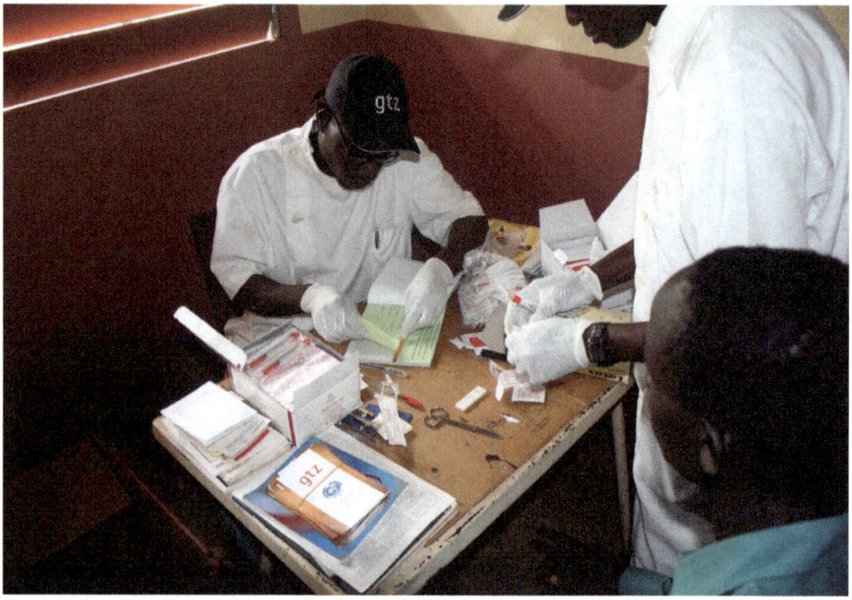

Photo 9. More than 3000 persons had been voluntarily, anonymously and free of charge HIV tested in the miners' village of Bossoui applying HEXAGON HIV rapid tests. Seroprevalence in mining communities is high and varies between 8% and 23%. Men and women working in the mining sector are both affected (18%) by the virus, whereas people working in other business like agriculture and trade reach lower figures of 14% in seroprevalence (J. Runge, October 2008).

Photo 10. Strengthening the role of women and of general gender aspects in the mining sector. The picture shows a gathering of women organized in 'groupements' in the diamond miners' village of Bossoui in the Central African Republic (J. Runge, August 2008).

PLATE 7

Photo 11. Example for sensitizing people in the framework of HIV/AIDS workplace programs. Here redwood truck drivers are addressed to use condoms ('K-Pote') when having casual sex with unknown partners. The apprentice inside the truck is remembering his boss to take a condom when joining the young girl. Similar posters can be used to focus on artisanal miners in the region (Source: CEMAC/GIZ 2008/2009).

PLATE 8

Photo 12. Oil drilling rig in front of the Cameroonian Atlantic coast at Limbe, west of Douala). Traditional fishing from dugouts meets modern exploitation of oil and gas (T. Bittiger, October 2010).

Photo 13. Group portrait with international delegates during the conference on 'Geological resources and good governance in Central Africa' in September 2009 in Cameroon's capital Yaoundé (www.yaounde2009.net). The editors are standing on the extreme right: James Shikwati; and on second position left from him, is Jürgen Runge (G. Zidro, September 2009).

For Product Safety Concerns and Information please contact our EU representative GPSR@taylorandfrancis.com Taylor & Francis Verlag GmbH, Kaufingerstraße 24, 80331 München, Germany

Batch number: 10726322

Printed by Printforce, the Netherlands